CHAOS IN LASER-MATTER INTERACTIONS

World Scientific Lecture Notes in Physics Vol. 6

CHAOS IN LASER-MATTER INTERACTIONS

Peter W Milonni
Mei-Li Shih
Jay R Ackerhalt

World Scientific

Published by

World Scientific Publishing Co Pte Ltd.
P.O. Box 128, Farrer Road, Singapore 9128

The authors and publisher would like to thank all authors and publishers for permission to reproduce or modify copyright materials (figure numbers within brackets refer to this publication). Credit goes to the following:

American Institute of Physics: H. G. Winful *et al.*, *Appl. Phys. Lett.* **48** (1986) 616 (Fig. 32.6); **American Physical Society**: T. Midavaine *et al.*, *Phys. Rev. Lett.* **55** (1985) 1989 (Fig. 32.1); C. O. Weiss *et al.*, *Phys. Rev.* **A28** (1983) 892 (Figs. 32.2 – 32.3); H. J. Carmichael *et al.*, *Phys. Rev.* **A26** (1982) 3408 (Fig. 34.1); H. Nakatsuka *et al.*, *Phys. Rev. Lett.* **50** (1983) 109; G. H. Walker *et al.*, *Phys. Rev.* **188** (1969) 416 (Figs. 41.1 – 41.6 & 42.1 – 42.5); S. W. McDonald *et al.*, *Phys. Rev. Lett.* **42** (1979) 1189 (Fig. 51.2); K. A. H. van Leeuwen *et al.*, *Phys. Rev. Lett.* **55** (1985) 2231 (Fig. 56.4); G. Casati *et al.*, *Phys. Rev. Lett.* **53** (1984) 2525 (Figs. 56.5 – 56.6); **The Institute of Electrical and Electronics Engineers, Inc.**: L. W. Casperson, *IEEE J. Quantum Electron* **QE-14** (1978) 756 (Figs. 26.1 – 26.2); **The Institute of Physics**: N Pomphrey, *J. Phys.* **B7** (1974) 1909 (Fig. 51.1); J. G. Leopold *et al.*, *J. Phys.* **B12** (1979) 709 (Figs. 56.2 – 56.3); **North Holland Physics Publishing**: A. Zardecki, *Phys. Lett.* **90A** (1982) 274 (Fig. 21.5); C. O. Weiss *et al.*, *Opt. Commun* **44** (1982) 59 (Fig. 32.4); **Optical Society of America**: L. W. Hillman *et al.*, *J. Opt. Soc. Am.* **B2** (1985) 211 (Figs. 32.7 – 32.8); **Kyoto University**: T. Kai *et al.*, *Prog. Theor. Phys.* **61** (1979) 54 (Fig. 32.5).

To those who have not granted us permission before publication, we have taken the liberty to reproduce their figures without consent. We will however acknowledge them in future editions of this work.

CHAOS IN LASER-MATTER INTERACTIONS

ISBN 9971-50-179-1
9971-50-180-5 pbk

Printed in Singapore by General Printing and Publishing Services Pte. Ltd.

For our parents

PREFACE

This volume is based on series of lectures given by the first author at the Universities of Arkansas, Puerto Rico, and Rochester, and evolved from the Ph.D thesis research of the second author. More importantly, it reflects the learning experience of all three authors over the past several years.

Our goal is to introduce the basic ideas of chaotic dynamics, as clearly and succinctly as we can, to graduate students and researchers in atomic and molecular physics, quantum optics, and laser physics. Much of this volume might also be useful for newcomers to chaos from other areas of physics.

Reviews and monographs on chaotic dynamics have usually been confined to either dissipative or Hamiltonian systems. Both are important to the theory of laser-matter interactions, and so we have given approximately equal emphasis to each, although most of the research on chaos in optical systems to date has focused on dissipative systems.

Our list of references is by no means exhaustive. Aside from some of the standard references on things like routes to chaos, the literature cited reflects mainly our own reading in the field. We apologize to the many authors whose work is not directly cited.

We thank Dr. K.K. Phua of World Scientific for the invitation to prepare this volume, Miss P.H. Tham for her editorial work and patience, and Michael E. Goggin for reading the manuscript and bringing relevant literature to our attention. We also benefitted from our collaboration with Hal Galbraith during the early stages of our interest in chaos, and from conversations with Doyne Farmer, Luigi Galgani, Gottfried Mayer-Kress, Howard Carmichael, and Beth Ann Reimer, among others. We thank the authors who gave us permission to use their figures, and the National Science Foundation for supporting the first two authors throughout their education in chaos. Peter Milonni wishes to express his gratitude to Professors Alfonso Rueda and Rafael Muller for an invitation to lecture at the Workshop on the Foundations of Physics in Puerto Rico in 1985, for which some of this material was first prepared. Talking about physics with them and the other lecturers, Professors Gordon Fleming (Pennsylvania State University) and Edward Nelson (Princeton University), made the Workshop unforgettable.

CONTENTS

PART ONE

DISSIPATIVE SYSTEMS

Call Pythagoras and bid
The sage to mark divine the laws which rule
Each planet's course; and when he reads and sees
Such harmony amidst the countless worlds,
Trembling with joy his heart will overflow
Before the sacred concert of high reason.

Hans Christian Oersted, The Soul in Nature.
Taken from Bern Dibner, Oersted (The Burndy
Library, Norwalk, Connecticut, 1961)

1. Introduction

The Greeks at the time of Ptolemy believed all motion could be decomposed into "perfect" circular motions. This idea was successfully applied in the description of planetary motions in terms of epicycles. With the usual acuity of hindsight we might phrase the Platonic ideal this way: all motion is quasiperiodic, such that the Fourier transform of any coordinate consists of sharp spikes. (Figure 1.1) Poincare', near the turn of the century, was perhaps the first person to understand that there are (bounded) systems whose spectra do not have this form, that there are so-called "nonintegrable" systems. Such systems have a broadband, continuous component in their spectra, as indicated in Figure 1.2. Einstein (1917) also recognized this possibility.

Spectra of the type shown in Figure 1.2 are associated with turbulent motion. Imagine two little corks placed near to each other in a flowing fluid, and suppose the flow is laminar (smooth and orderly). As time evolves the positions of the corks are well correlated and their separation grows linearly with time, at a rate proportional to the difference in their velocities. But if the flow is turbulent, the corks separate rapidly, typically exponentially with time. Their locations depend very sensitively on where they started out. We will call a system *chaotic* if it has this property of *very sensitive (exponential) dependence on initial conditions*.

Chaotic motion is non-quasiperiodic, having a spectrum like that sketched in Figure 1.2. Over the past few years many different kinds of physical systems have been found to evolve chaotically. Among the recent developments, two are especially exciting. One is the recognition that chaos may appear in systems described by relatively simple rules of evolution, such as three first-order differential equations. The other is the discovery that there are a few prevalent or "universal" ways by which a system can make the transition from regular, quasiperiodic behavior to chaos. These are called *routes to chaos*, and the same routes have been observed in systems as diverse as fluids, lasers, and semiconductor devices, to name but a few.

One of the seminal papers in the field was published by E.N. Lorenz in 1963. [1.1] Lorenz made a rather severe truncation of some hydrodynamical partial differential equations, ending up with a set of three ordinary, nonlinear differential equations. This Lorenz model exhibits, for certain parameter ranges, what is now called chaos, the property of very sensitive dependence on initial conditions. Lorenz offered a half-serious metaphor, which we can paraphrase as follows. Suppose you are trying to predict the weather using complicated hydrodynamical equations and a really super computer. You know initial conditions for the atmospheric pressure, temperature, etc., and feed all this information into your computer code. But you neglected to account for a butterfly fluttering about somewhere over Taipei! If your system is chaotic, this "uncertainty" makes detailed, long-term weather prediction impossible, because of the extreme sensitivity to initial conditions. What the butterfly does might affect the weather next month in New York.

Lorenz's work, which seemed to remain largely unnoticed for a long time, showed that we needn't study terribly complicated systems of equations to learn something about chaos: systems of just a few simple equations can have very sensitive dependence on initial conditions. This point has taken a long time to sink in.

At about the same time as Lorenz , Buley and Cummings [1.2] made a numerical study of essentially the same system of equations as Lorenz while modeling a single-mode laser. They remarked that "A case has been run...in which the output...appears as a series of almost random spikes." Evidently they were finding the chaotic behavior that Lorenz was wrestling with in a completely different context.

In modeling physical systems we usually work with differential equations (e.g., $F = ma$). Typically we specify initial conditions at some time $t = 0$, and try to predict the state of the system for times $t > 0$. From a computational standpoint things would be much easier if time evolved in discrete steps rather than continuously. Then a typical simulation would involve a discrete mapping like

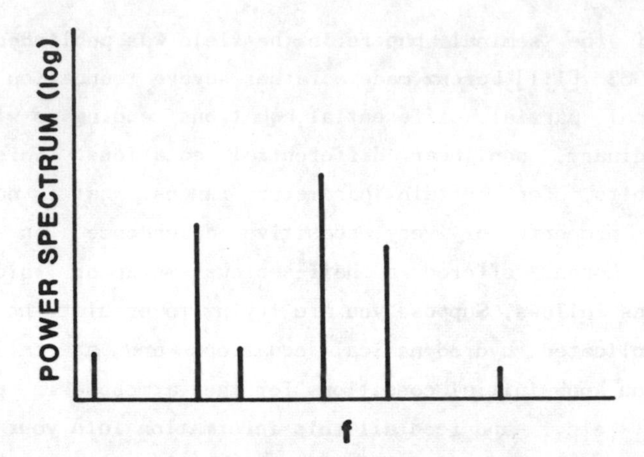

Figure 1.1 Typical frequency spectrum of some coordinate of a quasiperiodic system.

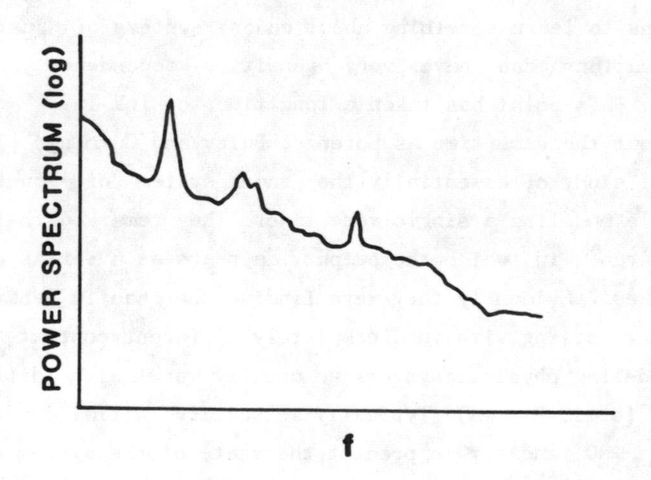

Figure 1.2 Typical frequency spectrum of some coordinate of a chaotic system.

$$x_{n+1} = f(x_n) \tag{1.1}$$

where f(x) is a prescribed function of x. Here the system at "time" n would be characterized by the number x_n, which is determined by (1.1) once the initial condition x_o is given. The mapping (1.1) is one-dimensional. We can easily conceive of higher-dimensional mappings, such as two-dimensional ones of the form

$$x_{n+1} = f(x_n, y_n) \tag{1.2a}$$

$$y_{n+1} = g(x_n, y_n) \tag{1.2b}$$

We might regard discrete mappings as contrived models for the kind of behavior exhibited by differential equations. Actually, however, it is possible, at least in principle, to construct discrete mappings from systems of differential equations. Such a technique was invented by Poincare' as part of his program to geometrize the theory of differential equations. Consider as an example a set of three differential equations of the form

$$\dot{x} = f(x,y,z), \quad \dot{y} = g(x,y,z),$$
$$\dot{z} = h(x,y,z) \tag{1.3}$$

Suppose we plot points $(x_n, y_n) \equiv (x(t_n), y(t_n))$ for times t_n for which $z(t_n) = 0$ and $\dot{z}(t_n) < 0$, as illustrated in Figure 1.3. Then the evolution $(x_n, y_n) \rightarrow (x_{n+1}, y_{n+1})$ defines a Poincare' map, which is a discrete mapping of the form (1.2). Although the only known general way of constructing a Poincare' map is by numerical integration of the differential equations, it is very useful to know that such a discrete mapping exists, for then we can study some aspects of "continuous flows" like (1.3) by looking at discrete mappings. Mappings are much easier to handle than differential equations (i.e., iteration is easier than integration).

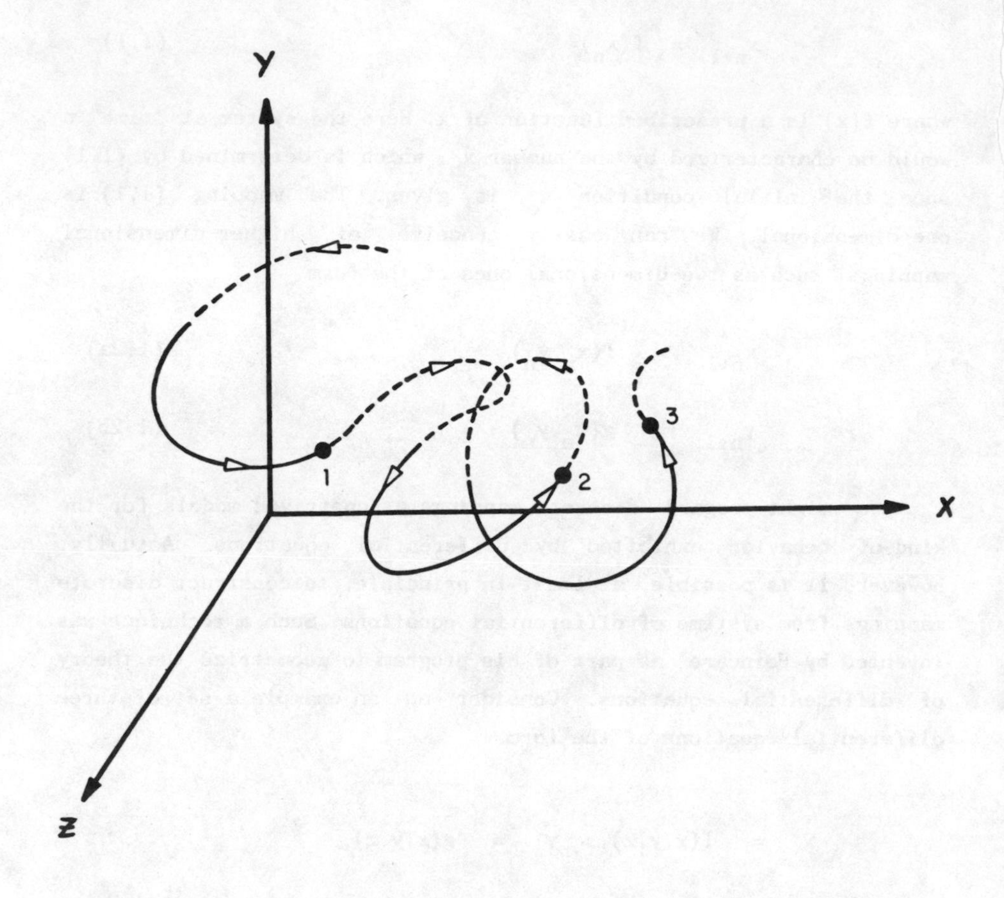

Figure 1.3 Construction of a Poincare' map with the xy plane as the surface of section. In this example point 1 is mapped into point 2, and 2 is mapped into 3.

In some instances the analysis of a physical system leads directly to a discrete mapping. Consider a particle constrained to move in one dimension under the influence of periodic impulses, the force being

$$F = A(x) \sum_{-\infty}^{\infty} \delta(t/T - n) \qquad (1.4)$$

Then the Hamilton equations of motion for the position and momentum are

$$\dot{x} = p/m \qquad (1.5a)$$

$$\dot{p} = A(x) \sum_{-\infty}^{\infty} \delta(t/T - n) \qquad (1.5b)$$

Let us integrate these equations from $t = nT-\epsilon$ to $(n+1)T-\epsilon$, where ϵ is an infinitesimally short time. Using the properties of the delta function, we easily obtain

$$x_{n+1} = x_n + T/m[p_n + TA(x_n)] \qquad (1.6a)$$

$$p_{n+1} = p_n + TA(x_n) \qquad (1.6b)$$

where x_n and p_n are the values of x and p just before the nth kick. This is a discrete mapping of the form (1.2). Periodically kicked systems of this type, because they are easy to handle numerically, have been useful in studies of "quantum chaos," as we will see later.

A one-dimensional mapping that has played an important role in the recent developments is the logistic map

$$x_{n+1} = 4\lambda x_n(1 - x_n), \quad 0 \le x_o,\lambda \le 1 \qquad (1.7)$$

Before ending our brief introduction, let us illustrate what we mean by very sensitive dependence on initial conditions by considering the case $\lambda = 1$ of the logistic map. In this case the transformation $x_n = \sin^2\pi\Theta_n$, plus the identity $\sin 2\pi\Theta_n = 2\sin\pi\Theta_n\cos\pi\Theta_n$, reduces the mapping to the form $\Theta_{n+1} = 2\Theta_n$, which has the explicit solution $\Theta_n = 2^n\Theta_o$. Since we can add any integer to Θ without changing the value of x, we can write

$$\Theta_n = 2^n\Theta_o \pmod 1 \tag{1.8}$$

as the solution of the logistic map for $\lambda = 1$.

It is easy to see that the mapping (1.8) has the property of very sensitive dependence on initial conditions: if we change the initial seed Θ_o to $\Theta_o + \epsilon$, then Θ_n changes by $2^n\epsilon = \epsilon e^{n\log 2}$. In other words, there is an exponential separation with "time" n of initially close "trajectories." The rate of exponential separation, namely log2, is called the Lyapunov exponent, and the fact that it is positive in this example means that we have very sensitive dependence on initial conditions, i.e., chaos.

We can understand this extreme sensitivity to initial conditions from another standpoint. Let us write Θ_o in base-2 notation. For instance, we can write the number 1/2 + 1/4 + 1/16 + 1/128 +...in base 2 as 0.1101001... In base 2 the algorithm (1.8) amounts to just shifting the "decimal" point. Thus if Θ_o = .1101001...then Θ_1 = .101001..., Θ_2 = .01001..., Θ_3 = .1001..., etc. Obviously then Θ_n will depend on the nth and higher digits of Θ_o, and when n is large the value of Θ_n depends extremely sensitively on the precise value of Θ_o. We can now begin to better appreciate the Lorenz butterfly metaphor!

This example illustrates another aspect of extreme sensitivity to initial conditions. If we iterate the map (1.8) on a digital computer, then after a relatively small number of iterations (typically $\approx$ 50 on 16-digit machines) we generate numerical "garbage," simply because in the digit-shifting process we eventually pick up digits representing round-off errors in the computer. (The reader unfamiliar with this sort of thing might find it amusing to program (1.8) on his computer, starting with $\theta_o = 1/7$. The sequence predicted by (1.8) is simply 1/7, 2/7, 4/7, 1/7, 2/7, 4/7, 1/7, ...,i.e., a "3-cycle." But this is <u>not</u> what is found by computer iteration, at least not if the iterations are carried out far enough.) For this reason it has been said that "Chaos will beat any computer!"

This example suggests that detailed, long-time predictions about a chaotic system are impossible in practical terms because (a) we don't know initial conditions with infinite precision, and (b) we can't handle an infinite string of digits in our computations. This practical indeterminability has obvious "philosophical" implications, some of which we will consider later on.

Chaos is often defined qualitatively as a more or less "random" or disorderly behavior that is <u>intrinsic</u> to a system and not due to any externally imposed "noise." That is, the chaotic behavior is described by completely <u>deterministic</u> equations of motion. In the Russian literature this intrinsic chaos is called <u>stochasticity</u>. Our goals are (a) to better understand this deterministic chaos, (b) to see how systems make the transition from orderly to chaotic behavior, (c) to study examples of how chaos arises in the interaction of light and matter, and (d) to consider some important aspects of "quantum chaos."

Aside from things like modular arithmetic or piecewise linearity, chaotic behavior is associated with <u>nonlinear</u> systems. For this reason we will devote the next section to a very brief introduction to nonlinear differential equations, emphasizing how they differ from linear differential equations. Of course the whole area of laser-matter interactions is replete with nonlinearities, and during these lectures we will see that even the simplest models in the field can be chaotic.

2. Nonlinearity

Any set of ordinary differential equations (ODE) can in principle be written as an autonomous set of equations - "autonomous" meaning there is no explicit dependence on the independent variable t. This is done by solving for t and differentiating once.

Example: $\ddot{x} + x = e^{-t}$ (non-autonomous)

$t = -\log(\ddot{x} + x)$

$1 = -\, d/dt \log(\ddot{x} + x) = -\,(\dddot{x} + \dot{x})/(\ddot{x} + x)$

$\dddot{x} + \ddot{x} + \dot{x} + x = 0$ (autonomous)

This example illustrates the fact than an nth order non-autonomous ODE is equivalent to an autonomous equation of order n + 1.

Furthermore an nth order autonomous ODE is equivalent to a set of n first-order ODE.

Example: $\dddot{x} + \ddot{x} + \dot{x} + x = 0$

Let $y_1 = x$, $y_2 = \dot{x}$, $y_3 = \ddot{x}$

Then the original ODE is equivalent to the three first-order ODE

$\dot{y}_1 = y_2$, $\dot{y}_2 = y_3$, $\dot{y}_3 = -\,(y_1 + y_2 + y_3)$

Thus any ODE, or set of ODE, is equivalent to a set of first-order ODE of the form

$$\dot{y}_j = F_j(y_1, y_2, \ldots, y_N)\ , \quad j = 1,2,3,\ldots,N \tag{2.1}$$

A solution $(y_1(t), y_2(t), \ldots, y_N(t))$ defines a trajectory in the N-dimensional phase space of (2.1). For a Hamiltonian system of n degrees of freedom, this phase space is the familiar 2n-dimensional phase space $(q_1, q_2, \ldots, q_n, p_1, p_2, \ldots, p_n)$. The definition of phase space based on (2.1), of course, applies regardless of whether the system is Hamiltonian.

We will assume the functions F_j in (2.1), as well as all their first derivatives $\delta F_j/\delta y_i$, to be bounded, continuous functions of their arguments. The existence and uniqueness theorem for ODE then ensures that any initial condition $(y_1(0), y_2(0), \ldots, y_N(0))$ - that is, any point in phase space - gives rise to a unique trajectory passing through that point. Another way to say this is that trajectories in phase space do not cross. This follows easily from geometrical considerations: (2.1) defines a unique tangent ("velocity") vector $(\dot{y}_1(t), \dot{y}_2(t), \ldots, \dot{y}_N(t))$ at every point on a trajectory, whereas if two trajectories did cross at some point this velocity would not be unique at the point of crossing.

Nonlinear systems can have certain properties that linear systems cannot have. For instance, it can be proved that a linear ODE can have a singular solution only at points where the coefficient functions themselves are singular. These are called fixed singularities, because they are independent of the initial conditions. But nonlinear ODE can have spontaneous and movable singularities.

Example: $\dot{y} = y/(t - 1)$ (linear ODE)

Coefficient function $1/(t - 1)$ has singular point $t = 1$

Solution $y(t) = 1/(1 - t)$ for $y(0) = 1$ has singular point $t = 1$

Solution $y(t) = 2/(1 - t)$ for $y(0) = 2$ also has singular point $t = 1$

Singular point is fixed, independent of initial condition

Example: $\dot{y} = y^2$ (nonlinear ODE)

No singularity for finite y in equation, but solution $y(t) = 1/(1 - t)$ for $y(0) = 1$ has a spontaneous singularity at $t = 1$

Solution $y(t) = 2/(1 - 2t)$ for $y(0) = 2$ has a spontaneous singularity at $t = 1/2$. The singularity of the solution is thus movable, depending on the initial condition.

It is rarely possible to find analytic solutions to nonlinear ODE. However, it is often possible to understand certain features of the solution just by a fixed-point analysis. A <u>fixed point</u> $(y_1^*, y_2^*, \ldots, y_N^*)$ of the system (2.1) is defined by the equations

$$F_j(y_1^*, y_2^*, \ldots, y_N^*) = 0, \qquad j = 1, 2, \ldots, N \tag{2.2}$$

In other words, a fixed point is a point in phase space at which the velocity vector $(\dot{y}_1, \dot{y}_2, \ldots, \dot{y}_N)$ vanishes. (Fixed points are also called critical points or equilibrium points.) For fixed-point analysis it is necessary to know whether the fixed points of a system are stable against small perturbations.

Consider the simple example of the harmonic oscillator defined by the Hamiltonian $H = (1/2)(p^2 + q^2)$. In this case the phase space is the plane (q,p). The equations of motion are $\dot{p} = -q$, $\dot{q} = p$, and so obviously the only fixed point is $(q^*, p^*) = 0$, corresponding to the oscillator at rest. If we perturb the solution about this fixed point, so that $q \to q^* + \epsilon_1$ and $p \to p^* + \epsilon_2$, we find that $\dot{\epsilon}_1 = \epsilon_2$ and $\dot{\epsilon}_2 = -\epsilon_1$, so that $\ddot{\epsilon}_1 + \epsilon_1 = \ddot{\epsilon}_2 + \epsilon_2 = 0$. Therefore the perturbations about the fixed point produce oscillations about the fixed point. If we start out close to the fixed point we remain close to it, and so the fixed point at the origin is called a <u>stable</u> fixed point. In fact, since $p^2 + q^2$ = constant, all trajectories in the neighborhood of the fixed point are closed loops. (In this example the loops are circles) The fixed point (0,0) is therefore called a <u>center</u>. Using Hamilton's equations, it may be shown that any stable fixed point of a Hamiltonian system is a center.

Consider next the less trivial example of the <u>van der Pol equation</u>:

$$\ddot{x} + b(x^2 - 1)\dot{x} + x = 0, \qquad b > 0 \tag{2.3}$$

With $y_1 = x$ and $y_2 = \dot{x}$, (2.3) is equivalent to the first-order system $\dot{y}_1 = y_2$, $\dot{y}_2 = -y_1 - b(y_1^2 - 1)y_2$. Again there is only one fixed point, $(y_1^*, y_2^*) = (0,0)$. To determine whether this fixed point is stable against small perturbations, we let $y_1 = y_1^* + \epsilon_1$ and $y_2 = y_2^* + \epsilon_2$, substitute into the equations for $\dot{y}_1$ and $\dot{y}_2$, and neglect powers of ϵ_1 and ϵ_2 higher than the first, because we are assuming small perturbations. This usual procedure of linear stability analysis leads in this example to the (linear) ODE

$$\dot{\epsilon}_1 = \epsilon_2 \tag{2.4a}$$

$$\dot{\epsilon}_2 = -\epsilon_1 - 2by_1^*\epsilon_1 y_2^* - b(y_2^{*2} - 1)\epsilon_2 = -\epsilon_1 + b\epsilon_2 \tag{2.4b}$$

It follows that $\ddot{\epsilon}_1 - b\dot{\epsilon}_1 + \epsilon_1 = \ddot{\epsilon}_2 - b\dot{\epsilon}_2 + \epsilon_2 = 0$ and, since $b > 0$, the perturbations grow exponentially with time. The fixed point (0,0) of the van der Pol equation for $b > 0$ is therefore <u>unstable</u>: all trajectories starting near the fixed point move away from it.

On the other hand it is not difficult to see from the van der Pol equation that trajectories starting out far from the origin will move toward it. We therefore expect that there exists at least one trajectory that neither goes toward nor away from the origin, but rather forms a closed loop about the origin. This is indeed the case. In fact it turns out that all initial conditions for the van der Pol equation give rise to trajectories that eventually settle on a closed loop about the origin. This is illustrated by the results shown in Figure 2.1, which were obtained by numerical integration of the van der Pol equation for the case $b = 0.1$. For obvious reasons the closed loop in Figure 2.1 may be called an <u>attractor</u>.

The closed-loop trajectory of Figure 2.1 is called a <u>limit</u>

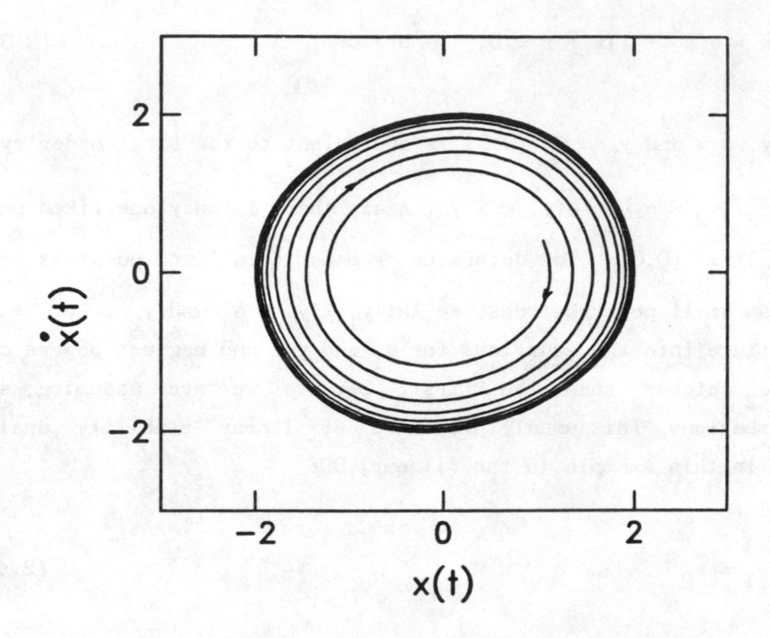

Figure 2.1 Limit cycle of the van der Pol equation.

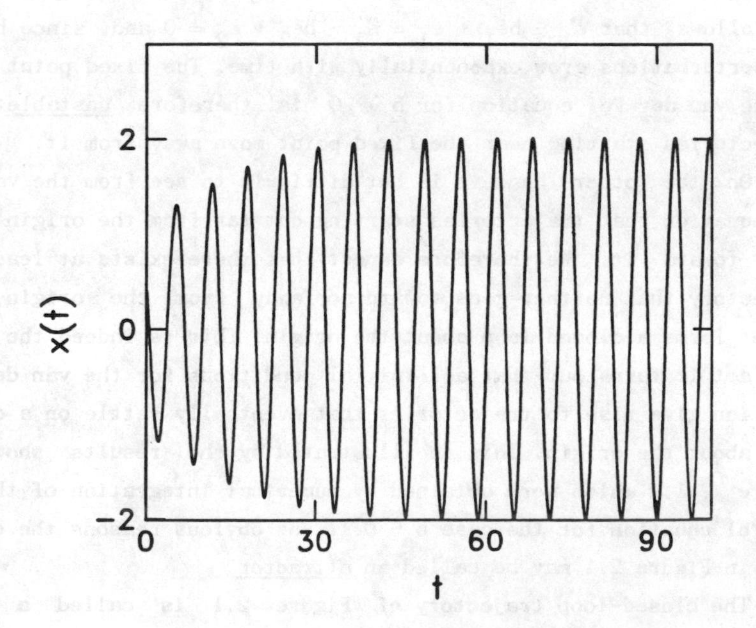

Figure 2.2 x(t) for the limit cycle of Figure 2.1.

cycle. A limit cycle is a closed, periodic trajectory, "isolated" in the sense that no nearby trajectory is also closed. The closed loops of the harmonic oscillator, for instance, are obviously not isolated, because we get many nearby closed loops just by changing the initial conditions slightly. The periodic character of a limit cycle is clear from the fact that it is a closed loop in phase space. In Figure 2.2 we plot x(t) for the case shown in Figure 2.1. After initial transients die out, x(t) settles into periodic oscillation.

Limit cycles appear only in nonlinear, dissipative systems, i.e., nonlinear systems with frictional forces. Like fixed points, limit cycles may be stable or unstable. Within its "basin of attraction" in phase space a stable limit cycle is independent of initial conditions, and is determined only by the parameters in the equations. A stable limit cycle may be thought of as an equilibrium state of oscillation of a nonlinear system, just as a stable fixed point represents a stationary equilibrium. There are no general theorems to tell us whether a limit cycle will exist for a given system, although some definite statements can be made for systems whose phase space is the plane (N = 2 in (2.1)). Thus Bendixson's Negative Criterion states that if $\delta F_1/\delta y_1 + \delta F_2/\delta y_2$ is of one sign in a simply-connected ("no holes") domain, there are no closed loops. And it is known that a limit cycle cannot enclose a region of phase space containing no fixed points.

Limit cycles obviously have Fourier spectra of the type sketched in Figure 1.1, and therefore represent a form of regular behavior. The Poincare'- Bendixson theorem for systems of the form (2.1) with N = 2 says that any bounded trajectory can either approach a fixed point as $t \to \infty$ or a closed curve, which may be a limit cycle. This is basically a consequence of the fact that trajectories do not cross. For N = 1 this non-crossing condition implies that any trajectory must approach a fixed point or go to $\pm\infty$ as $t \to \infty$. In other words, for $N \leq 2$ we can only have regular, quasiperiodic behavior, never chaos. Chaos in dynamical systems of the type (2.1) can only arise if $N \geq 3$, i.e., if the phase space is three-dimensional or larger.

Furthermore a linear system of the form (2.1) is, in principle, exactly solvable, with quasiperiodic (bounded) steady-state behavior, and so chaotic behavior can only occur in nonlinear sytems with three or more dimensions in phase space.

Obviously there are many nonlinear systems in laser physics with $N \geq 3$, and we will see many examples of chaotic behavior in such systems. But first we want to probe deeper into deterministic chaos, and in particular to investigate different pathways by which a system can go from quasiperiodic to chaotic behavior. We begin in the next section with one of the best known of such pathways, the period doubling route to chaos.

3. Period Doubling to Chaos

We have used the logistic map (1.7) with $\lambda = 1$ to illustrate the concept of very sensitive dependence on initial conditions (chaos). Now we want to see what happens for values of $\lambda \neq 1$, and in particular to see what happens as the "knob" λ is swept from 0 to 1. We will find that as λ is varied there is a transition from regular to chaotic behavior, and furthermore that this transition occurs according to a "universal" period doubling route to chaos.

We will first take an "experimental" approach and just see what happens as the knob is turned. In the next few sections we will then characterize the chaotic behavior in terms of Lyapunov exponents, power spectra, and decay of correlations. These are essential tools in the study of chaos, not only for discrete mappings but for differential equations as well. After these introductory discussions we will take up the notion of Feigenbaum universality, and we will see why the period doubling route to chaos appears so often in both numerical and laboratory experiments.

To help us understand the results of numerical experiments we define for the discrete mapping $x_{n+1} = f(x_n)$ a fixed point x^* as a solution to the equation

$$x^* = f(x^*) \tag{3.1}$$

A fixed point is just mapped into itself, as in the case of a fixed point of an ODE. As in the latter, it is important to know whether a fixed point is stable against small perturbations. If x^* is changed to $x^* + \epsilon$, then $f(x^*)$ is changed to

$$f(x^* + \epsilon) \cong f(x^*) + \epsilon f'(x^*) \tag{3.2}$$

which means that x^* is a stable fixed point if $|f'(x^*)| < 1$, unstable if $|f'(x^*)| > 1$. (If $|f'(x^*)| = 1$ the fixed point is said to be "marginally stable.")

For the logistic map, $f(x) = 4\lambda x(1-x)$, and the fixed points are the solutions of the equation

$$x^* = 4\lambda x^*(1 - x^*) \tag{3.3}$$

Thus there are two fixed points: $x^* = 0$ and $x^* = 1 - 1/4\lambda$. Since $f'(x) = 4\lambda(1-2x)$, we conclude that $x^* = 0$ is a stable fixed point when $\lambda < 1/4$, whereas $x^* = 1 - 1/4\lambda$ is stable for $1/4 < \lambda < 3/4$. With these results in hand, let us now begin our numerical experiments. The reader is encouraged to confirm the following results using a computer or just a pocket calculator (preferably programmable).

For $0 \leq \lambda \leq 1/4$ we find that whatever x we start out with between 0 and 1, the sequence of iterates $\{x_n\}$ generated by the logistic map (1.7) converges to $x^* = 0$. The stable fixed point $x^* = 0$ for $\lambda < 1/4$ is therefore an <u>attractor</u>. Similarly for $1/4 < \lambda < 3/4$ we find that, regardless of the value of x_0 ($\neq 0,1$), the sequence $\{x_n\}$ converges to the stable fixed point $x^* = 1 - 1/4\lambda$. Thus the stable fixed point $x^* = 1 - 1/4\lambda$ is an attractor for $1/4 < \lambda < 3/4$.

It is interesting now to see what happens for $3/4 < \lambda \leq 1$, since for such values of λ the logistic map has no fixed points. For λ = 0.76 we find, after some initial transients that depend on the initial seed x_0, that the sequence $\{x_n\}$ settles into a two-cycle oscillation: {.7306, .5984, .7306, .5984, ...}. This two-cycle is independent of x_0 and is thus an attractor of period two. If we let $x_1^* = .7306$ and $x_2^* = .5984$ we can write

$$x_2^* = f(x_1^*) = f(f(x_2^*)) \equiv f^2(x_2^*) \tag{3.4a}$$

$$x_1^* = f(x_2^*) = f(f(x_1^*)) \equiv f^2(x_1^*) \tag{3.4b}$$

where

$$f^2(x) = f(f(x)) = 16\lambda^2[x - (4\lambda+1)x^2 + 8\lambda x^3 - 4\lambda x^4] \tag{3.5}$$

is called the second iterate of f. According to (3.4), x_1^* and x_2^* are both (stable) fixed points of the second iterate mapping $x_{n+1} = f^2(x_n)$; this is easily confirmed by solving for the fixed points of the second iterate map and performing a linear stability analysis.

Thus, when λ is large enough that the fixed point $x^* = 1 - 1/4\lambda$ of the logistic map is unstable, a <u>period doubling bifurcation</u> occurs in which x^* "gives birth" to two stable fixed points of the second iterate map. The period has doubled in the sense that the new attractor for the map $x_{n+1} = f(x_n)$ is a two-cycle instead of a one-cycle.

<u>Exercise</u>: Show that the fixed points of the second iterate map become unstable when $\lambda = 1/4\ (1 + \sqrt{6}) \cong 0.862372...$

When the two fixed points of f^2 become unstable, they each give birth to two new fixed points of f^4; then we have four fixed points of the fourth iterate mapping $x_{n+1} = f^4(x_n)$, corresponding to a four-cycle, or an attractor of period four, of the logistic map. As the λ knob is turned further we find more and more period doublings, and the range of λ values associated with a 2^n cycle gets rapidly narrower as n increases. This is clear from Figure 3.1.

Let λ_n be the value of λ at which the nth period doubling bifurcation occurs. Feigenbaum [3.1] has established that the sequence $\{\lambda_n\}$ converges geometrically at a rate given by

$$\delta = \lim_{n\to\infty} (\lambda_n - \lambda_{n-1})/(\lambda_{n+1} - \lambda_n) = 4.6692016091\ldots \qquad (3.6)$$

The rapid convergence of the sequence of λ_n values (Figure 3.1) allows us to estimate λ_{n+1} fairly accurately from λ_n and λ_{n-1}. The sequence $\{\lambda_n\}$ has the limit point $\lambda_\infty = \lambda^* = 0.8924864\ldots$, beyond which the sequence $\{x_n\}$ of iterates of the logistic map appears to be a chaotic sequence without any periodicities (except for certain "windows" of λ values discussed later. For $\lambda = 0.959$, for instance, we find the 3-cycle $\{.9588, .1515, .4931, \ldots\}$).

As we will see later, Feigenbaum's δ is universal in the sense that (3.6) - and the period doubling route to chaos - applies to all maps with quadratic maxima, the logistic map being just one example. (We discuss later the conditions on the mapping function f for Feigenbaum universality.)

Exercise: Investigate experimentally the mapping $x_{n+1} = \lambda \sin \pi x_n$, with x_0 and λ between 0 and 1, as the λ knob is varied.

It is perhaps worth noting at this point that attractors - such as the stable n-cycles we find for the logistic map - occur only in dissipative systems. Since it is not immediately obvious that the logistic map has anything to do with dissipation, let us consider a dissipative mechanical system that leads directly to the logistic map. We consider the one-dimensional, periodically kicked system with

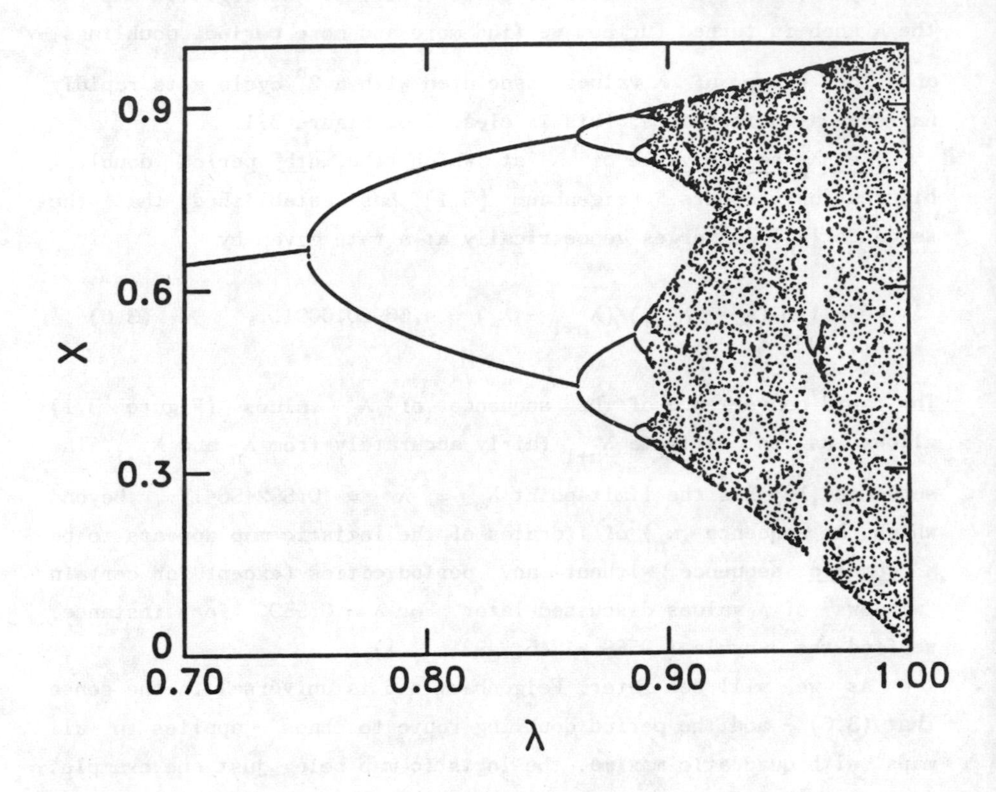

<u>Figure 3.1</u> Steady-state iterates of the logistic map as a function of the knob λ.

equations of motion

$$\dot{x} = p/m \tag{3.7a}$$

$$\dot{p} = -\beta p + A(x)\sum_{n=-\infty}^{\infty} \delta(t/T - n) \tag{3.7b}$$

which differs from the equations (1.5) only by the presence of the frictional term $-\beta p$ in the momentum equation. From (3.7) we have upon integration the mapping

$$x_{n+1} = x_n + (1 - e^{-\beta T})/m\beta \, [p_n + TA(x_n)] \tag{3.8a}$$

$$p_{n+1} = e^{-\beta T}[p_n + TA(x_n)] \tag{3.8b}$$

which reduces to the mapping (1.6) in the limit $\beta \to 0$ of no friction. Now let $A(x) = (m\beta/T)[4\lambda x(1 - x) - x]$ and consider the limit $\beta \to \infty$ of strong damping. In this limit (3.8) gives $p_{n+1} \to 0$ and $x_{n+1} = 4\lambda x_n(1 - x_n)$, and so we can regard the logistic map as the strong-damping limit of a periodically kicked system.

4. <u>Lyapunov</u> <u>Exponent</u>

In Section 1 we introduced the notion of the Lyapunov exponent as a measure of sensitivity to initial conditions. We computed this exponent for the case $\lambda = 1$ of the logistic map, and from the fact that this exponent is $\log 2 > 0$, we concluded that the sequence $\{x_n\}$ generated by the logistic map for $\lambda = 1$ is chaotic. Let us now see how to compute the Lyapunov exponent for a general one-dimensional mapping $x_{n+1} = f(x_n)$ on an interval of the real line.

We want to know how a small change in the initial seed x_0 affects x_n. Since

$$x_n = f^n(x_0) \tag{4.1}$$

a small change in x_0 by ϵ_o results in a change in x_n by

$$\epsilon_n = \epsilon_0 f^{n,}(x_0) \tag{4.2}$$

Now from the chain rule for derivatives we have $f^{2,}(x_0) = f'(x_1)f'(x_0)$ and in general

$$f^{n,}(x_0) = \prod_{j=0}^{n-1} f'(x_j) \tag{4.3}$$

It then follows from (4.2) that

$$|\epsilon_n| = |\epsilon_o| \prod_{j=0}^{n-1} |f'(x_j)| = |\epsilon_0| e^{n\chi_n} \tag{4.4}$$

where

$$\chi_n = (1/n)\log \prod_{j=0}^{n-1} |f'(x_j)| = (1/n) \sum_{j=0}^{n-1} \log|f'(x_j)| \tag{4.5}$$

The limit

$$\chi = \lim_{n\to\infty} \chi_n = \lim_{n\to\infty} (1/n) \sum_{j=0}^{n-1} \log|f'(x_j)| \tag{4.6}$$

defines the Lyapunov characteristic exponent (LCE) of the map. We see from (4.4) that if $\chi > 0$ we have exponential sensitivity to initial conditions, i.e., chaos. (We exclude unbounded systems like $x_{n+1} = 2x_n$, which have a trivial sort of exponential sensitivity to initial conditions.) If $\chi \leq 0$, on the other hand, then the iterates of the map are not very sensitive to the value of x_0, and we have regular (orderly) behavior. A computation of the LCE therefore tells us whether we have chaotic or regular behavior.

It is not difficult to show that if $\chi > 0$ then the sequence $\{x_n\}$ is aperiodic. Suppose there exists a stable n-cycle. Then the nth iterate map $x_{n+1} = f^n(x_n)$ has n stable fixed points x_i^*, i = 0, 1, ..., n − 1, so that $|f^{n\prime}(x_i^*)| < 1$ and

$$\chi = \lim_{m\to\infty} (1/m)\log \prod_{i=0}^{m} |f'(x_i)| = \lim_{m\to\infty} (1/m)\log \left(\prod_{i=0}^{n-1} |f'(x_i^*)|\right)^{m/n}$$

$$= (1/n)\log \prod_{i=0}^{n-1} |f'(x_i^*)| = (1/n)\log|f^{n\prime}(x_0^*)| < 0 \qquad (4.7)$$

Thus a stable n-cycle implies a negative LCE, and so a positive LCE implies the absence of any stable n-cycle. In other words, chaos implies the absence of any stable periodic motion.

It is found in numerical experiments that the limit (4.6) exists and is independent of the choice of x_0 (Figure 4.1), although a large value for n must usually be used to achieve convergence in these computations. Figure 4.1 shows the convergence of χ_n to $\chi = \log 2$ for the case $\lambda = 1$ of the logistic map. This case is atypical in that χ can be calculated without recourse to numerical computation, as we showed in Section 1; in general the LCE must be computed numerically, and this is usually a rather lengthy computation.

For the logistic map it is found that $\chi < 0$ for $\lambda < \lambda^*$, except at those values of λ at which the period doubling bifurcations occur, where $\chi = 0$. This is consistent with our assertion earlier that $\lambda < \lambda^*$ defines the regime of regular, periodic behavior of the map. For $\lambda > \lambda^*$, however, χ is mainly positive except for those values of λ at which periodic windows occur within this "chaotic regime." Plots of χ versus λ for the logistic map may be found in the literature. [4.1]

Frequently (4.6) is written in the form

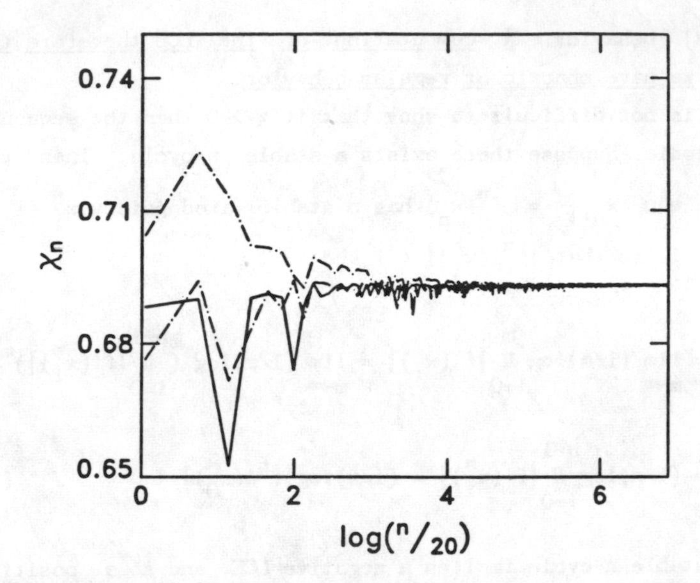

Figure 4.1 Convergence of the Lyapunov exponent to log2 for the logistic map with $\lambda = 1$, using three different values of the initial seed.

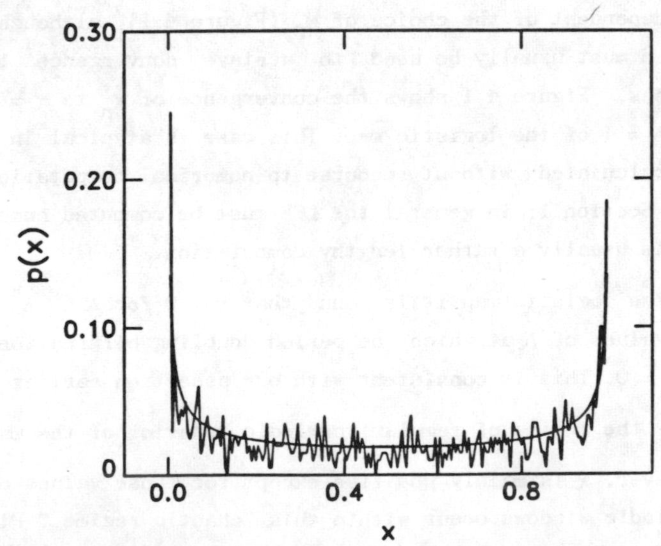

Figure 4.2 Distribution of iterates for the logistic map with $\lambda = 1$, compared with the theoretical distribution function $p(x) = [\pi\sqrt{y(1-y)}]^{-1}$.

$$\chi = \int dx p(x) \log|f'(x)| \tag{4.8}$$

where p(x) is the "probability distribution," or invariant measure, associated with the map. In other words, p(x)dx is the fraction of iterates lying in the interval [x, x+dx], and as such may be defined formally as

$$p(x) = \lim_{N\to\infty} (1/N) \sum_{n=0}^{N} \delta(x - x_n) = \lim_{N\to\infty} (1/N) \sum_{n=0}^{N} \delta[x - f^n(x_0)] \tag{4.9}$$

An equation for p(x) may be obtained as follows. The initial "distribution" is $\delta(x - x_0)$, and after one iteration this is mapped into $\delta[x - f(x_0)]$. In general a distribution $p_n(x)$ goes into

$$p_{n+1}(y) = \int dx p_n(x) \delta[y - f(x)] \tag{4.10}$$

after one iteration. The "steady-state," invariant distribution function p(x) must be unchanged by the mapping, and therefore satisfies the so-called Frobenius-Perron integral equation

$$p(y) = \int dx p(x) \delta[y - f(x)] \tag{4.11}$$

This equation can be solved exactly only in very special cases.

Exercise: For the logistic map with $\lambda = 1$ show that the Frobenius-Perron equation for the invariant measure is

$$4\sqrt{1-y}\, p(y) = p(\tfrac{1}{2} + \tfrac{1}{2}\sqrt{1-y}) + p(\tfrac{1}{2} - \tfrac{1}{2}\sqrt{1-y})$$

and show that

$$p(y) = [\pi\sqrt{y(1-y)}]^{-1}$$

is a (normalized) solution of this equation. This solution may be verified by numerical experiment. (Figure 4.2) Using (4.8), show that

$$\chi = \log 2.$$

Exercise: Consider the Bernoulli shift $x_{n+1} = 2x_n$ (mod 1). Show that the map in this case is ergodic, i.e., it covers uniformly the interval [0,1].

A positive LCE may be taken as a definition of chaos, whereas a negative or zero LCE is a signature of quasiperiodicity. A system of dimension greater than one has a spectrum of Lyapunov exponents, one for each dimension. If one of these LCE is positive, the system is chaotic. This means, loosely speaking, that in a certain (changing) direction there is an exponential stretching apart of initially close trajectories. Of course more than one of the LCE may be positive. For quasiperiodic motion, however, all the LCE are negative or zero.

It is not difficult to generalize the definition (4.6) beyond one dimension. Consider as an example the two-dimensional map (1.2). Let $x_0, y_0 \rightarrow x_0 + \epsilon_0, y_0 + \delta_0$. Then $x_1, y_1 \rightarrow x_1 + \epsilon_1, y_1 + \delta_1$, where (for small perturbations ϵ_0 and δ_0)

$$\begin{bmatrix} \epsilon_1 \\ \delta_1 \end{bmatrix} = J(x_0, y_0) \begin{bmatrix} \epsilon_0 \\ \delta_0 \end{bmatrix} \tag{4.12}$$

and

$$J(x,y) = \begin{bmatrix} \delta f/\delta x & \delta f/\delta y \\ \delta g/\delta x & \delta g/\delta y \end{bmatrix} \tag{4.13}$$

is the Jacobian matrix for the transformation. Then after n iterations the perturbations ϵ_n, δ_n in x_n, y_n are given by

$$\begin{bmatrix} \epsilon_n \\ \delta_n \end{bmatrix} = J(n) \begin{bmatrix} \epsilon_0 \\ \delta_0 \end{bmatrix} \tag{4.14}$$

where the matrix product

$$J(n) \equiv J(x_{n-1},y_{n-1})J(x_{n-2},y_{n-2})\ldots J(x_0,y_0) \tag{4.15}$$

In a coordinate system in which J(n) is diagonal we have

$$|\epsilon'_n| = |\lambda_1(n)\epsilon_0| = |\epsilon_0|e^{n\chi_n^{(1)}} \tag{4.16a}$$

$$|\delta'_n| = |\lambda_2(n)\delta_0| = |\delta_0|e^{n\chi_n^{(2)}} \tag{4.16b}$$

where $\lambda_1(n)$ and $\lambda_2(n)$ are the eigenvalues of the matrix J(n). The limits

$$\lambda^{(1,2)} = \lim_{n\to\infty} [\text{magnitude of eigenvalues of } J(n)]^{1/n} \tag{4.17}$$

are called the Lyapunov numbers of the mapping, and the Lyapunov characteristic exponents are defined by

$$\chi^{(1,2)} = \log\lambda^{(1,2)} \tag{4.18}$$

In order to determine whether a system is chaotic it is only necessary to determine the largest LCE; if this is positive, the system is chaotic. Fortunately the largest LCE is generally much easier to compute than the full spectrum of exponents. To see why this is so, consider the example above of a two-dimensional mapping. Suppose we iterate the maps (1.2) and (4.14) simultaneously, and compute at each iteration the quantity

$$d(n) = (1/n)\log[\epsilon_n^2 + \delta_n^2]^{1/2} \tag{4.19}$$

For large n this is the same as

$$d(n) = (1/n)\log[\mathrm{Max}(|\epsilon_n'|, |\delta_n'|)] \tag{4.20}$$

because only the direction of greatest stretching survives in the limit $n \to \infty$. Thus

$$\mathrm{Max}[\chi^{(1,2)}] = \lim_{n\to\infty} (1/n)\log[\epsilon_n^2 + \delta_n^2]^{1/2} \tag{4.21}$$

This procedure for computing the largest LCE is obviously much easier than having to compute the whole spectrum of LCE's; in particular, no matrix diagonalizations are required. The same procedure is easily extended to N-dimensional maps and, as we will see later, is applicable also to systems of ODE.

5. Power Spectra

In Section 1 we distinguished regular and chaotic behavior in terms of power spectra, and in our preceding discussion of Lyapunov exponents we showed that chaos implies there is no stable periodic cycle. Obviously spectral analysis is an essential tool in characterizing the temporal behavior of a system, and in this section we will discuss power spectra in more detail.

First let us consider in a bit more detail the relevance of Fourier spectra to the study of chaos. According to the terminology of Section 1, the temporal behavior of a function y(t) is quasiperiodic if its Fourier transform consists of sharp spikes, i.e., if

$$y(t) = \sum_{j=1}^{n} c_j e^{i\omega_j t} \tag{5.1}$$

The characteristic "recurrence" feature of quasiperiodicity may be stated formally as follows: for any $\epsilon > 0$ there exists a $T(\epsilon)$ such that any interval of length $T(\epsilon)$ of the real line contains at least one point t' such that $|y(t) - y(t')| < \epsilon$ for any t. [5.1] Given $y(t)$, we can always find a t' such that $y(t')$ is as close to $y(t)$ as we wish, and there are an infinite number of such times t'. Periodic functions are quasiperiodic, but of course quasiperiodicity does not imply periodicity. Thus the function

$$y(t) = c_1\cos\omega_1 t + c_2\cos\omega_2 t \tag{5.2}$$

is quasiperiodic, but it is not periodic unless ω_1 and ω_2 are commensurate frequencies, i.e., unless ω_1/ω_2 is a rational number.

Quasiperiodic motion is regular. That is, quasiperiodicity, like periodicity, is associated with a negative or zero Lyapunov exponent. Quasiperiodic motion can certainly look very complicated and seemingly irregular, but it cannot be truly chaotic in the sense of exponential sensitivity to initial conditions. In particular, the difference between two quasiperiodic trajectories is itself quasiperiodic, and so we cannot have the exponential separation of initially close trajectories that is the hallmark of chaos.

Since quasiperiodicity implies order, it follows that chaos implies non-quasiperiodic motion. Thus <u>chaotic motion does not have a purely discrete Fourier spectrum as in (5.1), but must have a broadband, continuous component in its spectrum, as in Figure 1.2.</u> Fourier analysis is therefore a very useful tool in distinguishing regular from chaotic motion, and furthermore it is generally much cheaper computationally than Lyapunov exponents.

In numerical computations we are dealing with discrete Fourier transforms of the form

$$Y(f_k) = \sum_{\ell=0}^{N-1} y(t_\ell)e^{-2\pi i\ell k/N} \quad , \ k = 0,1,\ldots,N-1 \tag{5.3}$$

where $t_\ell = \ell\Delta t$, $f_k = k/(N\Delta t)$, and N is the number of points, separated by Δt, sampled from the time series $y(t)$. The separation Δt of sampled points determines the maximum frequency component of $y(t)$ that can be computed with (5.3), and the total time span $N\Delta t$ determines the minimum frequency. For a discrete map the $y(t_\ell)$ are just iterates produced by the mapping, as in our examples below. In systems of ODE the $y(t_\ell)$ are obtained by numerical integration, and Δt may or may not be the same as the integration step size, depending on the range of frequencies of interest. We will refer to $|Y|^2$ as the <u>power spectrum</u> of y.

In practice it is convenient to use the Fast Fourier Transform (FFT) algorithm for the evaluation of (5.3). [5.2] To get some idea of what the FFT is, let us write (5.3) as

$$Y_k = \sum_{\ell=0}^{N-1} y_\ell W^{k\ell} , \quad W \equiv e^{-2\pi i/N} \tag{5.4}$$

For N = 8, for instance, we need W^0 through W^{49}. However, there is a lot of redundancy involved in the evaluation of the $W^{k\ell}$, because they can all be reduced to one of the 8 W's from W^0 to W^7, and furthermore $W^7 = -W^3$, $W^6 = -W^2$, $W^5 = -W^1$, and $W^4 = -W^0$. The FFT is basically just an algorithm that takes advantage of these redundancies to reduce substantially the number of operations required in the evaluation of (5.4). Useful FORTRAN listings are available in various places in the literature. [5.2]

To obtain an accurate FFT, a "cosine bell" or some other tapering function is applied to the sampled time series before the FFT is applied. This eliminates spurious frequency components associated with sharp ("diffraction") edges in the time series. For the cosine bell windowing, for example, the FFT is applied not to the time series y_ℓ itself, but rather to the time series

$$y'_\ell = (1/2)[1 - \cos(2\pi\ell/(N-1)]y_\ell , \quad \ell = 0,1,\ldots,N-1 \tag{5.5}$$

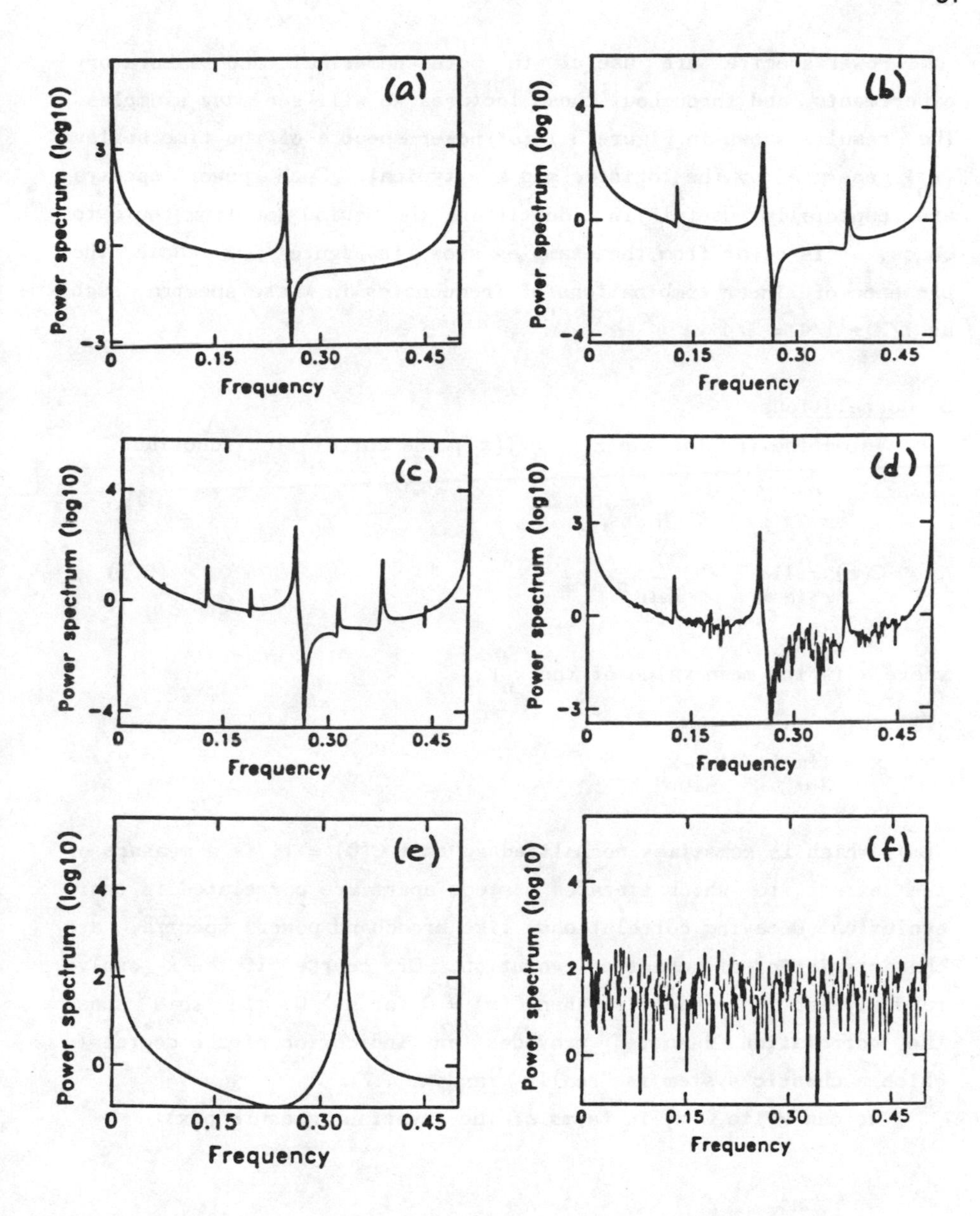

Figure 5.1 Power spectra obtained from the logistic map for (a) λ = .87, (b) λ = .89, (c) λ = .892, (d) λ = .893, (e) λ = .959, and (f) λ = 1.0

Power spectra are useful in both numerical and laboratory experiments, and throughout these lectures we will see many examples. The results shown in Figure 5.1 for power spectra of the time series $\{x_n\}$ generated by the logistic map are typical. Such power spectra are especially useful in identifying the period doubling route to chaos, as is clear from the examples shown in Figure 5.1. Note the presence of linear combinations of frequencies in these spectra, such as 1/2 + 1/4 = 3/4.

6. Correlations

We define for the map $x_{n+1} = f(x_n)$ the correlation function

$$C(m) = \lim_{N\to\infty} (1/N) \sum_{n=0}^{N} x_n x_{n+m} - \bar{x}^2 \tag{6.1}$$

where $\bar{x}$ is the mean value of the x_n:

$$\bar{x} = \lim_{N\to\infty} (1/N) \sum_{n=0}^{N} x_n \tag{6.2}$$

C(m), which is sometimes normalized so that C(0) = 1, is a measure of the extent to which iterates m steps apart are correlated in their evolution. Decaying correlations, like broadband power spectra, are characteristic of chaotic evolution. Of course if the x_n evolve randomly and independently, then C(m) = 0 for m > 0. In this sense the correlation function provides an indication of the degree to which a chaotic system is "really" random.

We can write C(m) in terms of the invariant measure p(x):

$$C(m) = \int_0^1 dx p(x) x f^m(x) - \left[\int_0^1 dx p(x) x\right]^2 \tag{6.3}$$

Consider, for instance, the correlation function C(1) for the logistic map with $\lambda = 1$. In this case $f(x) = 4x(1 - x)$ and we know that $p(x) = [\pi\sqrt{x(1 - x)}]^{-1}$, and so

$$C(1) = \frac{4}{\pi}\int_0^1 \frac{dx\, x^2(1 - x)}{\sqrt{x(1 - x)}} - \left[\frac{1}{\pi}\int_0^1 \frac{dx\, x}{\sqrt{x(1 - x)}}\right]^2 = 0 \quad (6.3)$$

whereas $C(0) = 1/8$.

It is easy to test these theoretical predictions by "experiment." Figure 6.1a shows numerical results for the correlation function C(m) in the case $\lambda = 1$ of the logistic map. These results confirm that the iterates of the logistic map are delta-correlated for $\lambda = 1$. In Figures 6.1b and 6.1c we show results for C(m) for the knob values $\lambda = .89$ and $\lambda = .92$, respectively. The oscillatory correlation function of Figure 6.1b is typical of quasiperiodic motion, whereas Figure 6.1c is typical of chaotic motion. The rapid decay of correlations shown in Figure 6.1a indicates that the $\lambda = 1$ case is very strongly chaotic. The more gentle decay shown in Figure 6.1c is perhaps more typical.

From the inverse Fourier transform

$$x_n = \sum_{k=0}^{N-1} X_k e^{2\pi ink/N} = \sum_{k=0}^{N-1} X_k^* e^{-2\pi ink/N} \quad (6.4)$$

we find that

$$(1/N)\sum_{n=0}^{N-1} x_n x_{n+m} = \sum_{k=0}^{N-1} |X_k|^2 e^{2\pi imk/N} \quad (6.5)$$

which allows us to relate power spectra and correlation functions by Fourier transformation:

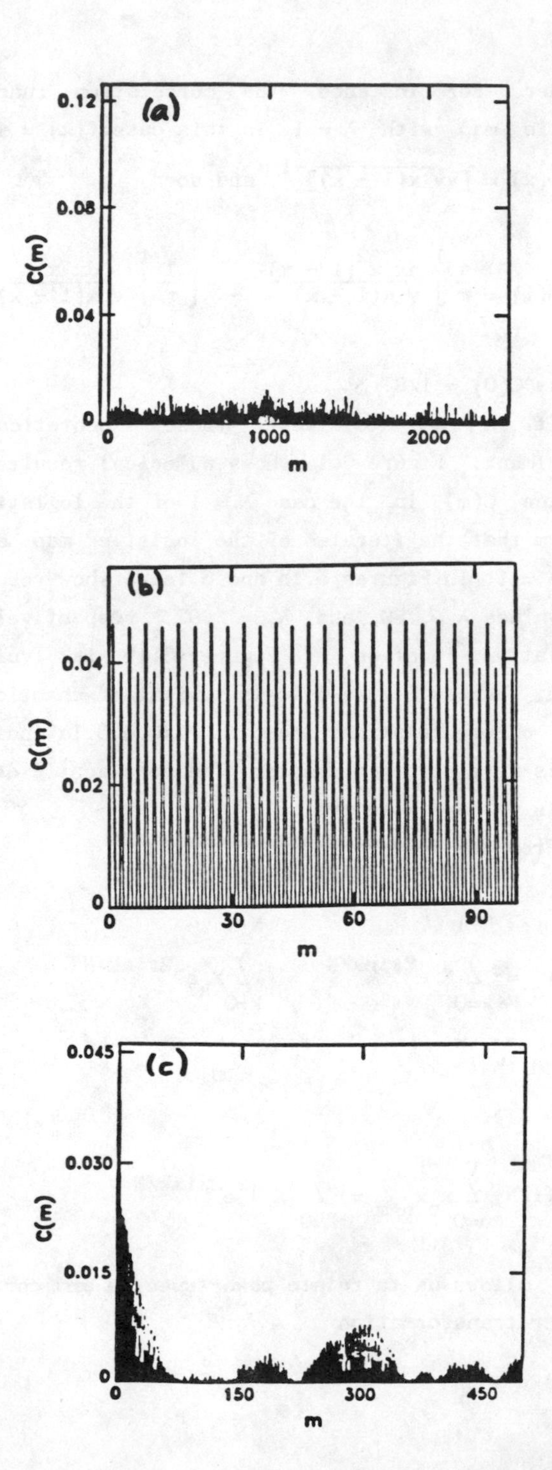

Figure 6.1 Correlation function C(m) for the logistic map with (a) $\lambda = 1.0$; (b) $\lambda = 0.89$; (c) $\lambda = 0.92$.

$$C(m) = \sum_{k=0}^{N-1} |X_k|^2 e^{2\pi imk/N} - \bar{x}^2 \tag{6.6a}$$

$$|X_k|^2 = (1/N) \sum_{m=0}^{N-1} C(m) e^{-2\pi ikm/N} + \bar{x}^2 \delta_{k,0} \tag{6.6b}$$

It is clear from these relations that quasiperiodicity implies that the correlation function C(m) is itself quasiperiodic, and furthermore that a decaying correlation function is associated with a broadband component in the power spectrum.

7. Remarks

Simple rules of evolution like the logistic mapping give chaotically varying outputs from perfectly prescribed inputs. The rules of evolution are perfectly deterministic, with no stochastic elements in either the equations of motion or the inputs. But how "random" is this chaos? How much is it like a game of coin flipping?

To address this question let us consider again the logistic map with $\lambda = 1$, which we know is equivalent to the Bernoulli shift $\theta_n = 2^n\theta_0 \pmod 1$. We can compare the iterates θ_n to the results of coin flipping by associating $0 < \theta_n < 1/2$ with heads (H) and $1/2 < \theta_n < 1$ with tails (T). As in Section 1 it is convenient to write the θ's in base 2, so that $\theta_n = .d_n d_{n+1} d_{n+2} \cdots$ with each d_i equal to either 0 or 1. Let us call θ_n "heads" if it lies in the left half of the unit interval, in which case $d_n = 0$, and "tails" if it is in the right half, in which case $d_n = 1$. Now suppose in tossing a coin we come up with some sequence of heads and tails like HTTHTHTT... It is easy to see that we can reproduce this same sequence from the Bernoulli shift by just choosing θ_0 in the right way, simply because $\theta_n = .d_n d_{n+1} d_{n+2} \cdots$ corresponds to heads or tails depending on whether $d_n = 0$ or 1, and so all we have to do is choose $\theta_0 = .d_0 d_1 d_2 \ldots d_n d_{n+1} d_{n+2} \cdots$ appropriately. (Recall that $\theta_1 = .d_1 d_2 d_3 \cdots$, $\theta_2 = .d_2 d_3 d_4 \cdots$, etc.) In other words, any arbitrary sequence of heads and tails corresponds to some choice of Θ_0.

Perhaps the coin flipping game involves some very complicated equations that reduce in some way to some sort of mapping that has chaotic behavior like the Bernoulli shift. To wit, are things we usually think of as "random" really examples of <u>deterministic</u> chaos, with underlying <u>deterministic</u> rules of evolution? Is the unpredictability just a consequence of extreme sensitivity to initial conditions? Very often devices like lasers and optical parametric oscillators have a "spiky," irregular sort of output. Is this deterministic chaos ?

Such questions are part of the reason for the great current interest in chaotic dynamics. However, it is probably not surprising that the large and complicated sets of equations needed to describe various "real world" phenomena can have very irregular and chaotic solutions. What <u>is</u> surprising to most people is that relatively simple-looking equations can have chaotic behavior, with solutions behaving as randomly as the toss of a coin. This is a relatively new idea, for which the major credit should probably go to E.N. Lorentz. [1.1] A more recent discovery is that there is a certain order in this chaos, as exemplified by the occurrence of certain "universal" routes to chaos. We next turn our attention to some universal features of the period doubling route to chaos.

But before going deeper into the subject, we would like to put things in a broader perspective by quoting from a paper by May:

"The elegant body of mathematical theory pertaining to linear systems (Fourier analysis, orthogonal functions, and so on), and its successful application to many fundamentally linear problems in the physical sciences, tends to dominate even moderately advanced University courses in mathematics and theoretical physics. The mathematical intuition so developed ill equips the student to confront the bizarre behavior exhibited by the simplest of discrete nonlinear systems, such as [the logistic map]. Yet such nonlinear systems are surely the rule, not the exception, outside the physical sciences.

"I would therefore urge that people be introduced to, say, [the logistic map] early in their mathematical education. This equation can be studied phenomenologically by iterating it on a calculator, or even by hand. Its study does not involve as much conceptual sophistication as does elementary calculus. Such study would greatly enrich the student's intuition about nonlinear systems.

"Not only in research, but also in the everyday world of politics and economics, we would all be better off if more people realized that simple nonlinear systems do not necessarily possess simple dynamical properties." [7.1]

And finally some words of Feynman are also worth quoting:

"If such variety is possible in a simple equation with only one parameter, how much more is possible with more complex equations! ... Unaware of the scope of simple equations, man has often concluded that nothing short of God, not mere equations, is required to explain the complexities of the world." [7.2]

8. <u>Feigenbaum Universality</u>

A quantitative theory of the period doubling sequence has been developed by Feigenbaum [3.1] and Collet and Eckmann [8.1] in terms of a renormalization procedure applicable to practically all unimodal (i.e., one hump) maps of interest. This theory yields, among other things, the number δ defined by equation (3.6). We will now present an <u>approximate</u> renormalization theory that yields an approximation to δ. [8.2] In the next section we will outline the exact renormalization theory.

We want to be fairly general and so we consider the general quadratic map

$$x_{n+1} = ax_n^2 + bx_n + c \qquad (8.1)$$

which can be linearly transformed to

$$x_{n+1} = x_n^2 + Ax_n = f(x_n) \tag{8.2}$$

Clearly then there is no generality lost in studying this map. Note that the logistic map $y_{n+1} = 4\lambda y_n(1-y_n)$ follows as a special case when we let

$$A = 4\lambda \quad \text{and} \quad x_n = -Ay_n \tag{8.3}$$

For generality we want to allow A to be positive or negative, which according to (8.3) implies the possibility of negative values for λ in the special case of the logistic map. To deal with negative λ's we note that the transformation

$$y_n = -(1 - 1/2\lambda)y'_n + (1 - 1/4\lambda) \tag{8.4}$$

yields the logistic map also for y'_n:

$$y'_{n+1} = 4\lambda' y'_n(1 - y'_n) \tag{8.5a}$$

where

$$\lambda' = 1/2 - \lambda \tag{8.5b}$$

All this says is that there is a mirror symmetry about $\lambda = 1/4$. Thus for both $\lambda = .80$ and $\lambda' = .50 - .80 = -.30$, for instance, there are stable two-cycles. This mirror symmetry is a general property of quadratic maps. In order to apply the results below to the special case of the logistic map when A is negative, we can use the formula (equations (8.3) and (8.5b))

$$\lambda = 1/2 - A/4 \tag{8.6a}$$

whereas when A is positive,

$$\lambda = A/4 \tag{8.6b}$$

We will consider the parameter range $-2 < A < 1$, for which (8.2) maps the interval $-1 < x < 1-A$ into itself. The only fixed point in this interval is $x^* = 0$, and it is stable for $-1 < A < 1$. At $A = -1$ this fixed point gives birth to a stable fixed point of the second iterate map f^2, i.e., to a two-cycle of f. The reason for this is the same as in the logistic map: as λ (or, in this case, A) is varied the humps in the second iterate map eventually allow for two stable fixed points (where $x^* = f^2(x^*)$ and $|f^{2\prime}(x^*)| < 1$) of f^2. (Figure 8.1)

Denote the two fixed points of f^2 by x_+^* and x_-^*, so that

$$f(x_+^*) = x_-^* \tag{8.7a}$$

and

$$f(x_-^*) = x_+^* \tag{8.7b}$$

x_+^* and x_-^* are found by solving $x^* = f^2(x^*)$:

$$x_{+,-}^* = -\tfrac{1}{2}(A + 1) \pm \tfrac{1}{2}\sqrt{(A + 1)(A - 3)} \tag{8.8}$$

For fixed A the map produces, according to (8.7), the sequence $\ldots, x_+^*, x_-^*, x_+^*, x_-^*, \ldots$ Writing

$$x_n = x_{+,-}^* + \Delta x_n \tag{8.9}$$

and using $x_{n+1} = f(x_n)$, we obtain the following expression for Δx_{n+1}, assuming $x_n = x_+^*$:

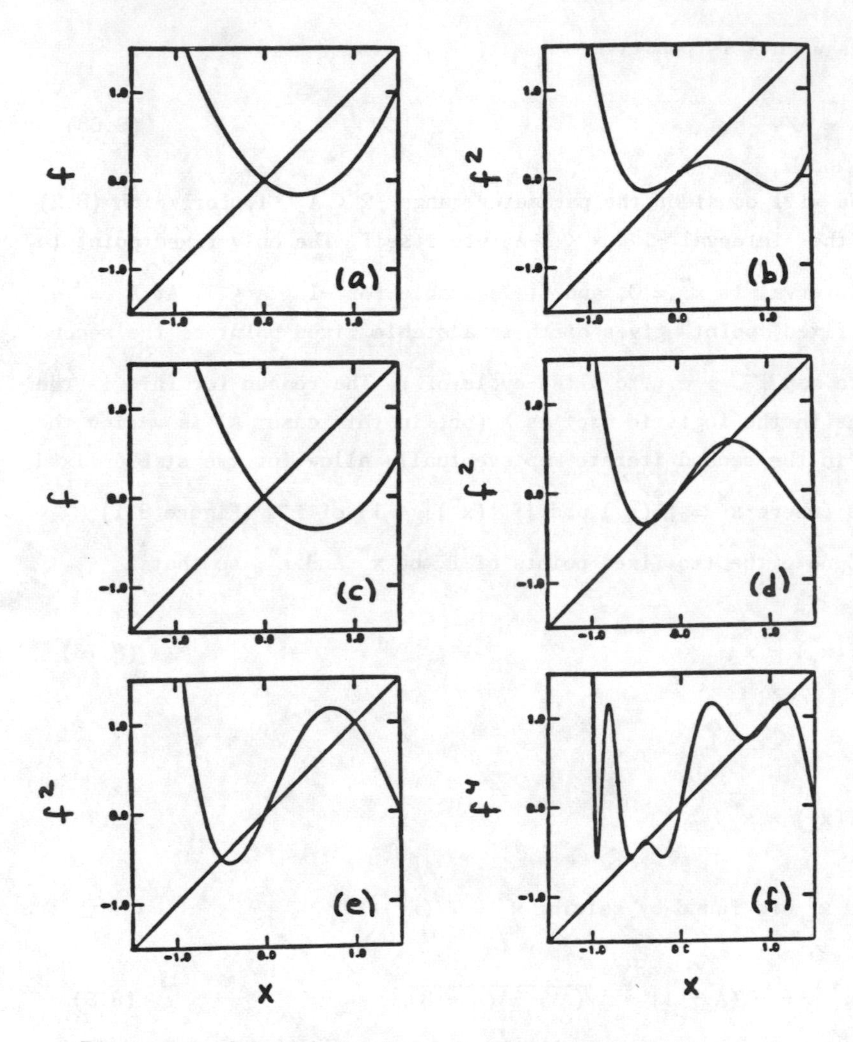

Figure 8.1 (a) $f(x)$ for $A = -.8$, showing the stable fixed point at the intersection with the line $y = x$. (b) $f^2(x)$ for $A = -.8$, showing that $x^* = 0$ is the only fixed point of f^2; (c) $f(x)$ for $A = -1.2$, at which the fixed point $x^* = 0$ is unstable ($|f'(0)| > 1$); (d) $f^2(x)$ for $A = -1.2$, showing two stable fixed points of f^2, corresponding to a two-cycle of f; (e) $f^2(x)$ for $A = -1.5$, at which f^2 has no stable fixed points; and (f) $f^4(x)$ for $A = -1.5$, showing four stable fixed points of f^4, corresponding to a four-cycle of f.

$$\Delta x_{n+1} = (A + 2x_+^*)\Delta x_n + \Delta x_n^2 \qquad (8.10)$$

If $x_n = x_+^*$ then $x_{n+1} = x_-^*$, and so for Δx_{n+2} we get

$$\Delta x_{n+2} = (A + 2x_-^*)\Delta x_{n+1} + \Delta x_{n+1}^2 \qquad (8.11)$$

Now we use (8.10) in (8.11) and keep only terms up to quadratic in the "perturbation" Δx_n. Then

$$\Delta x_{n+2} \cong (A + 2x_-^*)(A + 2x_+^*)\Delta x_n + [A + 2x_-^* + (A + 2x_+^*)^2]\Delta x_n^2 \qquad (8.12)$$

Of course if A is such that $x_{+,-}^*$ are stable fixed points of f^2, then the perturbations Δx_n generated by (8.12) will go to zero.

To understand the behavior of the map (8.12) it is convenient to rescale by defining

$$x_n' = \alpha \Delta x_n \qquad (8.13a)$$

with

$$\alpha \equiv A + 2x_-^* + (A + 2x_+^*)^2 \qquad (8.13b)$$

Then (8.12) becomes

$$x_{n+2}' = x_n'^2 + A'x_n' \qquad (8.14)$$

where

$$A' \equiv (A + 2x_-^*)(A + 2x_+^*) = -A^2 + 2A + 4 \qquad (8.15)$$

The map (8.14) has exactly the same form as the original map (8.2). Therefore we know that the fixed point $x'^* = 0$ of (8.14) is stable for $-1 < A' < 1$, or, using (8.15), for

$$1 - \sqrt{6} < A < -1 \tag{8.16}$$

If $x'^* = 0$ is a stable fixed point of (8.14), then $x^*_{+,-}$ must be stable fixed points of f^2, because (8.14) gives the perturbations about $x^*_{+,-}$. This tells us that $x^*_{+,-}$ become unstable fixed points of f^2 when

$$A < 1 - \sqrt{6} \cong -1.4495 \tag{8.17}$$

At $A = 1 - \sqrt{6}$ we then have the birth of a two-cycle of (8.14), corresponding to a four-cycle of f.

We can continue in this manner, introducing new perturbations Δx_n, rescaling, and recovering maps (8.14) of exactly the same form as the original map (8.2). Knowing that the first period doubling bifurcation of f occurs at $A = -1$, we can predict where successive bifurcations occur as in (8.17) above. The A values at which successive period doublings occur continue to get closer and closer together and they all pile up together at some value A_∞ given from (8.15) by

$$A_\infty = -A_\infty^2 + 2A_\infty + 4 \tag{8.18a}$$

or

$$A_\infty = \tfrac{1}{2}(1 - \sqrt{17}) \cong -1.5616 \tag{8.18b}$$

Let us pause to relate this result to the logistic map. Using

(8.6a) we have

$$\lambda_\infty = 1/2 - A_\infty/4 \cong 0.8904 \tag{8.19}$$

This is in good agreement with the exact result $\lambda_\infty = 0.8924864\ldots$ quoted in Section 3.

Let A_n be the value of A at which the nth bifurcation occurs in f. Since the (n+1)st bifurcation in the old map corresponds to the nth bifurcation in the new (rescaled) map, we can generalize (8.15) to read

$$A_n = -A_{n+1}^2 + 2A_{n+1} + 4 \tag{8.20}$$

Assume a geometric convergence of the sequence $\{A_n\}$ to the limit point A_∞, so that asymptotically

$$A_n \approx A_\infty + C\delta^{-n} \quad (\delta > 1) \tag{8.21}$$

Using this expression in (8.20), together with (8.18), we obtain

$$\delta = 2 - 2A_\infty = 1 + \sqrt{17} \cong 5.1231 \tag{8.22}$$

compared with the exact value $\delta = 4.6692\ldots$

Note that (8.21) can be expressed in the form

$$\lim_{n\to\infty} (A_n - A_{n-1})/(A_{n+1} - A_n) = (1 - \delta)/(\delta^{-1} - 1) = \delta \tag{8.23}$$

which is the form (3.6) we used to introduce δ in the special case of the logistic map.

Note also that the value of δ is independent of the specific parametrization used in writing a quadratic map, for if we write $x_{n+1} = x_n^2 + P(A)x_n$ instead of $x_{n+1} = x_n^2 + Ax_n$, where

$$P = g(A) \tag{8.24}$$

then, near $A = A_\infty$,

$$P_n - P_\infty = g(A_n) - g(A_\infty) \cong g'(A_\infty)(A_n - A_\infty) \tag{8.25}$$

and therefore

$$(P_n - P_{n-1})/(P_{n+1} - P_n) = (A_n - A_{n-1})/(A_{n+1} - A_n) \tag{8.26}$$

This means that we can replace A by some P(A) in (8.2), and the parameter P(A) will satisfy the same universal law (8.23) as A. Obviously this is a very important point in experimental studies of the period doubling route to chaos.

Finally let us consider the scaling parameter α defined by (8.13). At the accumulation point A_∞ we have, using (8.13), (8.8), and (8.18),

$$\alpha \cong -2.2372 \tag{8.27}$$

whereas the <u>exact</u> renormalization theory described in the next section gives

$$\alpha = -2.5029078750\ldots \tag{8.28}$$

9. Feigenbaum Universality: Outline of Exact Renormalization Theory

We now outline the theory leading to the exact values of the Feigenbaum numbers δ and α. We begin with the scaling parameter α, which in our approximate theory was introduced in equation (8.13) as a factor by which the perturbations Δx_n were magnified in order to recover a map (8.14) of the same form as the original map (8.2).

We will work again with the logistic map. The first part of the bifurcation diagram of Figure 3.1 is sketched in Figure 9.1, which is deliberately distorted for greater clarity. We indicate separations d_n of the pitchfork tines at those values Λ_n of λ for which x = 1/2 is a (stable) fixed point of f^{2^n}. α is given by

$$\alpha = -\lim_{n\to\infty}(d_n/d_{n+1}) \tag{9.1}$$

where $d_1 = f(\Lambda_1, 1/2) - 1/2$, $d_2 = f^2(\Lambda_2, 1/2) - 1/2$, and in general

$$d_n = f^{2^{n-1}}(\Lambda_n, 1/2) - 1/2 \tag{9.2}$$

where we have denoted $4\lambda x(1-x)$ by $f(\lambda, x)$.

At Λ_n the point x = 1/2 is a fixed point of $f^{2^n}(\Lambda_n, x)$, as is clear from Figure 9.1. From the chain rule (4.3) it is also clear that the derivatives of $f^{2^n}(\Lambda_n, x)$ at <u>all</u> its stable fixed points are equal to zero. For n = 1, for instance,

$$f^{2,}(\Lambda_1, x_1^*) = f^{2,}(\Lambda_1, x_2^*) = f'(\Lambda_1, x_2^*)f'(\Lambda_1, x_1^*) = 0 \tag{9.3}$$

since either x_1^* or x_2^* must be 1/2, where $f(\Lambda_1, x)$ has its maximum and its derivative vanishes. From (4.2) it follows that $\epsilon_n = 0$ to first order in the perturbation ϵ_0, and therefore the knob values Λ_n define what are called super-stable cycles, or simply <u>supercycles</u> of f.

Figure 9.2 shows d_1 and d_2 in plots of f and f^2 for $\lambda = \Lambda_1 \cong .8090$ and $\Lambda_2 \cong .8746$. The curve inside the dashed square of Figure 9.2c has nearly the same shape as the curve inside the entire square of Figure 9.2a when it is rotated and magnified by $|d_1/d_2|$. It is convenient to make a simple coordinate transformation displacing x = 1/2 to x = 0, so that the statement of self-similarity takes the form

$$f(\Lambda_1, x) \approx -\alpha f^2(\Lambda_2, -x/\alpha) = -\alpha f(\Lambda_2, f(\Lambda_2, -x/\alpha)) \tag{9.4}$$

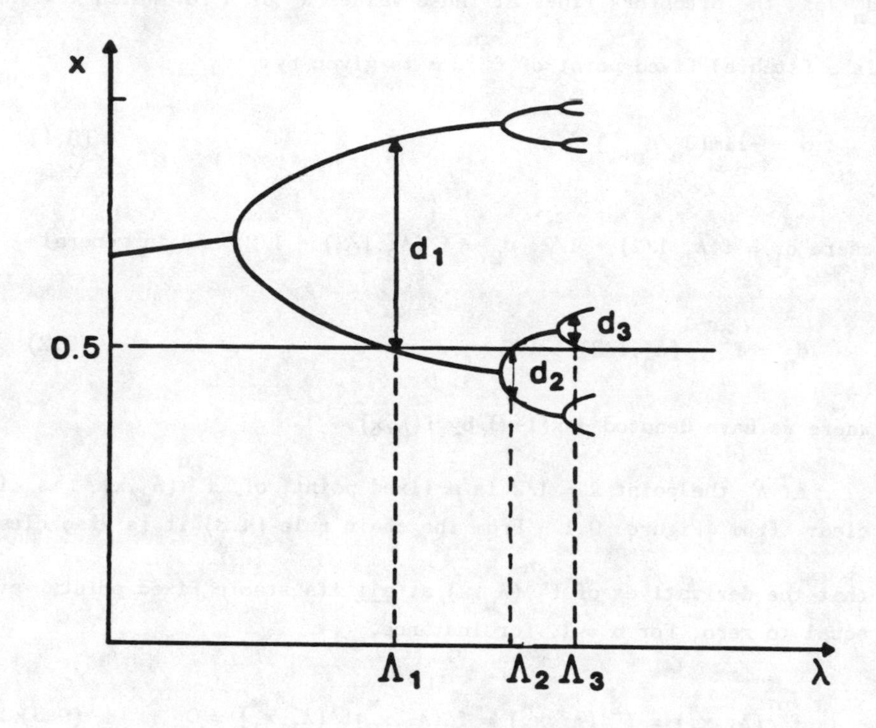

Figure 9.1 Sketch of the bifurcation diagram of Figure 3.1, showing the parameter values Λ_n at which supercycles occur.

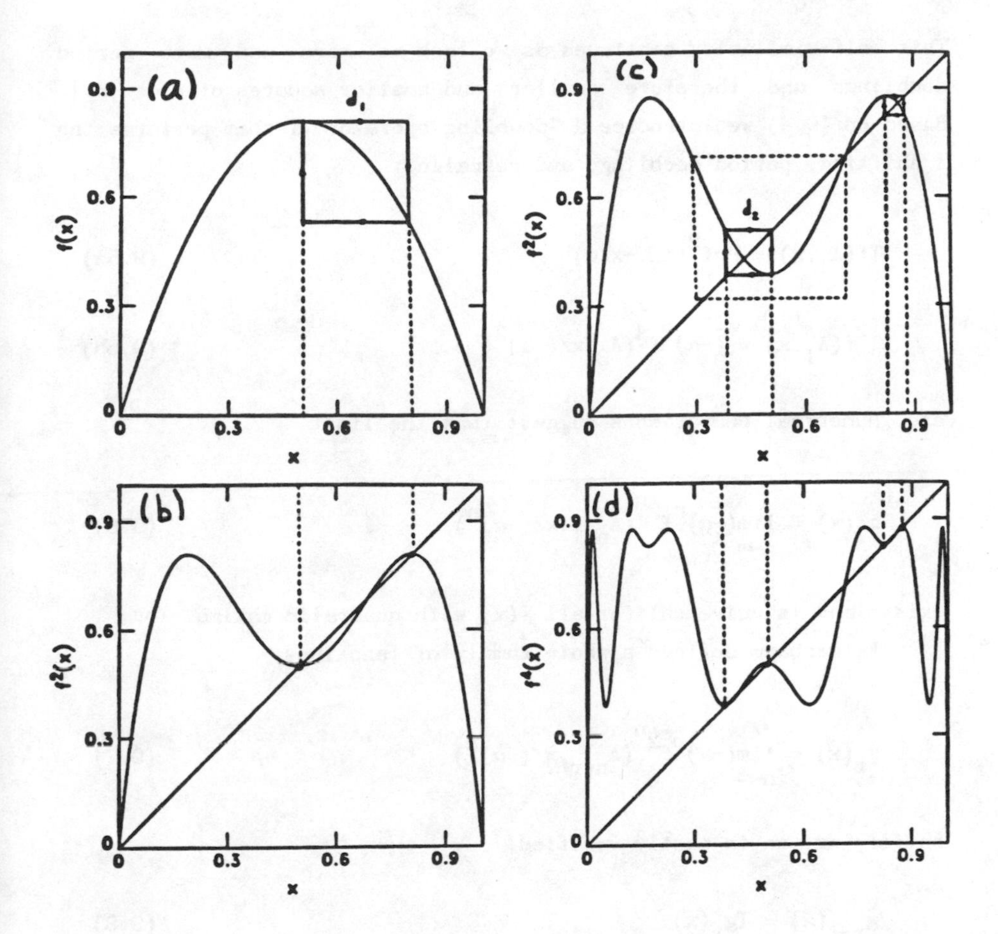

Figure 9.2 (a) f(x) for $\lambda = \Lambda_1$; (b) $f^2(x)$ for $\lambda = \Lambda_1$. For this value of λ f(x) has no stable fixed points, whereas $f^2(x)$ has the two stable fixed points indicated. The iterates generated by f form a two-cycle separated in x by the distance d_1. (c) $f^2(x)$ for $\lambda = \Lambda_2$; (d) $f^4(x)$ for $\lambda = \Lambda_2$. For this λ $f^2(x)$ has no stable fixed points, whereas $f^4(x)$ has the four stable fixed points indicated, corresponding to two stable two-cycles of f^2 as shown. The separation of the cycle element of f^2 closest to x = 1/2 from 1/2 is denoted d_2. The dashed square of (c) has approximately the same shape, when magnified and inverted, as the entire curve in (a).

This self-similarity continues as we look at more and more period doublings and therefore smaller and smaller squares of side $|d_n|$. Based on (9.4) we introduce a "doubling operator" T that performs the Λ shifting, period doubling, and rescaling:

$$Tf(\Lambda_1,x) = -\alpha f^2(\Lambda_2,-x/\alpha) \tag{9.5a}$$

$$T^2 f(\Lambda_1,x) = (-\alpha)^2 f^4(\Lambda_3,x/(-\alpha)^2) \tag{9.5b}$$

etc. Numerical comparisons suggest that the limit

$$g_1(x) \equiv \lim_{n\to\infty}(-\alpha)^n f^{2^n}(\Lambda_{n+1},x/(-\alpha)^n) \tag{9.6}$$

exists and is universal for all f(x) with quadratic maxima.

Feigenbaum defined a whole family of functions

$$g_r(x) = \lim_{n\to\infty}(-\alpha)^n f^{2^n}(\Lambda_{n+r},x/(-\alpha)^n) \tag{9.7}$$

satisfying, as is easily verified,

$$g_{r-1}(x) = Tg_r(x) \tag{9.8}$$

and conjectured the existence of a universal function

$$g(x) = \lim_{r\to\infty} g_r(x) = \lim_{n\to\infty}(-\alpha)^n f^{2^n}(\Lambda_\infty,x/(-\alpha)^n) \tag{9.9}$$

From (9.8) it follows that g(x) is a fixed point (in function space) of the doubling operator:

$$g(x) = Tg(x) = -\alpha g(g(-x/\alpha)) \tag{9.10}$$

and α is determined from the solution of this functional equation.

It may be checked that (9.10) holds also for $cg(x/c)$, i.e., it is invariant under magnification of g. In other words, the renormalization theory does not deal with absolute scales, and so we can arbitrarily fix a scale by setting $g(0) = 1$. Furthermore the supercycle condition $f^{2^n}{}'(\Lambda_n, 0) = 0$ implies $g'(0) = 0$. These boundary conditions are not sufficient to determine $g(x)$ satisfying (9.10): we must also specify the nature of $g(x)$ near $x = 0$, i.e.,

$$g(x) = 1 + Ax^z + Bx^{2z} + Cx^{3z} + \ldots \tag{9.11}$$

and different choices for z will produce different solutions of (9.10), and therefore different values of α and δ. We will focus on the "normal" case $z = 2$.

There are no general techniques for solving functional equations like (9.10). At a very crude level of approximation we assume, for the case of a quadratic maximum,

$$g(x) = 1 + Ax^2 \tag{9.12}$$

Then (9.10) becomes

$$1 + Ax^2 = -\alpha(1 + A) - 2A^2x^2/\alpha + O(x^4) \tag{9.13}$$

or

$$\alpha = 1 + \sqrt{3} = 2.732\ldots \tag{9.14a}$$

$$A = -\alpha/2 = -1.3666\ldots \tag{9.14b}$$

Even at this level of crudity we are in fairly good agreement with

Feigenbaum's numerical result (8.28) obtained by including terms up to x^{14} in an expansion like (9.12).

Using the renormalization equation (9.10), Feigenbaum obtained a (linearized) functional equation, in which δ is determined as the one "relevant" (> 1) eigenvalue of this equation. It is worth mentioning that the conjecture of the uniqueness of the appropriate solution of (9.10), and another conjecture relating to the spectral properties of the linear operator of which δ is an eigenvalue, were proved later by Collet, Eckmann, and Lanford. [9.1]

Not all unimodal (continuously differentiable, one hump) maps will show an infinite sequence of period doubling bifurcations, as we have assumed. What is required in addition is that the "Schwarzian derivative"

$$Sf(x) \equiv f'''(x)/f'(x) - 3[f''(x)/f'(x)]^2/2 \tag{9.15}$$

be negative over the whole mapping interval. Singer [9.2] proved in 1978 that $Sf(x) < 0$ is a necessary condition for $f(x)$ to have at most one stable period. (Obviously this cannot be a sufficient condition, for then we could not have chaos! Also there can be more than one stable n-cycle of the same period, as in Figure 9.2, where two distinct two-cycles of f^2 are seen for $\lambda = \Lambda_2$). Note that for the logistic map $Sf(x) < 0$ for all x in the interval [0,1]. Note also that $Sf(x) < 0$ implies $Sf^n f(x) < 0$, which may be verified with some algebra. Using simple examples and graphical constructions like Figure 8.1, the reader can convince himself that the condition $Sf(x) < 0$ for period doublings is at least plausible, although it is not at all obvious.

10. Remarks on Experimental Observations

Earlier we emphasized that the study of discrete maps is "relevant," because such maps can provide a sort of stroboscopic description of the behavior of ODEs. We will see many examples where the solution of an ODE system shows exactly the kind of period doubling to chaos we have described for the logistic map. At this point some remarks about experimental observations may be worthwhile. (Here again "experimental" includes numerical experimentation.)

First let us underscore the fact that the universality theory predicts certain <u>quantitative</u> features (such as α and δ) that are independent of the particular map we are dealing with. Thus if an experiment measures $\delta = 4.6692...$we are assured that the map has a quadratic hump. By the same token, the measurement of parameters like α and δ tells us <u>nothing</u> more specific about the map (e.g., we cannot determine from the measurements the precise form for $f(x)$).

We have discussed only one-dimensional maps. What about higher-dimensional systems? It turns out that dimensionality here is almost irrelevant. Whenever period doubling <u>ad infinitum</u> occurs, the basic process leading to universality is just functional composition, as in the renormalization equation (9.10). Consider, for instance, the Hénon map

$$x_{n+1} = y_n + 1 - Ax_n^2 \tag{10.1a}$$

$$y_{n+1} = 0.3x_n \tag{10.1b}$$

Depending on the initial seed (x_0, y_0), the sequence $\{x_n, y_n\}$ either settles onto an attracting set or diverges to infinity. (The set of all points (x_0, y_0) for which the sequence $\{x_n, y_n\}$ converges onto an attractor is called the <u>basin of attraction</u> of that attractor.) As the parameter A is varied a sequence of period doubling bifurcations is observed. Values A_n of A at which period doubling bifurcations occur are listed below [10.1]:

Period 2^n	A_n	$(A_n - A_{n-1})/(A_{n+1} - A_n)$
2	0.3675	
4	0.9125	4.844
8	1.026	4.3269
16	1.051	4.696
32	1.056536	4.636
64	1.05773083	4.7748
128	1.0579808931	4.6696
256	1.05803445215	4.6691

Note the evident convergence to the same universal number δ as in the case of a one-dimensional map with a quadratic maximum! This quantitative agreement with the one-dimensional case may be expected to occur in dissipative maps because there will be one direction of slowest contraction of "volumes," and this one direction will provide in effect a one-dimensional map for the system. The map (10.1), for instance, has one Lyapunov number greater than unity (corresponding to a positive Lyapunov exponent) and the other less than unity (corresponding to a negative Lyapunov exponent).

In Section 3 we noted that in the "chaotic regime" $\lambda_\infty < \lambda < 1$ there are periodic windows. This can be seen from the bifurcation diagram shown in Figure 3.1. Note in particular the rather wide period-3 window of λ values; narrower windows can also be discerned in Figure 3.1. The existence of such windows is consistent with a theorem given by Sarkovskii in 1964. [10.2] Define the two ordered sets $\{2^0 3,\ 2^0 5,\ 2^0 7, \ldots, 2^1 3,\ 2^1 5,\ 2^1 7, \ldots, 2^2 3,\ 2^2 5,\ 2^2 7, \ldots\} = \{3,\ 5,\ 7, \ldots, 6,\ 10,\ 14, \ldots, 12,\ 20,\ 28, \ldots\}$ and $\{\ldots, 2^5,\ 2^4,\ 2^3,\ 2^2,\ 2^1,\ 2^0\} = \{\ldots, 32,\ 16,\ 8,\ 4,\ 2,\ 1\}$. Sarkovskii proved that if a continuous, unimodal mapping f on an interval of the real axis has an n-cycle, then it must have an m-cycle for <u>every</u> m such that $n \leftarrow m$ in the ordering $3 \leftarrow 5 \leftarrow 7 \ldots 6 \leftarrow 10 \leftarrow 14 \ldots 12 \leftarrow 20 \leftarrow 28 \ldots 32 \leftarrow 16 \leftarrow 8 \leftarrow 4 \leftarrow 2$

$\leftarrow$ 1 between the two sets. Thus the existence of an 8-cycle implies the existence also of 1-, 2-, and 4-cycles. In particular, a 3-cycle implies the existence of all n-cycles, including aperiodic sequences associated with period doubling ad infinitum. ("Period Three Implies Chaos." [10.3]). However, it must be emphasized that Sarkovskii's theorem applies for a fixed value of a parameter λ, and that it says nothing about the stability of the m-cycles that are guaranteed to exist. In experiments, therefore, the predicted m-cycles may not actually be observed.

It is easy to see that in the "chaotic regime" there must be stable cycles of period $\neq 2^n$, i.e., stable cycles that do not arise from pitchfork bifurcations. Figure 10.1a shows the third iterate function $f^3(x)$ of the logistic map for $\lambda = .90$, where the only (unstable) fixed points of f^3 are the unstable fixed points $x^* = 0$ and $x^* = 1 - 1/4\lambda$ of f. As λ increases the hills and valleys of f^3 steepen until there are three tangencies with the line $y = x$; at this point a tangent bifurcation occurs in which six new fixed points of f^3 are born in stable and unstable pairs. This is clear from Figure 10.1b, which shows f^3 for $\lambda = .959$. The three stable fixed points, of course, correspond to a stable 3-cycle of f. When λ is increased further these stable fixed points eventually become unstable, and pitchfork (period doubling) bifurcations occur (not surprisingly, for f^3 is locally quadratic near its stable fixed points). At $\lambda = .961$ and .962, for instance, we find 6- and 12-cycles, respectively. There is a period doubling to chaos characterized by the universal Feigenbaum numbers α and δ.

Thus we have pitchfork bifurcations off the fundamental period 1 for $\lambda < \lambda_\infty$, whereas for $\lambda > \lambda_\infty$ there is a fundamental period 3 arising from a tangent bifurcation, followed by pitchfork bifurcations off this fundamental period. In fact in the chaotic regime there are infinitely many fundamental periods k arising from tangent bifurcations, and these all give rise to "harmonics" $2^n k$

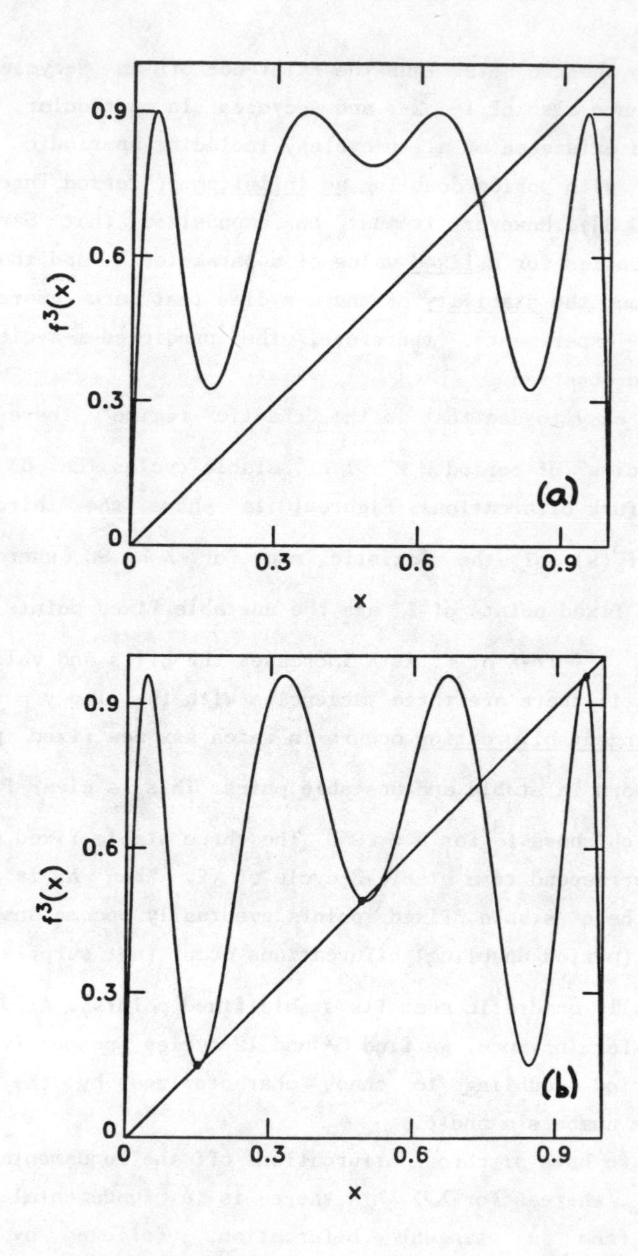

Figure 10.1 $f^3(x)$ **for the logistic map with (a)** λ **= .90 and (b)** λ **= .959, showing how a stable 3-cycle arises from a tangent bifurcation.**

produced by pitchfork bifurcations. The chaotic regime therefore has an enormous degree of "fine structure," with an infinity of stable n-cycles intermingled with chaotic bands.

In physical applications it is obviously of interest to know something about the order in which the fundamental periods k can be expected to occur as a parameter is varied. Remarkably, it is possible to make "universal" statements about the order in which various cycles appear, subject to certain constraints on f. This was shown in the "symbolic dynamics" of Metropolis, Stein, and Stein [10.4]. For the explicit example of the logistic map, for instance, the fundamental periods 1, 3, 4, 5, and 6 appear in the order indicated in the following table [7.1]:

Fundamental Period k	λ at which it first appears	λ at which it becomes unstable	λ at which all cycles $2^n k$ become unstable
1	.25	.75	.8925 (λ_∞)
6(a)	.9066	.9076	.9082
5(a)	.9346	.9353	.9358
3	.9571	.9604	.9624
5(b)	.9764	.9765	.9766
6(b)	.984379	.984399	.984412
4	.990025	.990200	.990300
6(c)	.994440	.994446	.994450
5(c)	.997565	.997575	.997580
6(d)	.9993958	.9993963	.9993965

As can be seen in these results, any particular stable cycle occurs typically over an extremely narrow range of parameter values, and therefore is unlikely or at least difficult to observe experimentally.

Probably the most useful experimental tool for observing the period doubling route to chaos is spectral analysis. (cf. Figure 5.1) If we have a 2^n-cycle the Fourier transform of $\{x_m\}$ is (equation (5.3))

$$X_k^{(n)} = \sum_{\ell=1}^{2^n} x_\ell e^{-2\pi ik\ell/2^n}, \quad k = 0,1,2,\ldots,2^n - 1 \tag{10.2}$$

After a pitchfork bifurcation to a 2^{n+1}-cycle we have

$$X_k^{(n+1)} = \sum_{\ell=1}^{2^{n+1}} x_\ell e^{-2\pi ik\ell/2^{n+1}}, \quad k = 0,1,2,\ldots,2^{n+1} - 1 \tag{10.3}$$

For instance,

$$X_k^{(1)} = x_1 e^{-2\pi ik/2} + x_2 e^{-2\pi ik}, \quad k = 0,1 \tag{10.4}$$

and

$$X_k^{(2)} = x_1 e^{-2\pi ik/4} + x_2 e^{-2\pi ik/2} + x_3 e^{-2\pi ik(3/4)} + x_4 e^{-2\pi ik}, \quad k = 0,1,2,3 \tag{10.5}$$

Thus the power spectrum for a 2-cycle has frequency components at 0 and 1/2, whereas for a 4-cycle there are peaks at 0, 1/4, 1/2, and 3/4. Similarly an 8-cycle has a power spectrum with spikes at 0, 1/8, 1/4, 3/8, 1/2, 5/8, 3/4, and 7/8. This is all pretty trivial, but it explains the various subharmonic peaks appearing in Figure 5.1, which are often not obvious to the beginner, who may expect to see a spike only at 1/n in the case of an n-cycle. It's all in the definition of the discrete Fourier transform.

There is a certain degree of universality in the form of the power spectra associated with the period doubling bifurcations. Let us approximate (10.2) and (10.3) by writing

$$X_k^{(n)} \cong \int_0^{T_n} dt x(t) e^{-2\pi ikt/T_n}, \quad T_n \equiv 2^n \tag{10.6}$$

and

$$X_k^{(n+1)} \cong \int_0^{2T_n} dt x(t) e^{-2\pi ikt/2T_n} = \int_0^{T_n} dt x(t) e^{-i\pi kt/T_n} \tag{10.7}$$

Furthermore we rewrite (10.7):

$$X_k^{(n+1)} = (1/2) \int_0^{T_n} dt[x(t) + (-1)^k x(t+T_n)] e^{-i\pi kt/T_n} \tag{10.8}$$

Now a 2^{n+1}-cycles looks a lot like a doubled version of a 2^n-cycle in the sense that $x(t+T_n) \cong x(t)$; for $\lambda = .88$, for instance, we find the 4-cycle {.879, .373, .823, .512}, and for $\lambda = .889$ the 8-cycle {.884, .363, .819, .525, .884, .364, .820, .522}. Thus for k <u>even</u> we have from (10.8)

$$X_{2k}^{(n+1)} \cong (1/2) \int_0^{T_n} dt x(t) e^{-2\pi ikt/T_n} = X_k^{(n)} \tag{10.9}$$

where we have written 2k instead of k to emphasize that this result applies for the even components of the spectrum. The result (10.9) says that the new even harmonics that appear after a period doubling bifurcation have their strengths determined (approximately) by the old spectrum (i.e., by the spectrum before the bifurcation). This explains, for example, why the peaks at frequencies 1/8, 1/4, and 3/8 in Figure 5.1c have about the same strengths as the peaks at 1/8,

1/4, and 3/8 in Figure 5.1b.

The same approximation $x(t) \cong x(t+T_n)$ applied to (10.8) for k odd would imply that $X_{2k+1}^{(n+1)} = 0$. However, we can go beyond this simple approximation and use the results of the renormalization theory; this leads to the conclusion that the new odd subharmonic amplitudes $|X_{2k+1}^{(n+1)}|$ are about a factor of 0.1525 times the old odd subharmonic amplitudes, corresponding to a reduction by about 8.17 dB ($= 10\log_{10}(.1525)^{-1}$). [8.1] This (approximately) universal behavior can be seen in Figure 5.1. As discussed later, this reduction imposes limits on the degree of external noise that can be tolerated when one is trying to observe a sequence of period doubling bifurcations.

11. The Duffing Oscillator

After all this discussion of discrete maps let us at last consider a "continuous" flow in the form of an ODE. Consider the anharmonic potential $V(x) = (1/2)x^2 - (1/4)\beta x^4$ shown in Figure 11.1. A particle of unit mass in this potential, subject to a frictional force $-\gamma\dot{x}$ and a driving force $\cos\omega t$, satisfies the Duffing equation

$$\ddot{x} + \gamma\dot{x} + x - \beta x^3 = \cos\mu t \tag{11.1}$$

Chaotic behavior of the Duffing oscillator was first studied by Huberman and Crutchfield. [11.1]

Let us begin by applying to (11.1) a well-known approximation of quantum optics, namely the rotating-wave approximation (RWA), or the method of slowly varying amplitudes. We write

$$x(t) = (1/2)[\alpha(t)e^{i\mu t} + \alpha(t)^* e^{-i\mu t}] \tag{11.2}$$

and assume $\alpha(t)$ is slowly varying in the sense that $|\ddot{\alpha}(t)| \ll \mu^2|\alpha(t)|$. Then (11.1) reduces to the approximate form

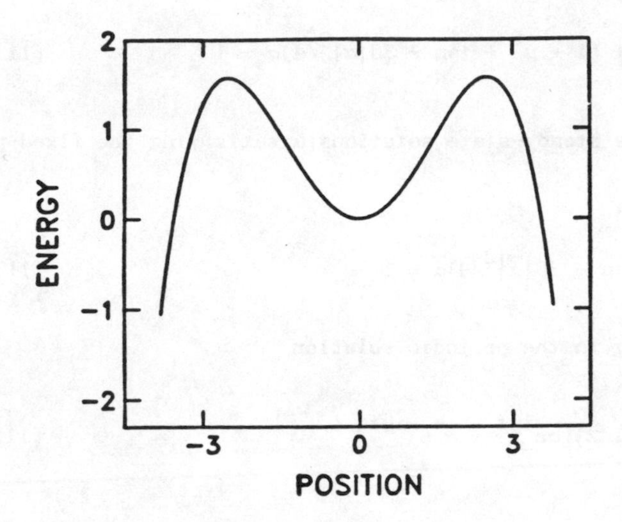

Figure 11.1 The anharmonic potential $V(x) = (1/2)x^2 - (1/4)\beta x^4$ for $\beta = .1587$.

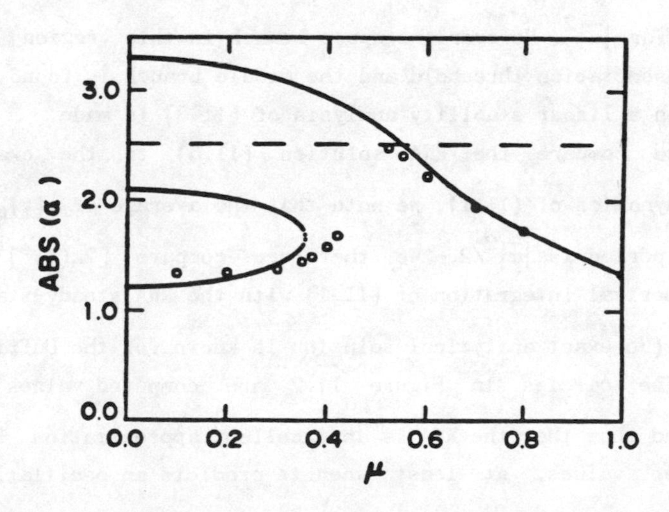

Figure 11.2 The solution of equation (11.4) for $|\bar{\alpha}|$ with $\beta = .1587$ and $\gamma = .72$. The dotted line shows the "dissociation threshold" for escape from the well. The circles were obtained by numerical integration of the Duffing equation, as discussed in the text.

$$(\gamma + 2i\mu)\dot{\alpha} + (1 - \mu^2 + i\gamma\mu - 3\beta|\alpha|^2/4)\alpha = 1 \tag{11.3}$$

This equation has steady-state solutions $\overline{\alpha}$ satisfying the fixed-point equation

$$(1 - \mu^2 + i\gamma\mu - 3\beta|\overline{\alpha}|^2/4)\overline{\alpha} = 1 \tag{11.4}$$

and corresponding to the periodic solution

$$x(t)_{RWA} = (1/2)[\overline{\alpha}e^{i\mu t} + \overline{\alpha}^{*}e^{-i\mu t}] \tag{11.5}$$

for x.

Equation (11.4) is equivalent to a (real) cubic equation for $|\overline{\alpha}|^2$. In Figure 11.2 we show the solution of this cubic for $\beta = .1587$ and $\gamma = .72$. (The dashed line indicates the "dissociation threshold" for escape from the potential well of Figure 11.1) Note that there is a region for $\mu < .4$ where there are three (real) solutions of the cubic equation for $|\overline{\alpha}|^2$. However the upper branch in this region is beyond the dissociation threshold and the middle branch is found to be unstable when a linear stability analysis of (11.3) is made.

In order to compare the RWA solution (11.5) to the exact steady-state dynamics of (11.1), we note that the average of $x(t)^2_{RWA}$ over a driving period is $|\overline{\alpha}|^2/2$. We therefore compare $[2x(t)^2]^{1/2}$ obtained by numerical integration of (11.1) with the RWA steady-state amplitude $|\overline{\alpha}|$. (No exact analytical solution is known for the Duffing oscillator.) The circles in Figure 11.2 are computed values of $[2x(t)^2]^{1/2}$, and show that the RWA is an excellent approximation for these parameter values, at least when it predicts an oscillation amplitude $|\overline{\alpha}|$ below the dissociation threshold.

However, the RWA obviously fails to account for harmonics (and subharmonics) of the driving frequency μ in x(t). Figure 11.3 shows

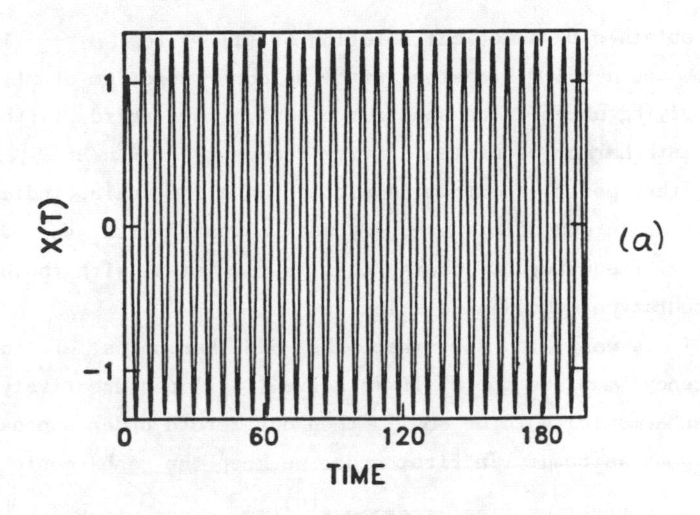

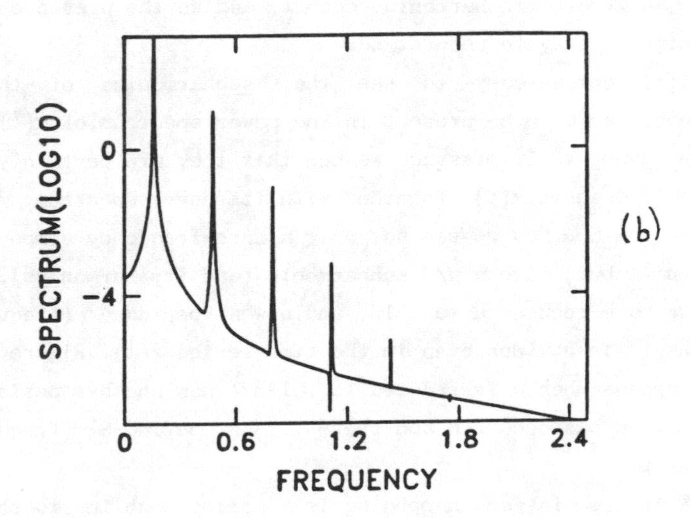

Figure 11.3 (a) $x(t)$ obtained by numerical integration of the Duffing equation for $\beta = .1587$, $\gamma = .728$, and $\mu = 1.0$; (b) power spectrum of the time series shown in (a).

x(t) obtained by numerical solution of (11.1) for β = .1587, γ = .728, and μ = 1.0, together with the power spectrum of x(t) obtained by applying an FFT. The spectrum clearly shows third, fifth, seventh, and ninth harmonics of the driving frequency ($f = \mu/2\pi \cong .16$). Note that the power spectrum is plotted on a log scale, indicating that the harmonic components are very weak compared with the fundamental driving frequency. Of course this is consistent with the high degree of accuracy of the RWA solution.

It is easy to see that the odd harmonics of the driving frequency must be there. If we solve (11.1) perturbatively, assuming the anharmonicity to be small, then our zeroth-order approximation to x(t) goes as $\cos\mu t$. In first order we keep the anharmonic term and get a correction that goes as $x^{(o)}(t)^3 \approx \cos^3\mu t$, which has a piece going as $\cos 3\mu t$. In the next order, similarly, we get a correction going as the fifth harmonic, $\cos 5\mu t$, and so the presence of the odd harmonics is easy to understand.

It is not as easy to see that <u>sub</u>harmonics of the driving frequency can also be present in the power spectrum of x(t), although if we <u>assume</u> their presence we see that they are certainly possible. Figure 11.4 shows x(t), together with its power spectrum, when μ is reduced to .53. Now we see not only a zero-frequency component in the spectrum, but also a $\mu/2$ subharmonic (and its harmonics). In Figure 11.5 we have reduced μ to .515, and now a period-4 (frequency $\mu/4$) component is obvious even in the time series x(t). Figure 11.6 shows what happens when μ is reduced to .5143 : now another period doubling bifurcation has occurred and there is a period-8 (frequency $\mu/8$) component.

What is in fact happening is a period doubling to chaos. If μ_n is the value of μ at which the n<u>th</u> period doubling occurs, and $\lambda = \mu_n^{-1}$, then it appears from numerical experiments that $(\lambda_n - \lambda_{n-1})/(\lambda_{n+1} - \lambda_n)$ is approaching the universal Feigenbaum number δ. In Figure 11.7 we show x(t) and its power spectrum for μ = .51, and observe the broadband power spectrum symptomatic of chaos.

To investigate as simply as possible the sensitivity to initial

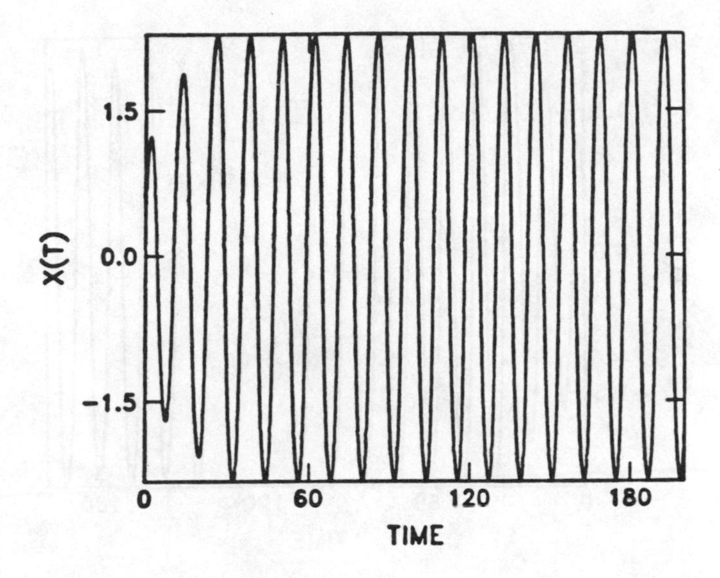

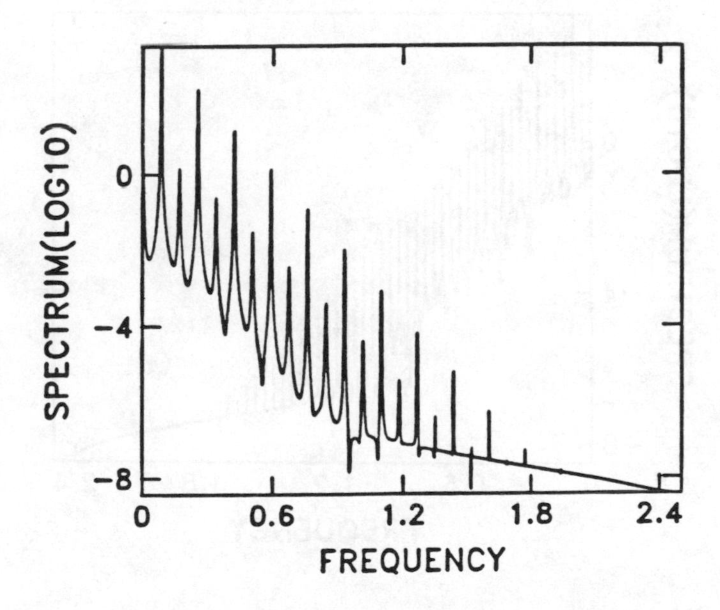

Figure 11.4 As in Figure 11.3, but with μ = .53.

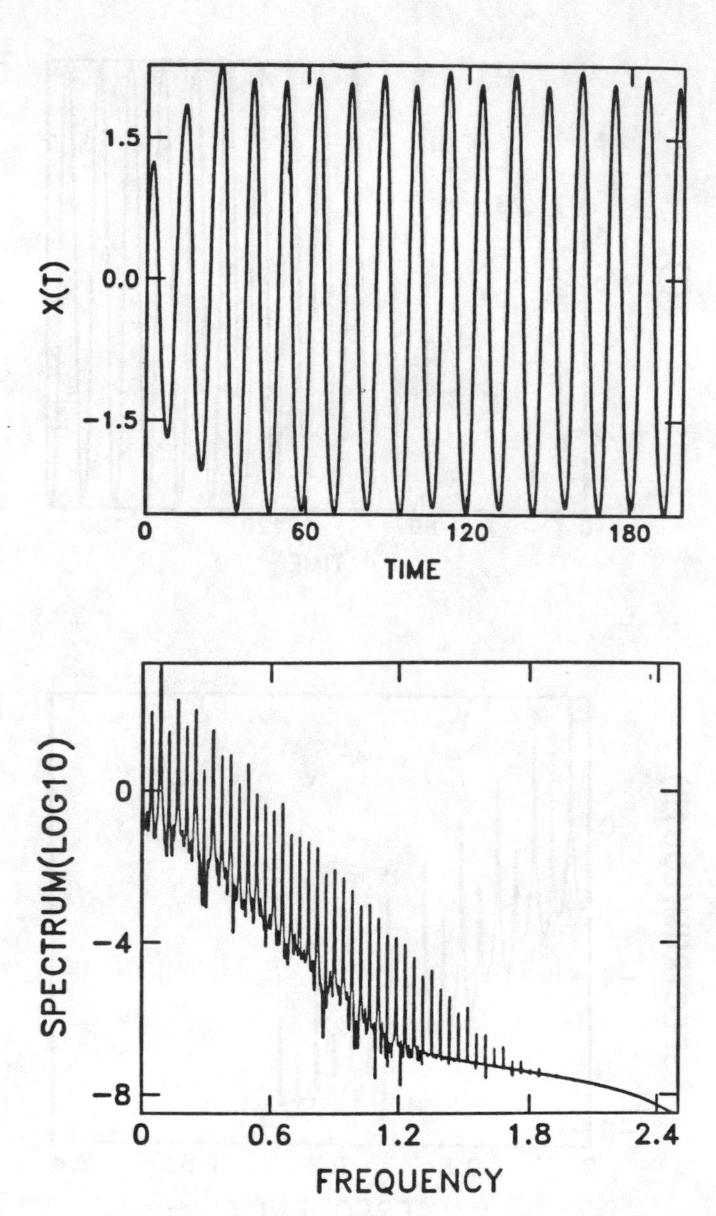

Figure 11.5 As in Figure 11.3, with μ = .515.

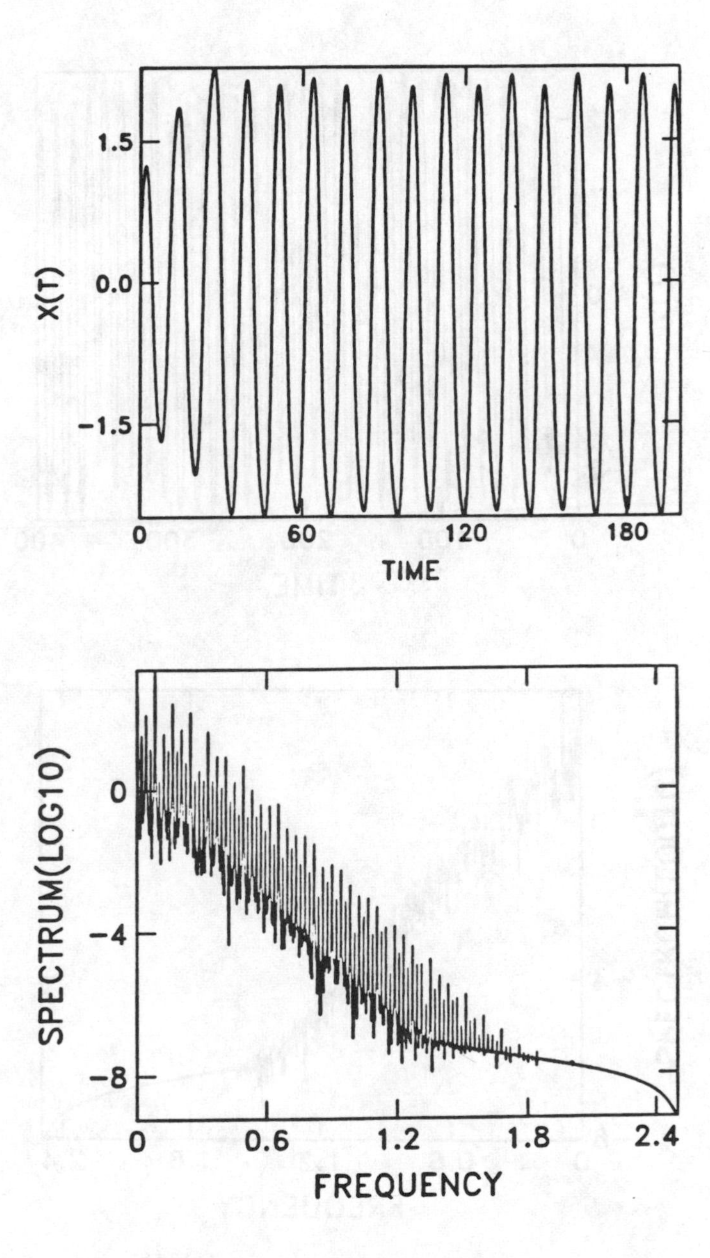

Figure 11.6 As in Figure 11.3, with μ = .5143.

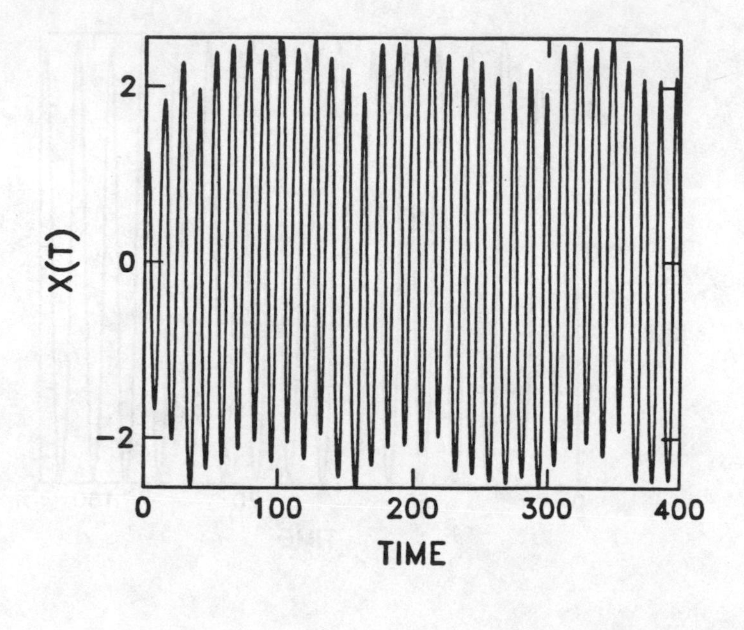

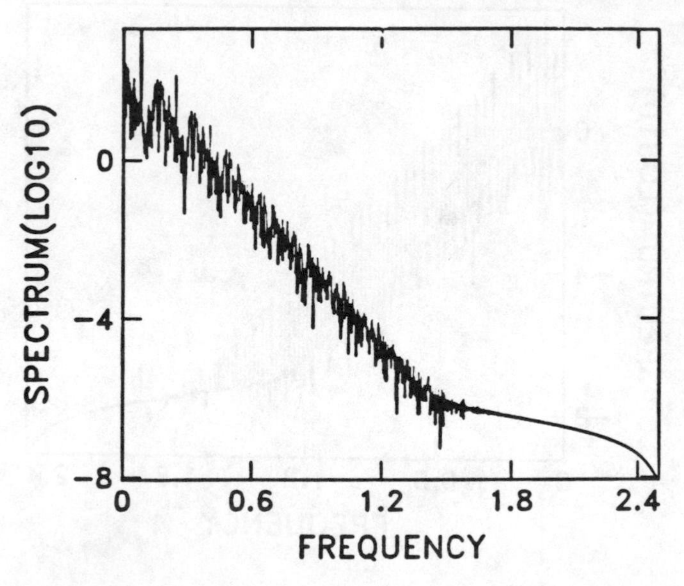

Figure 11.7 As in Figure 11.3, with μ = .51.

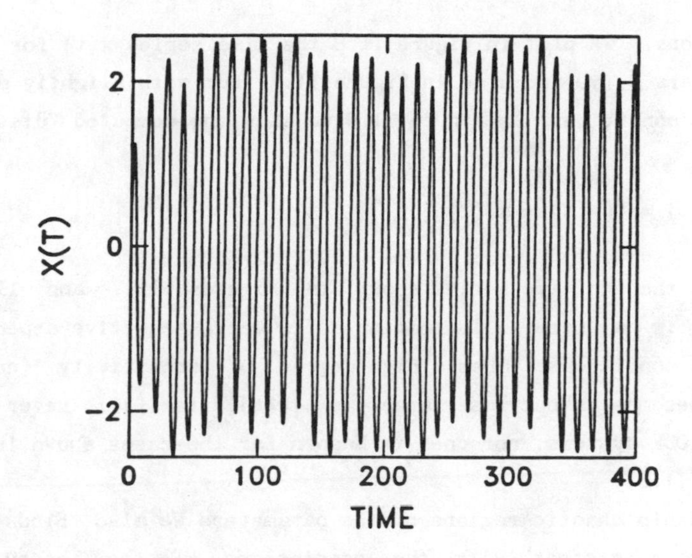

Figure 11.8 x(t) for the same parameters β, τ, and μ as in Figure 11.7, but with slightly different initial conditions.

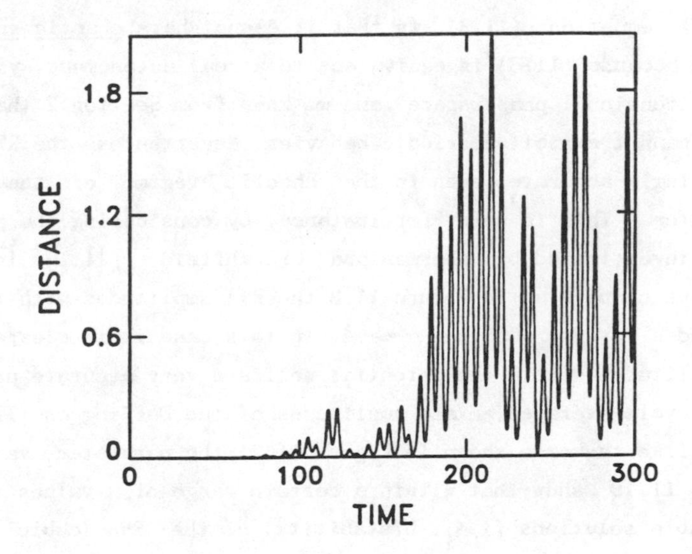

Figure 11.9 The distance (11.6) separating the two time series shown in Figures 11.7 and 11.8.

conditions, we plot in Figure 11.8 the time series $x(t)$ for the same parameters β, γ, and μ as in Figure 11.7, but with slightly different initial conditions. In Figure 11.9 we show the computed "distance"

$$d(t) = ([x(t) - x'(t)]^2)^{1/2} \tag{11.6}$$

between the two time series shown in Figures 11.7 and 11.8. The result is a nice illustration of "very sensitive dependence on initial conditions." Such a high degree of sensitivity (note that $d(t)$ becomes about as large as $|x(t)|$ itself) is never found in linear ODE systems, nor does it happen for the cases shown in Figures 11.3 - 11.6.

Within chaotic regions of the parameters we also find periodic windows consistent with the predictions of the theory based on one-dimensional maps with quadratic maxima. (Section 10) All this provides convincing confirmation that the theory is indeed relevant to ODE systems.

Finally let us return to the RWA. One thing that is clear from the RWA equation (11.3) is that *it cannot have chaotic solutions*. This is because (11.3) is equivalent to a real autonomous system with a two-dimensional phase space, and we know from Section 2 that such a system cannot exhibit aperiodic behavior. Nevertheless the RWA can be surprisingly accurate, even in the chaotic regime of the Duffing oscillator. This is seen, for instance, by considering the parameter regime investigated by Huberman and Crutchfield. [11.1] In Figure 11.10 we compare as in Figure 11.2 the RWA amplitudes with the exact amplitudes for $\beta = .055$ and $\gamma = .4$. In this case it is clear that the RWA amplitudes within the potential well are very accurate predictors of the cycle-averaged, exact amplitudes of the Duffing oscillator.

Unlike the case shown in Figure 11.2, the parameter values for Figure 11.10 show that within a certain range of μ values there are two stable solutions (i.e., bistability) of the RWA cubic equation for $|\overline{\alpha}|^2$. Which stable branch the system follows depends on its past history, as indicated by the arrows in Figure 11.10. In particular,

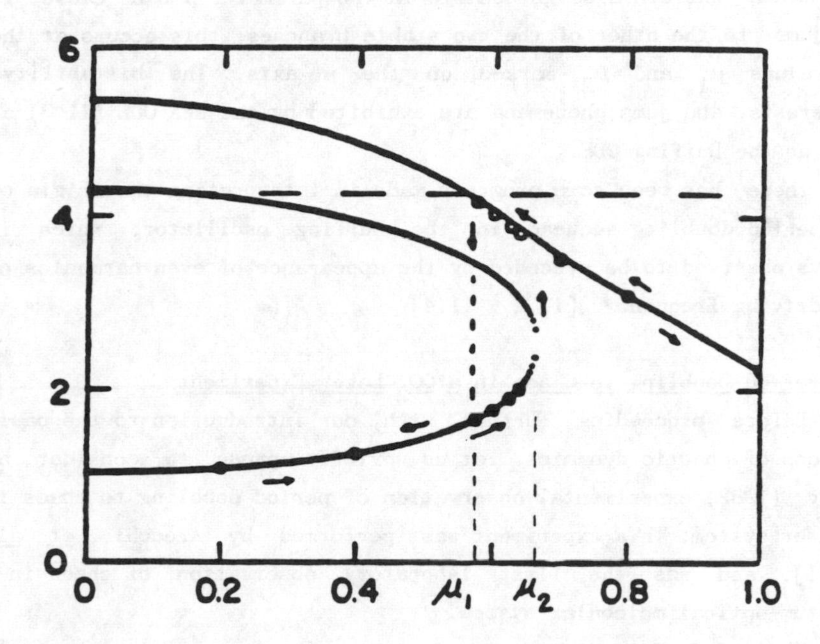

<u>Figure 11.10</u> **As in Figure 11.2, but with β = .055 and τ = 0.4.**

there is a hysteresis loop as indicated. The system exhibits a "jump phenomenon" wherein a slight change in the parameter μ can cause it to jump to the other of the two stable branches; this occurs at the two values μ_1 and μ_2 marked on the μ axis. The bistability, hysteresis, and jump phenomena are exhibited by the RWA ODE (11.3) as well as the Duffing ODE.

There has been some progress made in interpreting the origin of the period doubling sequence in the Duffing oscillator, which is always observed to be preceded by the appearance of even harmonics of the driving frequency. [11.2 - 11.4]

12. Period Doubling to Chaos in a CO_2 Laser Experiment

Before proceeding further with our introduction to the basic notions of chaotic dynamics, let us briefly pause to consider an early (1982) experimental observation of period doubling to chaos in a laser system. This experiment was performed by Arecchi, *et al.* [12.1], and was the first laboratory observation of chaos in a quantum-optical molecular system.

The standard rate equations for a two-level system in a resonant field of photon number n are

$$\dot{N}_2 = R - \gamma_2 N_2 - Gn(N_2 - N_1) \tag{12.1a}$$

$$\dot{N}_1 = -\gamma_1 N_1 + Gn(N_2 - N_1) \tag{12.1b}$$

where R is a pump rate to the upper level, G is proportional to the stimulated emission cross section ($G = \sigma/\hbar\omega$), and γ_2 and γ_1 are the relaxation rates of levels 2 and 1, respectively. N_2 and N_1 are respectively the populations of the upper and lower levels of the laser transition. Subtracting (12.1b) from (12.1a), we have

$$\dot{\Delta} = R - \gamma\Delta - 2Gn\Delta \tag{12.2a}$$

where $\Delta \equiv N_2 - N_1$ is the population inversion, and for simplicity we have assumed $\gamma_1 = \gamma_2 \equiv \gamma$. This equation describes a homogeneously broadened, two-level system in a single-mode resonant field. For a self-consistent description we also write a rate equation for the cavity photon number:

$$\dot{n} = Gn\Delta - Kn \tag{12.2b}$$

where K is the cavity damping rate, which is inversely proportional to the cavity Q factor. Equations (12.2) are used by Arecchi, et al. to model their experiments.

The system (12.2) has the single fixed point

$$\bar{\Delta} = K/G \tag{12.3a}$$

$$\bar{n} = (R/2K - \gamma/2G) \tag{12.3b}$$

Let $\Delta \to \bar{\Delta} + \epsilon$ and $n \to \bar{n} + \eta$, and assume ϵ and η are small perturbations. Then from (12.2) we obtain, in the usual fashion of linear stability analysis,

$$\dot{\epsilon} = -(GR/K)\epsilon - 2K\eta \tag{12.4a}$$

$$\dot{\eta} = (GR/2K - \gamma/2)\epsilon \tag{12.4b}$$

From (12.3) we have the threshold condition $GR > \gamma K$ (i.e., $\bar{n} > 0$) for laser oscillation. This condition implies that the solutions of (12.4) are always decaying in time, as is easily shown. In other words, the fixed point (12.3) is a stable fixed point of the laser rate equations.

At the crudest level of analysis this would suggest that cw

laser oscillation is always stable. But of course the two-level rate equations (12.2) are much too simple for most lasers. For one thing, they are valid only in the "rate equation approximation" in which the homogeneous linewidth of the transition is large compared with the Rabi frequency, which is not always the case (although this condition is well satisfied in the Arecchi experiment). For another, they do not describe a multimode situation. And as regards chaos, the two-dimensional phase space of (12.2) is too small to allow anything but regular behavior.

Arecchi, et al. used an electro-optical modulator in a stabilized CO_2 laser to obtain an oscillatory cavity loss: $K \rightarrow K_1(1 + m\cos\Omega t)$ in (12.2b). In this case, in somewhat of an analogy to the Duffing oscillator, there is the possibility of chaotic behavior. In the experiments the modulation index $m \approx .01$ and the modulation frequency $f = \Omega/2\pi$ ranged in the neighborhood from 40 to 150 kHz. Numerical simulation of the system (12.2) with $K = K_1(1 + m\cos\Omega t)$, and for parameters appropriate to the laser, predicted a period doubling route to chaos as Ω is varied over a certain range. This behavior was observed experimentally. For instance, a period-2 oscillation of the output intensity was found for f = 62.75 kHz, a period-4 for f = 63.80 kHz, and chaos for f = 64.00 kHz. For f = 64.13 kHz, presumably within the "chaotic regime," a period-3 window was found.

It is difficult experimentally to see more than a few period doubling bifurcations, as discussed later. The important point of the present section is simply to emphasize that the period doubling route to chaos has been observed - by now in a rather large number of systems - in "real-world" laboratory experiments. So once again we can emphasize the relevance of the results obtained for "simple" one-dimensional maps!

13. Bifurcations

We have described the Feigenbaum scenario in which chaos develops as the result of an infinite sequence of period doubling or pitchfork bifurcations. There are various types of bifurcation, i.e., various ways in which a system can change as some parameter is varied. The pitchfork bifurcation is indicated pictorially in Figure 13.1a: as a parameter is varied a stable fixed point becomes unstable and bifurcates into a stable two-cycle. (The dotted line indicates that the fixed point exists but is unstable, whereas solid lines denote stability.) In connection with the supercycles of the logistic map we also mentioned the tangent bifurcation, which is illustrated in Figure 13.1b. Here two fixed points are created as a parameter is varied, one of them stable and the other unstable. We can also have an exchange of stability bifurcation as shown in Figure 13.1c, or a reverse pitchfork bifurcation as in Figure 13.1d.

Different scenarios for the onset of chaos are distinguished by the different kinds of bifurcation that are supposed to occur en route to chaos. An especially important type of bifurcation in this regard is the Hopf bifurcation, as illustrated in Figure 13.1e. A Hopf bifurcation is a change from a stable fixed point to a limit cycle. As should be clear from our discussion near the end of Section 2, a limit cycle requires at least two dimensions of phase space, and so a Hopf bifurcation can occur in two-dimensional but not one-dimensional flows in phase space. We can also have an inverted Hopf bifurcation, as illustrated in Figure 13.1f.

14. The Intermittency (Pomeau-Manneville) Route to Chaos

The intermittency route to chaos was proposed by Pomeau and Manneville in 1979. [14.1] Consider the Lorenz model equations

$$\dot{x} = \sigma(y - x) \tag{14.1a}$$

$$\dot{y} = -xz + rx - y \tag{14.1b}$$

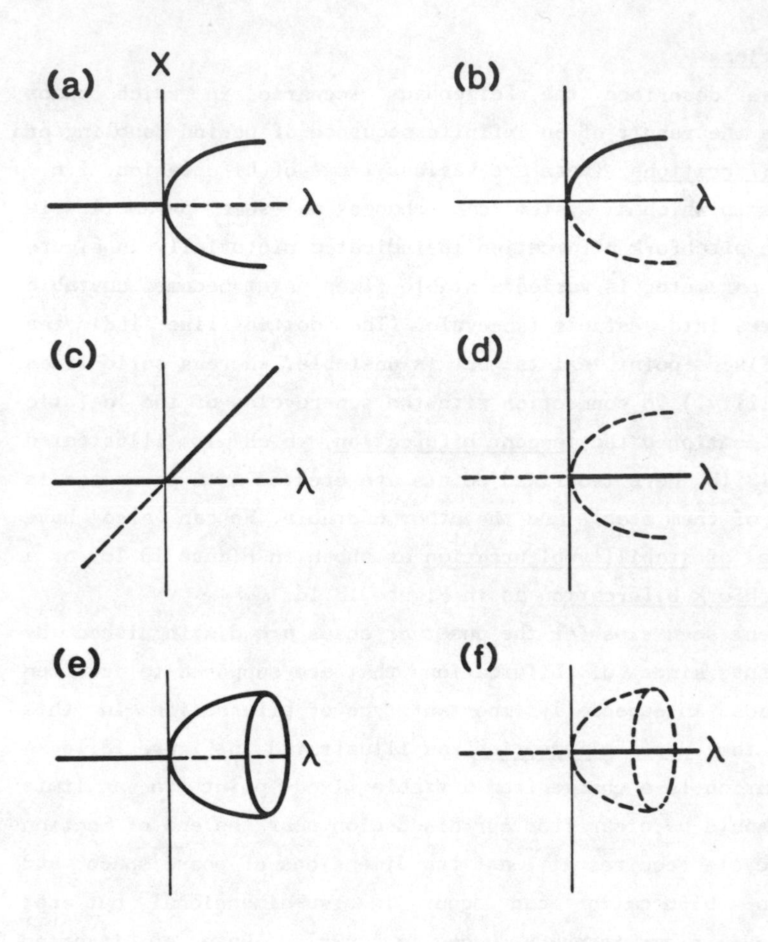

Figure 13.1 **Schematic illustration of various types of bifurcation. (a) pitchfork bifurcation, in which a stable fixed point gives way to a stable two-cycle; (b) tangent bifurcation, in which two fixed points are created, one stable and the other unstable; (c) exchange of stability bifurcation, in which one stable fixed point gives way to another; (d) reverse pitchfork bifurcation; (e) Hopf bifurcation, in which a stable fixed point gives way to a limit cycle; (f) inverted Hopf bifurcation. In each case the solid lines imply stability, whereas the dotted lines imply instability.**

$$\dot{z} = xy - bz \qquad (14.1c)$$

(This system is discussed further in Section 23.) Following Pomeau and Manneville we plot in Figure 14.1 the time series z(t) obtained by numerical integration of (14.1) for $\sigma = 10$, $b = 8/3$, and several values of r around 166. We see that as r is increased slightly there are increasingly frequent, irregular bursts interrupting the regular oscillations. This illustrates the <u>intermittency route to chaos</u>, a more or less continuous transition from regular to chaotic behavior.

Pomeau and Manneville interpreted the behavior shown in Figure 14.1 as follows. For r less than some critical value r_c there is a stable fixed point in a Poincare' map for (14.1), corresponding to the stable oscillation of Figure 14.1a. For $r > r_c$ the map has no fixed points, and its iterates can wander around erratically. But near the original fixed point the trajectory slows down and is temporarily trapped in the vicinity of the fixed point (which of course is not an attractor because it is unstable). This is illustrated in Figure 14.2, which shows that a large number of iterations of the map may be required to escape the "channel" of regular motion. This figure also indicates a particular way in which the fixed point becomes unstable. Namely, it shows an <u>inverse tangent bifurcation</u> in which a pair of fixed points, one stable and the other unstable, "collide" and disappear at $r = r_c$.

The intermittency route to chaos portrayed in Figures 14.1 and 14.2 is called <u>type-I intermittency</u>. Here a real eigenvalue of the linearized Poincare' map crosses the unit circle at $x = +1$, as shown in Figure 14.3a. In <u>type-II intermittency</u>, two complex conjugate eigenvalues cross the unit circle, as shown in Figure 14.3b. We can also have the situation shown in Figure 14.3c, in which a real eigenvalue crosses the unit circle at $x = -1$. This is called <u>type-III intermittency</u>.

Type-I intermittency can also be found in the logistic map. For $\lambda = \lambda_c = (1/4)(1 + \sqrt{8}) = .9571$ a period-3 cycle is born in the "chaotic regime." (Section 10) We show in Figure 14.4 the iterates

Figure 14.1 Intermittent behavior in the Lorenz model for $\sigma = 10$, $b = 8/3$, and (a) $r = 166.1$, (b) $r = 166.4$, (c) $r = 167$.

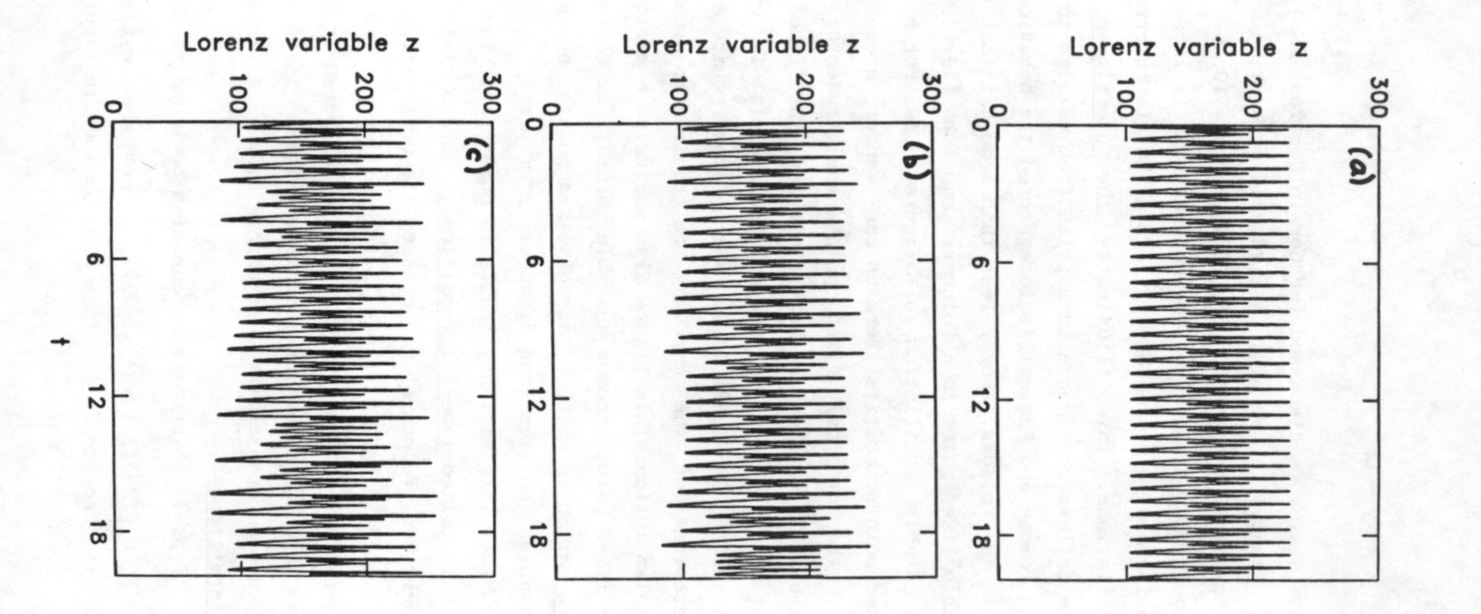

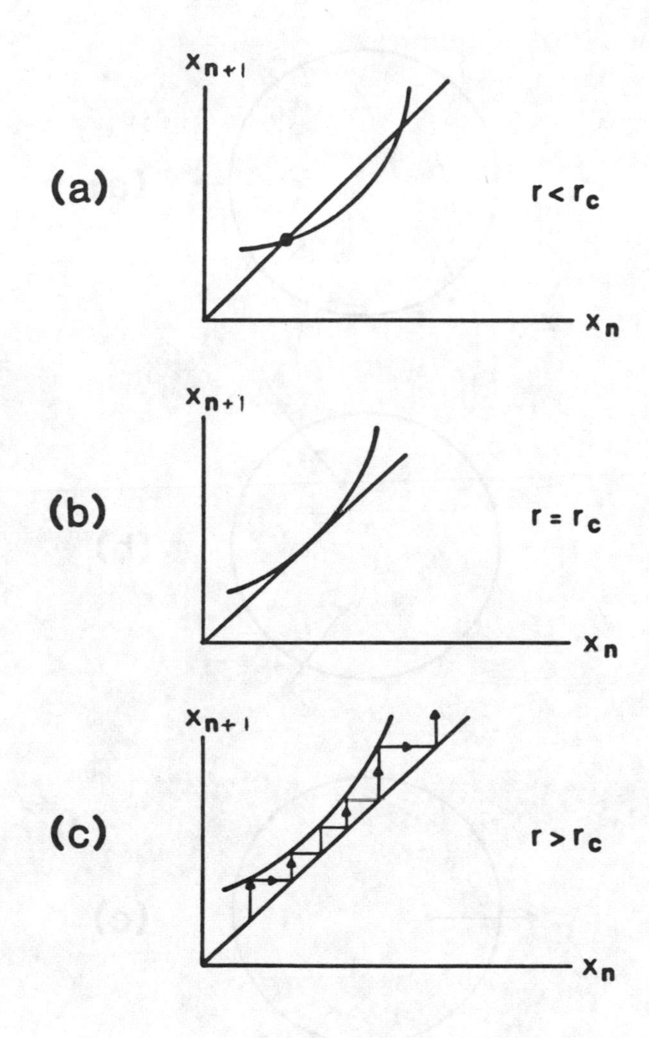

Figure 14.2 (a) and (b) An inverse tangent bifurcation as a knob r reaches a critical value r_c from below. For r slightly above r_c, many iterates of the map may be required to escape the channel near the point of tangency, as shown in part (c).

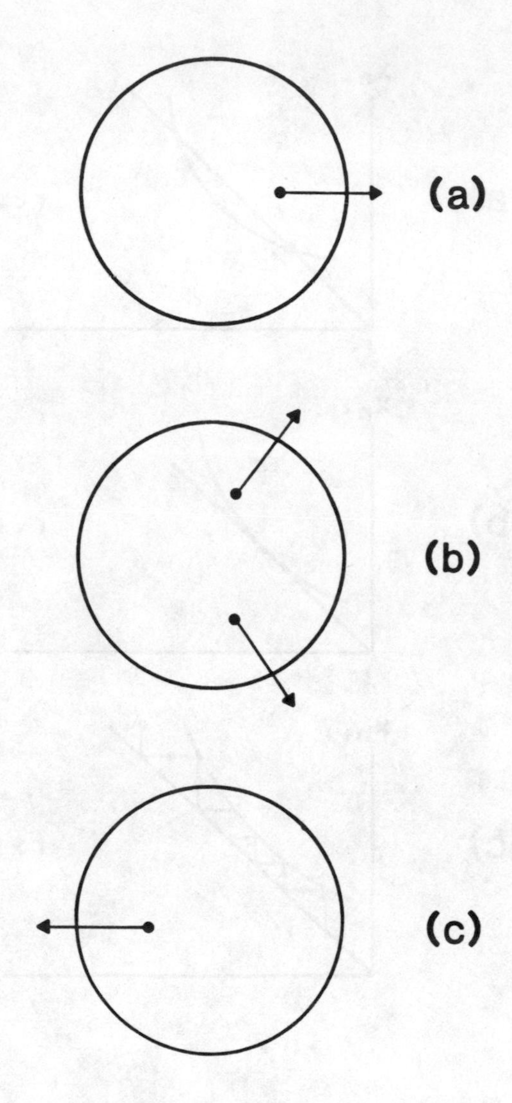

Figure 14.3 (a) A real eigenvalue of a linearized map crosses the unit circle at $x = +1$; (b) two complex conjugate eigenvalues cross the unit circle; and (c) a real eigenvalue crosses the unit circle at $x = -1$. These three ways by which stability is lost are associated with intermittency of types I, II, and III, respectively.

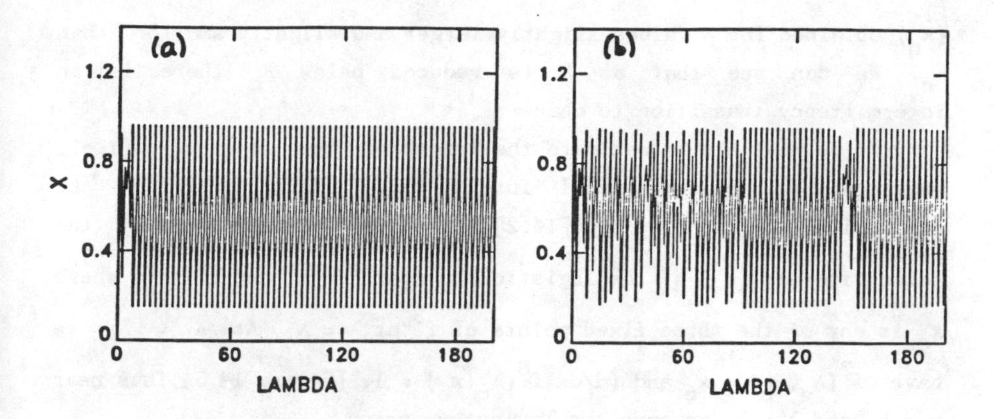

Figure 14.4 Intermittency in the logistic map. (a) $\lambda = \lambda_c + .002$, (b) $\lambda = \lambda_c - .0001$.

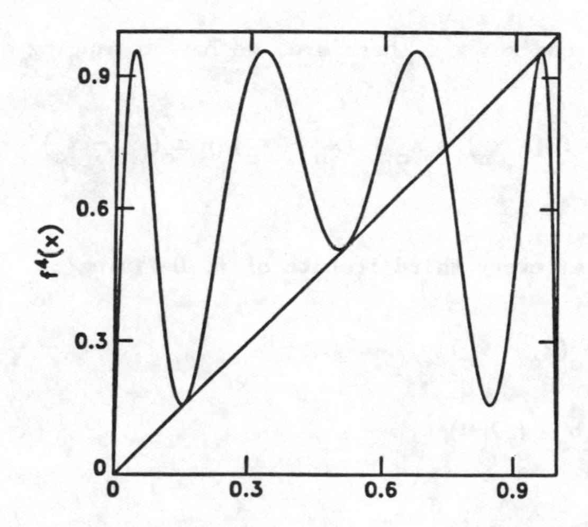

Figure 14.5 At $\lambda = \lambda_c$ in the logistic map the derivative of the third iterate map becomes unity at each of the three fixed points.

$\{x_n\}$ obtained for λ values slightly larger and slightly smaller than λ_c. We can see that as λ is reduced below λ_c there is an intermittency transition to chaos.

It is possible to estimate the length of the "laminar" (i.e., regular) regions for type-I intermittency. Here we follow the derivation by Hirsch, et al. [14.2] We consider (as an example) the third iterate (f^3) of the logistic map near $\lambda = \lambda_c$ and $x = x_c$, where x_c is any of the three fixed points of f^3 at $\lambda = \lambda_c$. At $\lambda = \lambda_c$ we have $f^3(\lambda_c,x_c) = x_c$ and $(d/dx)f^3(\lambda_c,x_c) = 1$. (Figure 14.5) Thus near $x = x_c$ and $\lambda = \lambda_c$ we have the Taylor expansion

$$f^3(\lambda,x) \cong x_c + (x - x_c) + a_c(x - x_c)^2 + b_c(\lambda_c - \lambda) \qquad (14.2)$$

Near $\lambda = \lambda_c$ and $x = x_c$, therefore, we have a mapping of the form

$$x_{n+1} = f^3(\lambda,x_n) = x_c + (x_n - x_c) + a_c(x_n - x_c)^2 + b_c(\lambda_c - \lambda) \qquad (14.3)$$

if we look at every third iterate of f. Defining

$$y_n = b_c(x_n - x_c) \qquad (14.4a)$$

$$a = a_c b_c \quad (> 0) \qquad (14.4b)$$

$$\epsilon = \lambda_c - \lambda \qquad (14.4c)$$

we can write (14.3) as

$$y_{n+1} = y_n + ay_n^2 + \epsilon \qquad (14.5)$$

The laminar regions are defined by the condition that $|y_n|$ is small, say $|y_n| < y_0$. Then we can replace (14.5) by the ODE

$$dy/dt = ay^2 + \epsilon \qquad (14.6)$$

to a good degree of approximation. Outside the range $|y_n| < y_0$, y_n is assumed to vary chaotically and then to re-enter the laminar region from some point y_{in}. Thus we solve (14.6) for the time $t = \ell(y_{in})$ it takes to reach $y = y_0$ starting out at y_{in}. This gives the amount of time spent in the laminar region:

$$\ell(y_{in}) = (1/\sqrt{a\epsilon})[\tan^{-1}y_0\sqrt{a/\epsilon} - \tan^{-1}y_{in}\sqrt{a/\epsilon}] \qquad (14.7)$$

The average length $\langle\ell\rangle$ of the laminar regions is therefore given by

$$\langle\ell\rangle = \int dy_{in}P(y_{in})\ell(y_{in}) = (1/\sqrt{a\epsilon})\tan^{-1}y_0\sqrt{a/\epsilon} \cong (\pi/2)/\sqrt{a\epsilon}$$

$$(14.8)$$

for $y_0\sqrt{a/\epsilon} >> 1$. Here $P(y_{in})$ is the probability distribution for the points y_{in} from which re-entry into the laminar regime occurs; (14.8) applies for any $P(y_{in})$ such that $P(y_{in}) = P(-y_{in})$ (i.e., for P symmetric about x_c). In other words, we have the scaling law

$$\langle\ell\rangle \approx (\lambda_c - \lambda)^{-1/2} \qquad (14.9)$$

for the average length of the laminar regimes. Numerical experiments with the logistic map have confirmed this scaling. Hirsch, et al. [14.2] have shown that for a mapping with maximum of order z, the scaling law (14.9) generalizes to

$$\langle\ell\rangle \approx (\lambda_c - \lambda)^{-(z-1)/z} \qquad (14.10)$$

The intermittency and period doubling routes to chaos are rather closely related, as might be surmised from the fact that both routes are found in the logistic map. The period doubling route occurs as λ increases toward λ_∞ from the left, and in the chaotic regime ($\lambda > \lambda_\infty$) as the periodic windows lose their stability as λ is increased. (Recall Section 10) However, if we decrease λ within a periodic window, the intermittency route to chaos occurs in the neighborhood of an inverse tangent bifurcation.

There are infinitely many periodic windows born of tangent bifurcations in the chaotic regime after period doubling _ad infinitum_, and so _we can expect intermittency to be a rather typical phenomenon, like period doubling_. In fact intermittent behavior has been observed for some time in fluid flows (before the theory of Pomeau and Manneville), and we will see that it has also been observed in lasers.

A Poincare' map of the form (14.5) near λ_c typifies type-I intermittency, whereas maps like

$$r_{n+1} = (1 + \epsilon)r_n + ar_n^3 \tag{14.11a}$$

$$\theta_{n+1} = \theta_n + \Omega \tag{14.11b}$$

(in polar coordinates) and

$$x_{n+1} = -(1 + \epsilon)x_n - ax_n^3 \tag{14.13}$$

typify type-II and type-III intermittency, respectively. In type-II and type-III intermittency the average length of a laminar region scales as ϵ^{-1} rather than $\epsilon^{-1/2}$.

Intermittency may also be treated by renormalization theory. [14.3, 14.4] Consider the map (14.5) with $\epsilon = 0$ (i.e., $\lambda = \lambda_c$), or a generalization of the form

$$x_{n+1} = f(x_n) = x_n + ax_n^z \tag{14.13}$$

Because of the term linear in x, the second iterate f^2 has the same behavior near $x = 0$. This suggests the possibility of finding a (functional) fixed point of the doubling operator T (cf. equation (9.10)):

$$Tg(x) = \alpha g(g(x/\alpha)) = g(x) \tag{14.14}$$

Instead of the boundary conditions $g(0) = 1$ and $g'(0) = 0$ used in the renormalization theory of the period doubling sequence, we now impose the conditions $g(0) = 0$ and $g'(0) = 1$. (In the period doubling case we chose $g'(0) = 0$ because we were looking at supercycles. Now we choose $g'(0) = 1$ because we are interested in a tangent bifurcation when an eigenvalue crosses the unit circle at +1.) With the present boundary conditions we have a rare situation where a renormalization group equation can be solved exactly. For $z = 2$, for instance, the solution of (14.14) is

$$g(x) = x/(1 - ax), \quad \alpha = 2 \tag{14.15}$$

Using such renormalization group results it is possible, for instance, to rederive scaling relations for the lengths of laminar regions.

15. From Quasiperiodicity to Chaos: The Ruelle-Takens-Newhouse Scenario

Perhaps the oldest scenario for the transition from laminar to turbulent flow is the one proposed by Landau in 1944. [15.1] It is supposed that as some parameter or "knob" characterizing the flow is increased, the number of spikes in the Fourier spectrum grows. It is clear that we can get behavior that looks quite complicated by adding just a few sine waves of different frequency, and so in the limit of an infinite number of (incommensurate) frequencies we might call the motion turbulent or "chaotic." In other words, we would have a transition to chaos by frequency proliferation. Another way to say this is that chaos would develop via an infinite sequence of Hopf bifurcations.

One difficulty with the Landau scenario, we now realize, is that it does not really describe a route to chaos per se: a spectrum with discrete frequencies (the spikes of Figure 1.1), however many, is characteristic of quasiperiodic motion, not chaos. The motion might certainly look very complicated, and we might even suspect at a glance that it is aperiodic, but it is not chaotic in the sense of very sensitive dependence on initial conditions, nor of having a broadband spectrum like that shown in Figure 1.2.

Aside from this, the Landau scenario is even "unlikely." In 1971 Ruelle and Takens [15.2] argued that, after the appearance of a few (incommensurate) frequencies in its spectrum, a system is more likely to go directly to chaos, without first undergoing more Hopf bifurcations. Originally Ruelle and Takens argued that after the onset of four incommensurate frequencies the motion would generally be unstable and give way to chaos under the action of small perturbations. Later Newhouse, Ruelle, and Takens [15.3] argued that small perturbations would cause the motion to become unstable after the appearance of just three incommensurate frequencies. In other words, after the appearance of two incommensurate frequencies a system is likely to become chaotic, because small perturbations would destroy the three-frequency motion. This route to chaos is therefore referred to sometimes as the two-frequency route.

In the Ruelle-Takens-Newhouse scenario a system becomes chaotic after two Hopf bifurcations, whereas the Landau scenario required an infinite sequence of Hopf bifurcations. It is possible to go from two-frequency to three-frequency motion without the onset of chaos, as discussed later. However, the two-frequency route does seem generic, in the sense that it has by now been observed in many different types of experiment. Before focusing more attention on the two-frequency (or Ruelle-Takens-Newhouse) scenario for the transition from order to chaos, we will summarize some background material on quasi-periodicity and N-tori. [15.4] This background will be useful also in our discussion later of chaos in Hamiltonian systems.

Consider an ODE system in some M-dimensional phase sapce. When we have an <u>N-frequency</u> <u>quasiperiodic</u> <u>orbit</u>, all the dependent variables $y_i(t)$ may be written in the form

$$y_i(t) = F_i(f_1 t, f_2 t, \ldots, f_N t), \; i = 1,2,3,\ldots,M \geq N \qquad (15.1)$$

where F_i is a function that is periodic with period one in all its arguments:

$$F_i(\psi_1 \pm 1, \psi_2 \pm 1, \ldots, \psi_N \pm 1) = F_i(\psi_1, \psi_2, \ldots, \psi_N) \qquad (15.2)$$

The frequencies $f_1, f_2, \ldots, f_N$ are incommensurate, meaning that the equation

$$\ell_1 f_1 + \ell_2 f_2 + \ldots + \ell_N f_N = 0 \qquad (15.3)$$

admits no solutions for integer ℓ_j other than the trivial one with all $\ell_j = 0$. Because of (15.2) we may express any of the $y_i(t)$ as a Fourier series of the form

$$y(t) = \sum Y_{n_1 n_2 \ldots n_N} e^{-2\pi i[n_1 f_1 + n_2 f_2 + \ldots + n_N f_N]t} \qquad (15.4)$$

i.e., N-frequency quasiperiodic motion has a discrete Fourier spectrum made up of spikes at linear combinations of N incommensurate frequencies. (Figure 1.1)

Another consequence of (15.2) is that the ψ_j may be defined modulo one and so we can think of them as angles. Thus for N = 2 we have two such angles, ψ_1 and ψ_2, and (ψ_1, ψ_2) may be regarded as a point on a 2-torus. (Figure 15.1) And because of (15.1) the motion of the variables $y_i(t)$ is also confined to a 2-torus. This motion may be parametrized by equations like

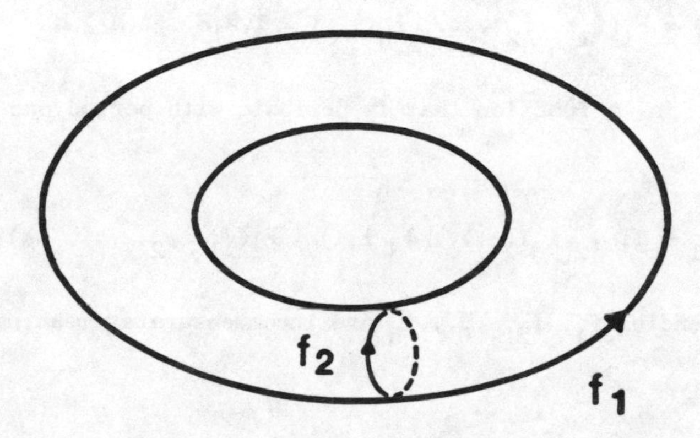

Figure 15.1 **A 2-torus for 2-frequency quasiperiodic motion.**

$$y_i = F_i(\psi_1, \psi_2), \quad i = 1,2 \tag{15.5}$$

More generally, N-frequency quasiperiodic motion is confined to an N-torus. Suppose we take a Poincare' surface of section cutting the N-torus at $\psi_N = 0$. The Poincare' map then characterizes the motion of the system at the discrete times $t_N = n/f_N$ and, from (15.1),

$$y_i(t_N) = F_i(f_1 n/f_N, f_2 n/f_N, \ldots, f_{N-1} n/f_N, 0)$$

$$= F_i(n\omega_1, n\omega_2, \ldots, n\omega_{N-1}, 0) \tag{15.6}$$

Thus the surface of section corresponds to an (N-1)-torus. Since F_i is periodic with period one in each of its arguments, it follows that

$$\overline{y}(f_N) = \overline{F}(\overline{\psi}_n) \tag{15.7}$$

in vector notation, where

$$\overline{\psi}_{n+1} = (\overline{\psi}_n + \overline{\omega}) \bmod 1 \tag{15.8}$$

and $\overline{\omega} = (\omega_1, \omega_2, \ldots, \omega_{N-1}) = (f_1/f_N, f_2/f_N, \ldots, f_{N-1}/f_N)$.

In the 2-frequency quasiperiodic case we have from (15.8) the map

$$\Theta_{n+1} = (\Theta_n + \Omega) \bmod 1 \tag{15.9}$$

The frequency ratio $\Omega = f_1/f_2$ is called the winding number. It is easy to see that for rational values $\Omega = p/q$ (p and q integers) of the winding number, the trajectories on the 2-torus close after q iterations of the Poincare' map (15.9). This defines what is called a <u>mode-locked</u> state in which the motion is periodic. For irrational winding numbers, however, the motion on the 2-torus is quasiperiodic.

In this case (15.9) covers the circle uniformly, and in fact the (quasiperiodic) motion on the 2-torus covers the whole torus. This ergodic behavior on the 2-torus for irrational Ω is not, of course, chaotic - remember that (15.9) is associated with a Poincare' map for 2-frequency quasiperiodic motion. To study the transition to chaos we must add nonlinearity. Following Arnold [15.5] it is conventional to work with the circle map

$$\theta_{n+1} = [\theta_n + \Omega - (K/2\pi)\sin 2\pi\theta_n] \bmod 1 \equiv T(\theta_n) \tag{15.10}$$

where $K > 0$ and $0 \leq \Omega < 1$. Note that $T(\theta + 1) = T(\theta)$ and, for $K < 1$, the map (15.10) is invertible, and so we can still associate it with a Poincare' map for motion on a 2-torus. (A Poinare' map must be invertible. This is a consequence of the uniqueness theorem for ODE.)

For $K > 1$ the circle map is found to exhibit chaotic behavior. The value $K = 1$ is therefore the transition point between order and chaos. In other words, it is at this point that the 2-torus is destroyed. (Motion on a 2-torus cannot, according to the Poincare'-Bendixson theorem mentioned in Section 2, be chaotic.)

In general the winding number is defined by

$$w(K,\Omega) = \lim_{n\to\infty}(1/n)[T^n(\theta) - \theta] \tag{15.11}$$

which obviously reduces to Ω when $K = 0$. In order to study the transition from quasiperiodicity to chaos as $K \to 1$, one would like to keep the winding number (15.11) fixed at a particular irrational. [15.6 - 15.8] This irrational has been chosen in renormalization group approaches to be the "most irrational" number, i.e., the Golden Mean $\overline{w} = (\sqrt{5} - 1)/2$, which has a continued fraction representation consisting of all ones:

$$\overline{w} = 1/(1 + 1/(1 + \ldots \tag{15.12}$$

The first few truncation approximations to $\overline{w}$ are 1/1, 1/2, 2/3, ..., and in general $w_n = F_n/F_{n+1}$, where the Fibonacci numbers F_n are defined by $F_0 = 0$, $F_1 = 1$, and $F_{n+1} = F_n + F_{n-1}$. Thus

$$\overline{w} = \lim_{n\to\infty} w_n = \lim_{n\to\infty}(F_n/F_{n+1}) \tag{15.13}$$

(This is made clear by writing $w_n = F_n/F_{n+1} = 1/(1 + F_{n-1}/F_n) = 1/(1 + w_{n-1})$. Then as $n \to \infty$, $\overline{w} = 1/(1 + \overline{w})$, and the positive solution of this equation is the Golden Mean.) Thus the w_n furnish a succession of rational approximations to the Golden Mean.

Let $\Omega_n(K)$ be the value of Ω, for fixed K, such that

$$T^{q_n}(0) = p_n \tag{15.14}$$

That is, $\Omega_n(K)$ generates a winding number of the circle map equal to $w_n = p_n/q_n$, and allows $\Theta = 0$ as a cycle element. (A winding number p_n/q_n implies a q_n-cycle.) It is assumed that the $\Omega_n(K)$ converge to that value of Ω, denoted $\overline{\Omega}(K)$, that gives $\overline{w}$ as a winding number:

$$\lim_{n\to\infty}\Omega_n(K) = \overline{\Omega}(K) \tag{15.15}$$

Shenker [15.6] obtained the following numerical results. First, the $\Omega_n(K)$ converge geometrically:

$$[\Omega_n(K) - \Omega_{n-1}(K)]/[\Omega_{n+1}(K) - \Omega_n(K)] \to \delta \tag{15.16}$$

where $\delta = -2.6180339... = -1/\overline{w}^2$ for $K < 1$, and $\delta = -2.83362...$ for K = 1. Second, the distances d_n from $\Theta = 0$ to the nearest element of a cycle belonging to $\Omega_n(K)$,

$$d_n = T^{F_n}(0) - F_{n-1} \tag{15.17}$$

scale like

$$d_n/d_{n+1} \to \alpha \tag{15.18}$$

where $\alpha = -1.618\ldots = -1/\bar{w}$ for $K < 1$ and $\alpha = -1.28857\ldots$ for $K = 1$.

There is an obvious resemblance of these results to those of the Feigenbaum universality theory for the period doubling route to chaos. In fact these results are universal for maps T such that (a) $T(\Theta + 1) = T(\Theta)$; (b) $T(\Theta)$ is invertible and differentiable (i.e., T is a diffeomorphism) for $K < 1$; and (c) $T^{-1}(\Theta)$ is nondifferentiable for $K = 1$, and does not exist for $K > 1$. In this case the numbers α and δ are universal (although they depend on the choice of the irrational $\bar{w}$), and an exact renormalization theory exists for the two-frequency transition to chaos. [15.7, 15.8]

16. Remarks

Experimental corroboration of universal scaling features in the two-frequency transition to chaos is complicated by the need to keep the winding number fixed while a nonlinear coupling parameter (K) is increased. Evidence for universal scaling behavior has nevertheless been obtained in a fluid flow experiment by Fein, et al., [16.1] who were able to obtain two-frequency quasiperiodic flow with frequency ratio $\bar{w}$ by a thermal modulation technique. The evidence is found in a power spectrum. According to Shenker [15.6] and Rand, et al. [15.8] the power spectrum at the transition to chaos has its main peaks at frequencies given by the Fibonacci numbers, and these peaks divide the spectrum into bands of similar structure at the low-frequency end. Such a power spectrum was measured by Fein, et al. at (or near) the critical Rayleigh number.

Another complication in the two-frequency route to chaos concerns mode locking (also known as frequency locking), which occurs when the winding number $w(K,\Omega)$ defined by (15.11) is a rational number p/q. In this case the motion prescribed by the circle map is periodic, being in fact just a q-cycle. This is obvious for $K = 0$ ($w(K,\Omega) = \Omega$). For $K > 0$ it is found that, for Ω values in an interval of length of order K^q near $\Omega = p/q$, the circle map has a unique q-cycle for each (positive) integer p for which p/q cannot be further reduced. If K is small a random choice for Ω is likely to give quasiperiodic (rather than periodic) motion. But as the nonlinear coupling parameter K increases the measure of the set of Ω values corresponding to periodic motion increases. This seems consistent with results of both numerical and laboratory experiments. As $K \to 1$ there are quasiperiodic "gaps" of Ω values from which the transition from 2-frequency quasiperiodicity to chaos can occur.

As noted earlier, the transition from 2-frequency quasiperiodicity to chaos has been observed in a variety of experiments, especially in fluid flows [16.2] and, as we will see later, in lasers. However, it is also possible to have a 1-frequency → 2-frequency → 3-frequency → chaos transition. Such a three-frequency route to chaos has been observed experimentally. [16.3 - 16.5]

In the last few years the genericness of the Ruelle-Takens-Newhouse scenario has come into question. [15.4, 16.6] The arbitrarily small perturbations that Ruelle, Takens, and Newhouse proved would destroy a 3-frequency quasiperiodic attractor (3-torus), and produce a structurally stable chaotic attractor, were assumed to have small first and second derivatives, but not necessarily small higher derivatives. Numerical experiments of Grebogi, Ott, and Yorke [16.6] suggest that, for N = 3 and 4, the likelihoods of occurrence of quasiperiodic motions for small (and smooth) perturbations are ordered as follows: N-frequency quasiperiodicity most likely, followed by (N-1)-frequency quasiperiodicity, ..., followed by periodicity. As the nonlinear coupling is increased, N-frequency quasiperiodicity becomes less likely, and does not occur at all when the map becomes noninvertible.

In any case we have now described three types of transition from order to chaos: period doubling, intermittency, and quasiperiodicity → chaos. The first two are better understood at present than the third. Period doubling may be diagnosed experimentally by spectral analysis, whereas intermittency is characterized by the scaling relations for the lengths of the laminar regions. The quasiperiodicity → chaos scenario is also investigated by spectral analysis, although the transition to chaos may be interrupted by mode locking or additional Hopf bifurcations.

Before finally getting down to the relevance of these ideas to laser-matter interactions, we will discuss some aspects of the chaotic attractors whose onset these various scenarios are meant to describe.

17. Strange Attractors, Dimensions, and Fractals

Our discussion thus far has been restricted by and large to dissipative dynamical systems. Consider a continuous flow described by (2.1), and let S be some surface in the N-dimensional phase space. Each point on S may be regarded as a possible initial condition for the system, and S will thus change its shape under the action of the flow defined by (2.1). The volume enclosed by S then evolves according to the equation

$$dV/dt = \int_S dS\vec{v}\cdot\hat{n} = \int_V d^N y \sum_{j=1}^{N} \delta F_j/\delta y_j \qquad (17.1)$$

where $\vec{v}$ is the velocity vector $(\dot{y}_1, \dot{y}_2, \ldots, \dot{y}_N)$, $\hat{n}$ is a unit normal to S, and the last step follows from the divergence theorem in N dimensions. The system is said to be dissipative if $dV/dt < 0$, so that $V \to 0$ as $t \to \infty$. Similarly a discrete mapping is said to be dissipative if the determinant of the Jacobian matrix has modulus less than unity. (Section 4)

Within some "basin of attraction" in phase space, a dissipative system will settle onto an attracting set or attractor. Lanford

[17.1] has given the following definition: A subset X of phase space is an attractor if (1) X is invariant under the flow; (2) there is a neighborhood around X that shrinks down to X under the flow; (3) no part of X is transient; and (4) X cannot be decomposed into two non-overlapping pieces. We can define the basin of attraction of X as the set of points in phase space that approach X as $t \to \infty$.

An attractor may be a simple fixed point or a limit cycle or an N-torus, but it may also be a chaotic attractor with the property of very sensitive dependence on initial conditions (at least one positive Lyapunov exponent). In other words, the contraction of volume characteristic of dissipative systems does not necessarily imply contraction of all lengths. Consider the following example of a two-dimensional map, called the baker's transformation (because it is reminiscent of a baker kneading dough by stretching, cutting, and stacking):

$$x_{n+1} = 2x_n \bmod 1 \tag{17.1a}$$

$$\begin{aligned} y_{n+1} &= ay_n \text{ for } 0 \leq x_n < 1/2 \\ &= ay_n + 1/2 \text{ for } 1/2 \leq x_n \leq 1 \text{ , } a < 1 \end{aligned} \tag{17.1b}$$

The x part of the transformation is the familiar Bernoulli shift (Section 1) and represents a stretching along x followed by a folding over back onto the unit interval if $x > 1$. The y part is a contraction for $x < 1/2$, and a contraction and displacement for $x > 1/2$. In Figure 17.1 we show what successive applications of the transformation with $a = .4$ do to a circle centered at $x = y = 1/2$. The mapping is certainly contracting volume (i.e., area in this two-dimensional example), but at the same time we know that (17.1a) leads to an exponential separation of initially close points. How does the system manage to contract volume while stretching lengths? It does so by forming an attractor that stretches but then folds over on itself. This produces a ribbon-like pattern characteristic of chaotic attractors; note the "structure within structure" developing

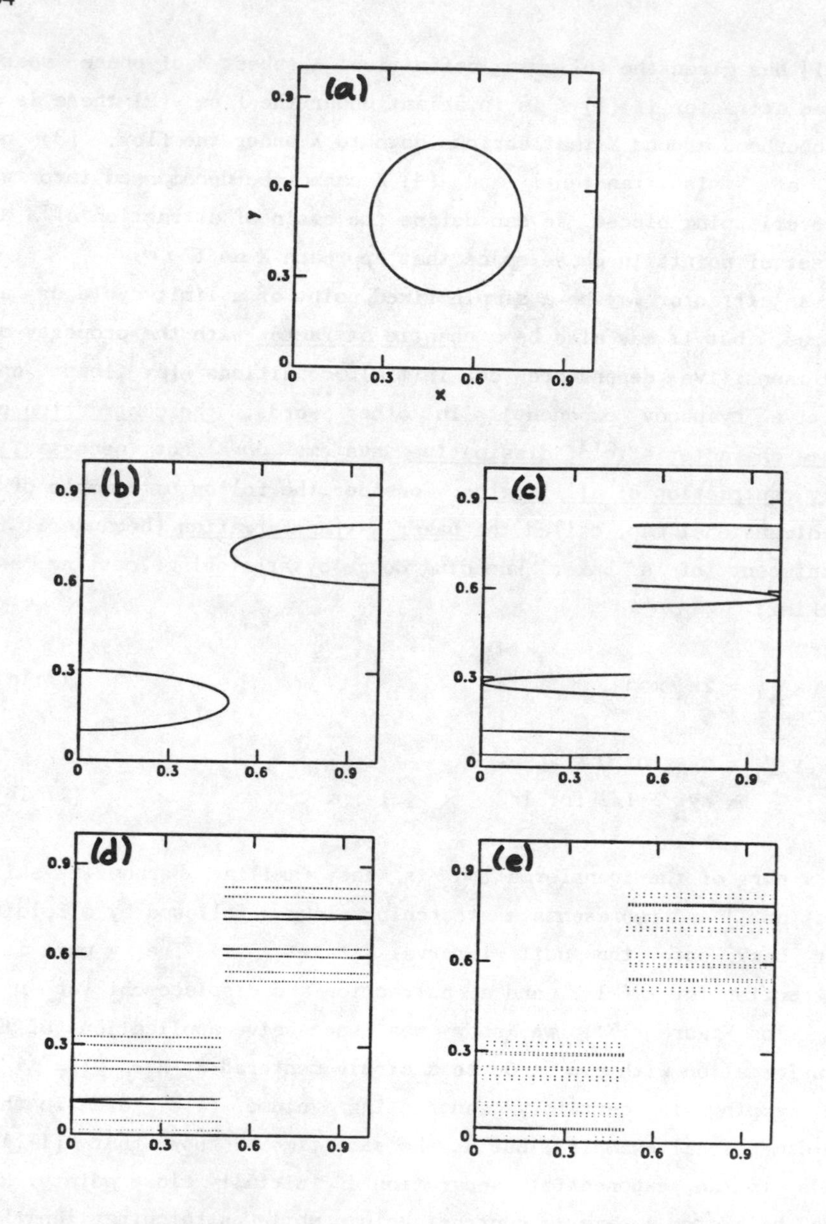

<u>Figure 17.1</u> **Effect of the dissipative baker's transformation with a = 0.4 on a circle (a) after one (b), two (c), three (d), and four (e) iterations.**

in Figure 17.1. In fact the actual attractor for the dissipative baker's transformation consists of an infinite set of horizontal lines. Its basin of attraction is the whole unit square.

This sort of structure inside structure means we have a strange attractor. The chaotic (strange) attractor developing in Figure 17.1 has one positive Lyapunov exponent (λ_x = log2 - recall Section 1) and one negative Lyapunov exponent (λ_y = loga). Thus we have stretching in the x-direction and contraction in the y-direction, and the net effect is area contraction.

A strange attractor has zero volume, but it does not necessarily have an integer dimension corresponding to a point or a line or some other obvious zero-volume set in the N-dimensional phase space. Here the generalized notion of dimension introduced by Hausdorff around 1919 is useful. If we have an object of dimension D, and increase each linear dimension by a factor ℓ, its volume is increased by a factor $k = \ell^D$, or in other words

$$D = \log k/\log \ell \tag{17.2}$$

If we take this as a definition of , then D need not have integer dimension, as we will see shortly. Equation (17.2) provides a (loose) definition of the Hausdorff dimension.

A classic example of a set with fractional Hausdorff dimension is the Cantor set, which is defined by the following construction. Divide the line segment [0,1] into thirds and remove the middle (open) third. Do the same with the remaining two segments, and continue the process of removing middle thirds. The set obtained by continuing *ad infinitum* is the Cantor set (or an example of a Cantor set). The Hausdorff dimension of the Cantor set is easily calculated. If we multiply the segment [0,1/3] by $\ell = 3$, for instance, we produce two such segments, so that k = 2 and D = log2/log3 = 0.6309...

Another definition of dimension is the fractal dimension or capacity. To motivate this definition, consider that a unit cube contains $N(\epsilon) = \epsilon^{-3}$ cubes of side ϵ, a unit square contains $N(\epsilon)$ =

ϵ^{-2} squares of side ϵ, and a unit line segment contains $N(\epsilon) = \epsilon^{-1}$ segments of length ϵ. In each case the dimension of the object is $\log N(\epsilon)/\log(1/\epsilon)$. The capacity is defined as

$$D_c = \lim_{\epsilon \to 0} \log N(\epsilon)/\log(1/\epsilon) \tag{17.3}$$

where ϵ is in general a hypercube in a phase space of arbitrary dimension. D_c is often called the Hausdorff dimension, because D and D_c are conjectured to be equal. Equation (17.3) is convenient because it furnishes an explicit, box-counting algorithm for computing dimension, although in general this counting procedure is computationally expensive. Mandelbrot [17.2] defines a _fractal_ as an object having a fractal dimension greater than its topological dimension. The Cantor set, for instance, has topological dimension 0 and so it is an example of a fractal.

Chaotic attractors almost always seem to be strange attractors, i.e., they have non-integer Hausdorff dimension. The structure of the strange attractor, however, does not tell us anything about the frequency with which different points on the attractor are visited in the course of the flow. It is therefore useful to modify the definition (17.3) by covering the attractor with cells and weighting the _jth_ cell by its visitation probability P_j. This gives the _information dimension_

$$D_I = \lim_{\epsilon \to 0} I(\epsilon)/\log(1/\epsilon) \tag{17.4}$$

where

$$I(\epsilon) \equiv -\sum_{j=1}^{N(\epsilon)} P_j \log P_j \tag{17.5}$$

Note that if each cell has the same visitation probability, so that $P_j = 1/N(\epsilon)$ for all j, then $I(\epsilon) = N(\epsilon)$ and D_I reduces to D_c. In

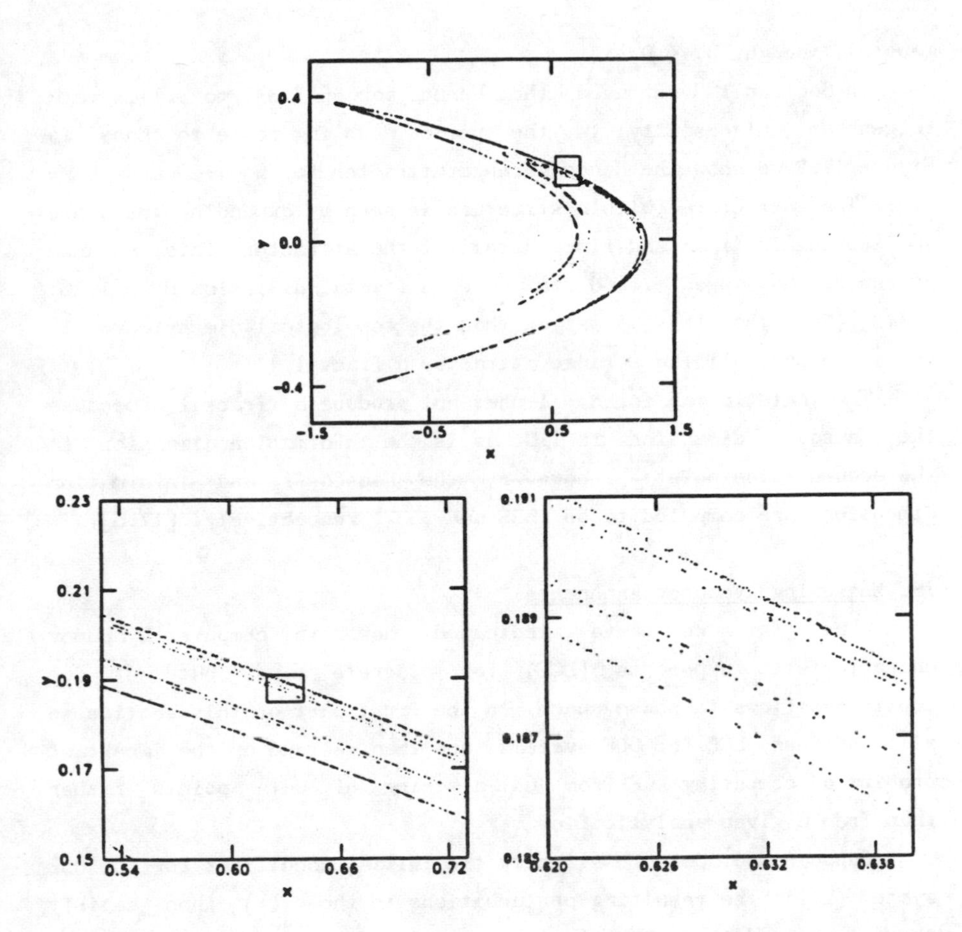

Figure 17.2 Structure within structure of the Hénon strange attractor, obtained by successively expanding the scale of resolution.

general, though, $D_I \lesssim D_c$.

In Section 10 we used the Hénon map [17.3] to illustrate Feigenbaum universality in the period doubling route to chaos. In Figure 17.2 we show the Hénon strange attractor for a =1.4 and b = 0.3. The structure within structure is seen by expanding the scale and looking in finer and finer detail at the attractor. This strange attractor has been found to have a fractal dimension $D_c \cong 1.26$. [17.4] (Note that this is larger than the topological dimension of 1, and so we can call the strange attractor a fractal.)

The logistic map for $\lambda = 1$ does not produce a fractal, because the Hausdorf dimension is 1.0, as is the information dimension. At the accumulation point λ_∞, however, the Hausdorff and information dimensions are computed to be .538 and .537, respectively. [17.5]

18. Measuring Lyapunov Exponents

Thus far we have indicated how to compute Lyapunov characteristic exponents (LCE) for discrete maps, but not for continuous flows in phase space. In the first part of this section we will discuss LCE for ODE systems, and then we take up the important problem of computing LCE from just a string of data points rather than from a given analytic flow.

Suppose we perturb slightly the initial conditions for the ODE system (2.1). The resulting perturbations in the $y_j(t)$ then satisfy the (linearized) equations

$$\delta\dot{y}_j(t) = \sum_{i=1}^{N} (\delta F_j/\delta y_i)\delta y_i(t) \tag{18.1}$$

or, in matrix form,

$$\delta\dot{y}(t) = J(t)\delta y(t) \tag{18.2}$$

The vector δy is said to belong to the tangent space of the original system (2.1). The formal solution of (18.2) is

$$\delta y(t) = \exp[\int_0^t dt' J(t')]\delta y(0) \equiv \Lambda(t)\delta(0) \tag{18.3}$$

where the N × N matrix Λ satisfies the equation

$$\dot{\Lambda}(t) = J(t)\Lambda(t) \tag{18.4}$$

We may define the N LCE

$$\chi_j = \lim_{t\to\infty}(1/t)\log|\lambda_j(t)| \tag{18.5}$$

where the $\lambda_j(t)$ are the eigenvalues of the matrix $\Lambda(t)$.

If one of the χ_j is greater than zero, we have very sensitive dependence on initial conditions. To determine whether a system is chaotic, we therefore need to compute only the largest LCE. In Section 4 we described how to compute the largest LCE for a discrete map. An analogous procedure is applicable to continuous flows: solve the 2N equations (2.1) and (18.1) simultaneously, assuming "random" values of the initial perturbations $\delta y_j(0)$. Compute along the way the quantity

$$\chi(t) = (1/t)\log||\delta y(t)|| \tag{18.6}$$

where $||\ldots||$ denotes the Euclidean norm:

$$||\delta y|| = [\sum_{j=1}^{N} (\delta y_j)^2]^{1/2} \tag{18.7}$$

(A different definition of the norm may be used.) Then

$$\chi = \lim_{t\to\infty}\chi(t) \qquad (18.8)$$

(or more rigorously, the limit superior, lim sup) is the largest LCE of the system. The system is chaotic if and only if $\chi > 0$. This technique of computing the largest LCE, which is described in detail by Benettin, et al. [18.1], works for the same reason as the analogous procedure for maps: only the "direction" of greatest stretching survives the $t \to \infty$ limit. (If we imagine each of the $\delta y_j(t)$ to grow like $\exp(\chi_j t)$, then the procedure using (18.7) and (18.8) yields the largest χ_j, as the reader may easily show.)

Benettin, et al. [18.1] show how to compute the full spectrum of N LCE. Since Wolf, et al. [18.2] have given a clear description of the method, together with a FORTRAN listing that may be used for computing the LCE spectrum of an ODE system, we will not go into the general method here.

In practice we might have a string of experimental data points and we want to know whether this time series is chaotic. [18.2, 18.3] That is, we would like to be able to compute the largest LCE associated with a single time series rather than a given system of equations. A method for making such a computation has been described by Wolf, et al., [18.2] and we summarize here the basic idea for computing the largest (non-negative) LCE. (Note: It may be shown that at least one LCE is zero if the system has no fixed point, [18.4] and so a computation of the largest non-negative LCE is enough to determine whether the system is chaotic.)

The first step is to "construct" the attractor from the time series, as follows. From the experimental time series x(t), construct a set of points in an m-dimensional phase space:

$$\vec{x}_i = (x(t_i), x(t_i + \tau), \ldots, x(t_i + (m-1)\tau) \qquad (18.9)$$

where τ is some fixed time delay and m is called the embedding dimension. It is assumed that the set of points so constructed will accurately represent the properties of the actual attractor of the system, if m is large enough. [18.5] Figure 18.1 illustrates the Wolf algorithm for computing χ, the largest LCE. Begin with the first data point $\vec{x}(t_0)$ and its nearest neighbor (in the sense of Euclidean norm) $\vec{y}_0(t_0)$, which is at some distance L_0 from $\vec{x}(t_0)$. Let both points evolve until their separation

$$L_0' = ||\vec{x}(t_1) - \vec{y}_0(t_1)|| \tag{18.10}$$

exceeds some pre-chosen number ϵ. Retain $\vec{x}(t_1)$ and find a near neighbor $\vec{y}_1(t_1)$ such that its distance L_1 from $\vec{x}(t_1)$ is less than ϵ and such that it is as nearly as possible (from the data) in the same direction as $\vec{y}_0(t_1)$. Continue this process until the end of the time series is reached. The largest LCE is then estimated from the formula

$$\chi = \frac{1}{t_N - t_0} \sum_{i=0}^{N-1} \log(L_i'/L_i) \tag{18.11}$$

This algorithm has been found in numerical experiments to give good approximations to χ with relatively small numbers of points characterizing the attractor. [18.2] In their paper Wolf, et al. [18.2] also provide a FORTRAN listing of a code for estimating χ by the procedure just described.

An alternative approach has been described by Eckmann and Ruelle. [18.3] Basically it involves a least-squares estimate of the Jacobian of the system. Based on a few numerical experiments it has been argued that the Wolf algorithm may be preferable. [18.6]

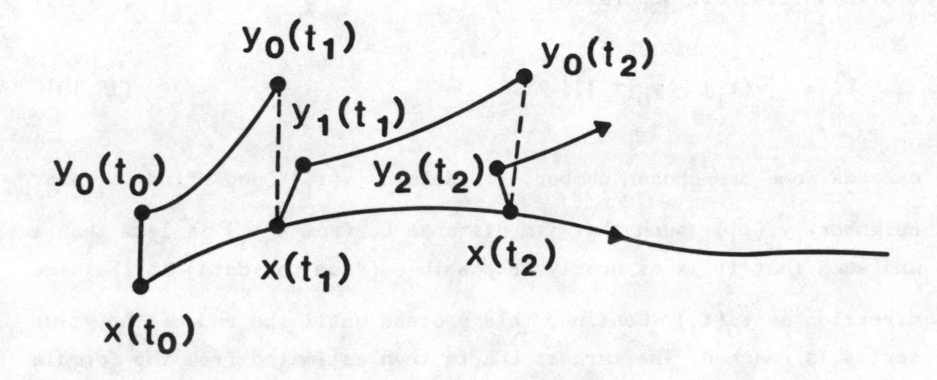

<u>Figure 18.1</u> **Illustrating the Wolf algorithm for computing the maximal LCE from a data set. (See text)**

19. Measuring Dimensions

It is convenient to define an infinite set of dimensions. Let $\vec{y}(t) = (y_1(t), y_2(t), \ldots, y_N(t))$ denote a trajectory in our N-dimensional phase space and construct from the trajectory a sequence of points $\vec{y}(0)$, $\vec{y}(\tau)$, $\vec{y}(2\tau)$, ..., $\vec{y}(M\tau)$ in phase space, where M is large. Partition the phase space into $N(\epsilon)$ cells of side ϵ and let M_i be the number of points appearing in cell i. Define $P_i = M_i/M$ as the probability of finding a point in cell i. Then the dimensions D_n are defined by

$$D_n = \lim_{\epsilon \to 0} \left(\frac{1}{n-1}\right) \log\left(\sum_{i=1}^{N(\epsilon)} P_i^n\right) / \log\epsilon \tag{19.1}$$

Note that

$$D_0 = -\lim_{\epsilon \to 0} \log N(\epsilon)/\log\epsilon = D_c \tag{19.2}$$

and

$$D_1 = \lim_{\epsilon \to 0} \left(\sum_{i=1}^{N(\epsilon)} P_i \log P_i\right) / \log\epsilon = D_I \tag{19.3}$$

That is, D_0 and D_1 reduce to the fractal and information dimensions defined earlier.

If $P_i = 1/N(\epsilon)$ for each cell (i.e., if the cells are uniformly filled) then $D_0 = D_1$ and the Hausdorff (fractal) and information dimensions are the same. In general $D_1 \le D_0$ and in fact $D_n \le D_m$ for $n > m$. For the logistic map with $\lambda = 1$ it is found that $D_0 = D_1 = D_\infty = 1$, whereas at $\lambda = \lambda_\infty$, $D_0 = .538$, $D_1 = .537$, $D_2 = .500$, and $D_\infty = .394$. [17.5] The difference $D_0 - D_1$ provides some measure of the nonuniformity with which different parts of the attractor are covered by the flow.

The dimension D_2 is related to the <u>correlation</u> <u>integral</u>

$$C(\epsilon) \equiv \lim_{M\to\infty}(1/M^2) \sum_{i,j} \theta[\epsilon - |\vec{y}_i-\vec{y}_j|] \tag{19.4}$$

where θ is the unit step function. To see this, note that $\sum P_i^2$ is the probability that any two points lie in the same cell. This is approximately the probability that two points on the attractor are within a distance ϵ of each other, which is given by (19.4). Thus the <u>correlation</u> <u>dimension</u>

$$D_2 = \log C(\epsilon)/\log\epsilon \tag{19.5}$$

This relation is useful because it circumvents the use of a box-counting algorithm for computing D_2.

In practice we might have only an experimental data set to work with, as discussed in the preceding section. In this case we again assume that a single time series $y_j(t)$ (such as $x(t)$ or $y(t)$ or $z(t)$ in the Lorenz model) will provide us with an accurate representation of the motion on the attractor if we use a large enough embedding dimension. For the Lorenz system, for instance, it has been found that the same value D_2 is obtained regardless of whether points $(x(t_i), y(t_i), z(t_i))$ or $(x(t_i), x(t_i+\tau), x(t_i+2\tau))$ are used to reconstruct the attractor. [19.1] Another example is provided by the Mackey-Glass system [19.2]

$$\dot{x}(t) = \frac{ax(t-\tau)}{1 + [x(t-\tau)]^{10}} - bx(t) \tag{19.6}$$

which provides a model for the regeneration of blood cells. Of course this time-delay system is infinite dimensional, but it is found that essentially the same value $D_2 \cong 1.95$ (for $a = .2$, $b = .1$, $\tau = 17$) is obtained for embedding dimensions $m = 3$, 4, and 5. [17.5] Thus we can define the embedding dimension to be the smallest m (Equation (18.9)) above which the computed correlation dimension D_2 does not change.

One suspects there might be a connection between the dimensions of an attractor and its LCE spectrum. [19.3] In a three-dimensional phase space, for instance, three negative LCE imply a fixed point; two negative and one zero LCE imply a limit cycle; two zero and one negative LCE imply a 2-torus; and one positive, one negative, and one zero LCE imply a chaotic (presumably strange) attractor. The dissipative baker's transformation considered in Section 17 suggests there may be a more quantitative connection. In this case the LCE are $\chi_1 = \log 2$ (for the x-direction) and $\chi_2 = \log|a|$ (for the y-direction). The Haussdorff dimension may be shown to be $D_0 = \log|2a|/\log|a|$, and so we have the relation

$$D_0 = 1 + \chi_1/|\chi_2| \tag{19.7}$$

Kaplan and Yorke (See [17.4]) have conjectured that more generally

$$D_0 = n + \sum_{j=1}^{n} \chi_j/|\chi_{n+1}| \tag{19.8}$$

for a strange attractor. Here the LCE are ordered as $\chi_1 > \chi_2 > \chi_3 \cdots$ and n is the largest integer for which

$$\sum_{j=1}^{n} \chi_j > 0. \tag{19.9}$$

The Kaplan-Yorke conjecture has been checked numerically by Russell, <u>et al</u>., [17.4] and it appears to have precise validity only for uniform attractors. For nonuniform attractors it may be a good approximation, [17.4] but we do not know of any rigorous statements that can be made.

The Grassberger-Procaccia algorithm based on (19.5) has been used by various experimentalists to measure dimension. A small value of measured dimension, compared with a presumably large-dimensional phase space, is an indication that the dynamics is confined to a low-dimensional attractor embedded in the phase space. The idea that deterministic chaos in dissipative systems might be associated with motion on a low-dimensional (strange) attractor appears to have originated in the paper by Ruelle and Takens in 1971. [15.2]

20. Kolmogorov Entropy

The Kolmogorov entropy of a system is defined in such a way that, in most instances, it is equal to the sum of the positive Lyapunov exponents. It provides an estimate of the average time over which accurate predictions can be made about a chaotic system, before long-term predictability is lost due to the sensitivity to initial conditions. Since the basic concept is tied to the notion of information-theoretic entropy, we begin by briefly summarizing the latter.

Suppose a system has two possible states (like heads or tails of a coin). Then its maximum information content is 1 binary digit (bit), since the answer to one question (heads or tails?) fully specifies the state of the system. Similarly if the system has four states, its maximum information content $I = 2$ bits, since the answer to <u>two</u> questions with "yes" or "no" answers determines which of the four possible states the system is in. (If the states are represented by four slots in a line, we can ask first whether the system is in one of the left two slots, and then one more question nails down the state.) In general $I = \log_2 N$, where N is the number of states of the system.

Information is gained when we make a measurement on a system. Thus a coin flip yields 1 bit of information. Similarly the answer (yes or no) to a question about the state of a four-state system provides 1 bit of information, because it reduces by 1 the number of questions we have to ask in order to determine the state the system is in.

Suppose a system has N states, and the <u>ith</u> state is known <u>a</u>

priori to occur with probability P_i. We define the entropy

$$S = -\sum_{i=1}^{N} P_i \log_2 P_i \tag{20.1}$$

Entropy is a measure of the amount of information necessary to determine the state of the system. For example, one bit of information is necessary to determine the state of a coin with equal a priori probabilities of heads or tails ($S = 1$), but if the coin were biased so that it always comes up heads, then no more information is necessary to know its state ($S = 0$). Anytime one of the $P_i = 1$, of course, $S = 0$, because a measurement (or the answer to a question with a yes or no answer) doesn't tell us anything we didn't already know. And whenever all the P_i are equal, then $S = \log_2 N$, the maximum for a system with N states. This is because we didn't know anything about the system to begin with (i.e., we had no a priori knowledge of which states were more likely than others) and so a determination of the state of the system increases our knowledge by the maximum possible amount. As with the thermodynamical notion of entropy, S here is also a measure of "disorder." Thus in the "orderly" case ($P_i = 1$ for some i) $S = 0$, whereas in the "disorderly" case ($P_i = 1/N$ for all i) S has its maximum value.

Consider a trajectory $\vec{y}(t) = (y_1(t), y_2(t), \ldots, y_N(t))$ and partition the phase space into n hypercubes of side ϵ. Let $P(i_0, i_1, \ldots, i_n)$ be the joint probability that the point $\vec{y}(0)$ lies in the i_0th cell, $\vec{y}(\tau)$ lies in the i_1th cell, ..., and $\vec{y}(n\tau)$ lies in the i_nth cell. Then, from the above discussion,

$$K_n \equiv -\sum_{i_0 \ldots i_n} P(i_0, i_1, \ldots, i_n) \log_2 P(i_0, i_1, \ldots, i_n) \tag{20.2}$$

is a measure of the amount of information necessary to specify the trajectory to within a precision ϵ, assuming only the probabilities $P(i_0, i_1, \ldots, i_n)$ are known a priori. It follows that $K_{n+1} - K_n$ is the additional amount of information required to specify which cell $\vec{y}(n\tau+\tau)$ will fall in. The Kolmogorov entropy may be defined as [20.1]

$$K = \lim_{\tau\to 0} \lim_{\epsilon\to 0} \lim_{N\to\infty} (1/N\tau) \sum_{n=0}^{N-1} (K_{n+1} - K_n)$$

$$= - \lim_{\tau\to 0} \lim_{\epsilon\to 0} \lim_{N\to\infty} (1/N\tau) \sum_{i_0 .. i_N} P(i_0, i_1, \ldots, i_N) \log_2 P(i_0, i_1, \ldots, i_N) \qquad (20.3)$$

We see that K is the average rate of information loss.

For non-chaotic systems, $K = 0$, i.e., there is no loss of information because initially close points on a trajectory remain close together as time evolves. For chaotic systems, however, initially close points separate exponentially on average, and therefore joint probabilities for cell occupations decrease exponentially with time. Thus $K > 0$ for chaotic systems. For truly (non-deterministic) random systems, initially close points take on a statistical distribution over all the allowed new cells. Thus if $P(i_0) \approx \epsilon$, then $P(i_0, i_1) \approx \epsilon^2$, etc., and so $K \to \infty$ as $\epsilon \to 0$ for pure randomness. The K-entropy is therefore useful not only for distinguishing regular from chaotic behavior, but also for distinguishing deterministic chaos from noise.

For one-dimensional chaotic maps the K entropy and the LCE are identical (except for possible differences depending on whether $\log_{10}$ or $\log_2$ is used). For higher-dimensional chaotic systems information is lost as a consequence of the stretching apart of initially close points due to all the positive LCE. It is therefore reasonable that the K entropy should be just the sum of the positive LCE. Actually K can be defined as [20.2]

$$K = \int d^N y \; p(\vec{y}) \sum_i \chi_i^{(+)}(\vec{y}) \tag{20.4}$$

where $p(\vec{y})$ is the invariant distribution for the attractor (Section 4) and the $\chi_i^{(+)}(\vec{y})$ are the positive LCE. In many cases in dissipative systems the $\chi_i^{(+)}$ are found to be independent of $\vec{y}$, and we have just

$$K = \sum_i \chi_i^{(+)} \tag{20.5}$$

(Actually the K-entropy is typically defined for a given region in which $\chi_i^{(+)}(\vec{y})$ is independent of $\vec{y}$.)

Consider the logistic map in the chaotic regime. If the initial state (x_0) is specified with a precision ϵ, then after n iterations the uncertainty in the state (x_n) is $\approx \epsilon\exp(\chi n)$, where χ is the LCE. If this uncertainty is larger than 1 we cannot make precise estimates of the state; we can only make statistical statements based on the invariant distribution. In other words, we can only make accurate predictions about x_n for

$$n < (1/\chi)\log(1/\epsilon) \tag{20.5}$$

More generally the LCE should be replaced by the K-entropy, [20.1] and for a continuous flow n is replaced by the time t. Thus

$$T \approx (1/K)\log(1/\epsilon) \tag{20.6}$$

is about the time over which the state of a chaotic system may be predicted, given an initial-state specification within an accuracy ϵ (i.e., within a hypercube of side ϵ).

21. Noise

An important consideration in connection with the experimental study of bifurcations and the transition to chaos is the influence of noise. We now discuss briefly some results of numerical experiments on the influence of noise in model systems.

Mayer-Kress and Haken [21.1] have studied the influence of noise on the logistic map by adding to each x_n a random number ξ_n:

$$x_n = f(x_{n-1}) + \xi_n = ax_{n-1}(1-x_{n-1}) + \xi_n \tag{21.1}$$

(We use a instead of 4λ to conform to the notation of Mayer-Kress and Haken.) The ξ_n were taken to be uniformly distributed on an interval $[-\beta, \beta]$, and therefore had zero mean and standard deviation $\sigma = \beta/\sqrt{3}$. (The ξ_n are actually pseudo-random, being generated by a random-number generator on a computer. That is, they are produced by a deterministically chaotic mapping with a large, positive LCE.)

Mayer-Kress and Haken observed that their numerically generated probability distributions for $\{x_n\}$ became stationary after $\approx 10^6$ iterations and appeared to be independent of x_0 and ξ_1. Figure 21.1 shows results for $a = a_4 = 3.498$, in which case the logistic map without noise ($\sigma = 0$) has a super-stable 4-cycle. In (a) the noise parameter $\sigma = .001$. What is plotted is the numerically determined probability distribution for 10^7 iterates and a partition of the unit interval into 1000 subintervals. The distribution is broadened but it is still concentrated around the periodic points, i.e., there is a "semiperiodic" orbit of period 4. At a noise level $\sigma = .011$ two of the peaks overlap and we have a semiperiodic orbit of period 2, as shown in (b). For increasing σ the two humps merge together and the distribution becomes flatter, as shown in (c) and (d).

Figure 21.2 shows what happens for $a = a_3 = 3.831...$, where the unperturbed logistic map has a super-stable 3-cycle. In (a) the noise level $\sigma = .00035$, but in (b) and (c) $\sigma = .001$ and $.004$, respectively. The situation here is different from the case $a = a_4$. (Compare Figures 21.1a and 21.2b.) Evidently the orbit of iterates can escape from the period-3 "islands" and remain for a while on the aperiodic

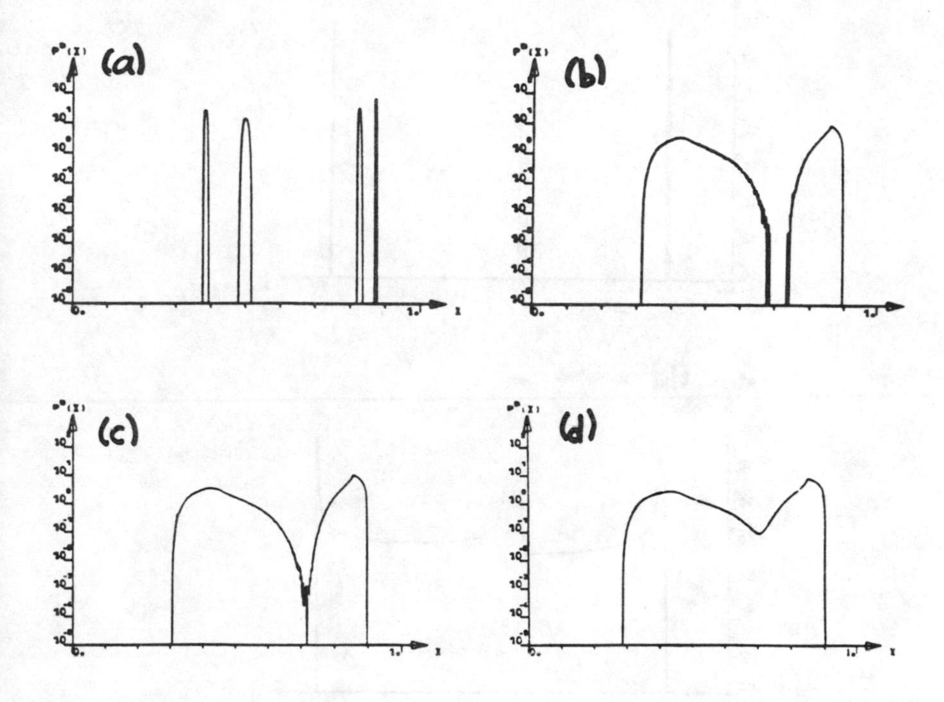

Figure 21.1 Probability distribution for the logistic map with $a = a_4 \cong 3.498$ and noise level (a) $\sigma = .001$, (b) $\sigma = .011$, (c) $\sigma = .012$, (d) $\sigma = .015$. From G. Mayer-Kress and H. Haken, Reference [21.1], with permission.

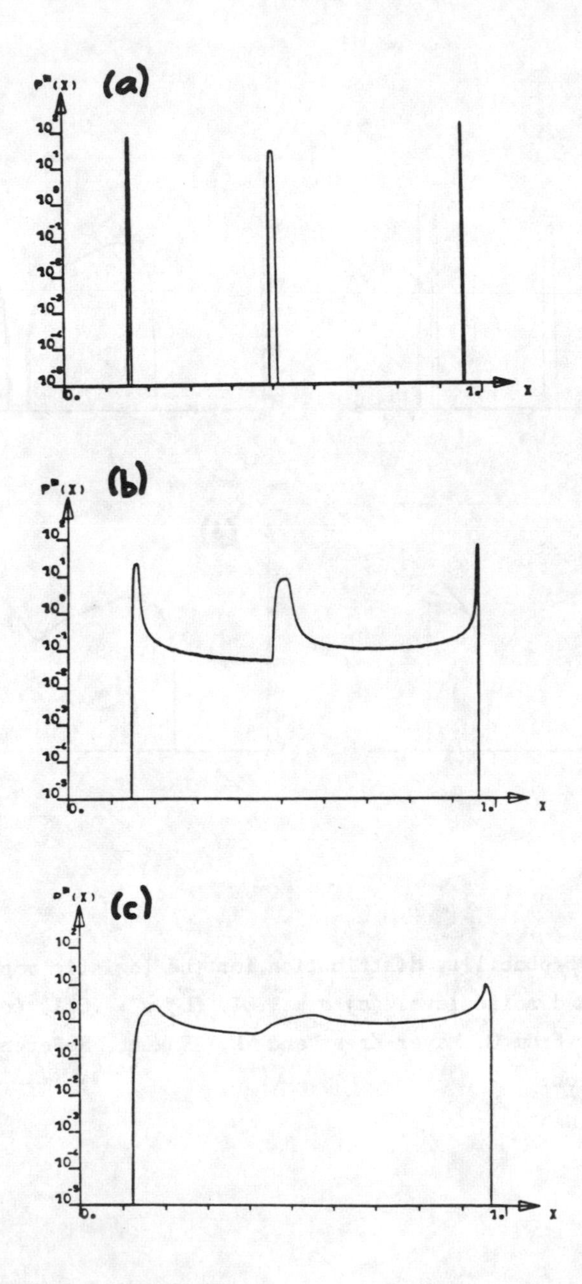

Figure 21.2 Probability distribution for the logistic map with $a = a_3 \cong 3.831$ and (a) $\sigma = .00035$, (b) $\sigma = .001$, (c) $\sigma = .004$. From G. Mayer-Kress and H. Haken, Reference [21.1], with permission.

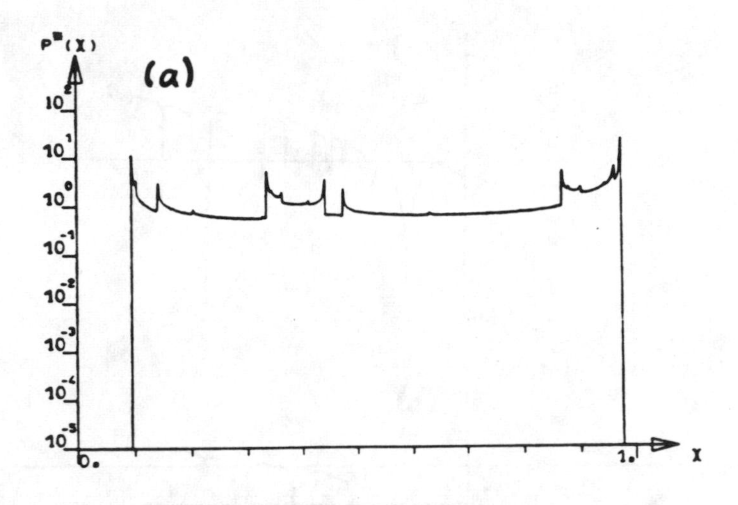

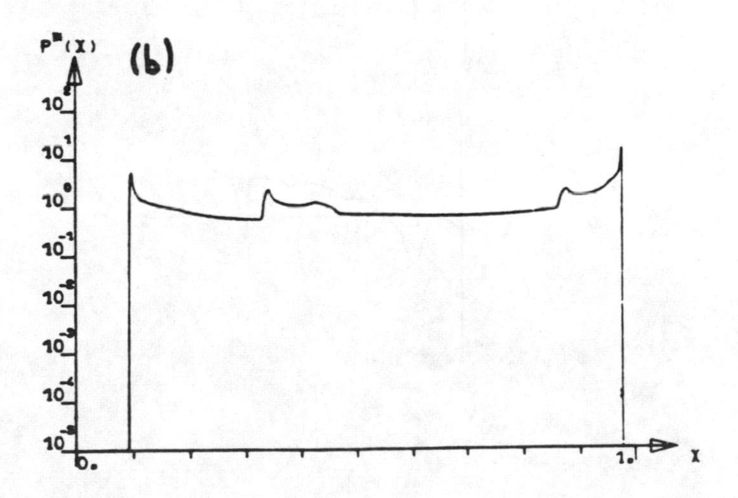

Figure 21.3 Probability distribution for the logistic map with a = 3.9 and (a) σ = 0.0, (b) σ = .001. From G. Mayer-Kress and H. Haken, Reference [21.1], with permission.

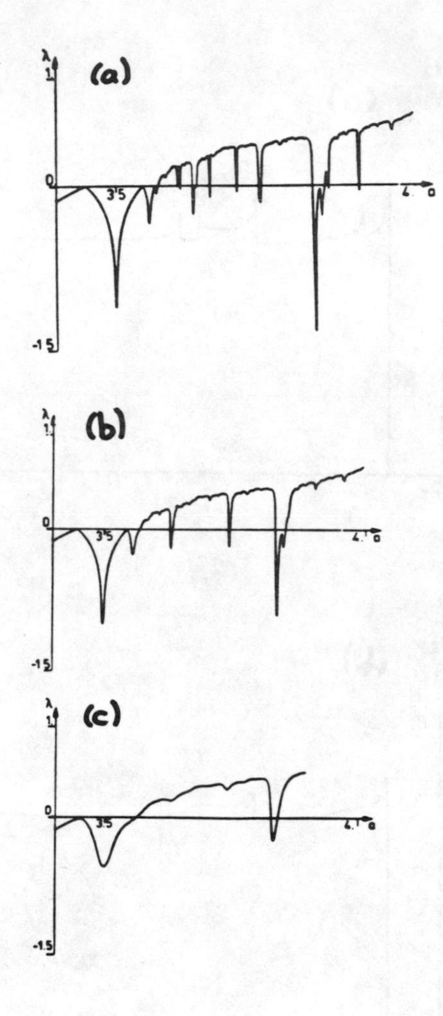

Figure 21.4 Effect of noise on the LCE for the logistic map. In (a) the noise level $\sigma = 0.0$. In (b) and (c) σ is .00015 and .001, respectively. From G. Mayer-Kress and H. Haken, Reference [21.1], with permission.

sets (which are not present for $a = a_4$ but are present for $a = a_3$, albeit they are unstable) before getting caught again on the periodic islands. [21.1] Figure 21.3 shows the probability distribution for a chaotic case (a = 3.9) with and without noise.

Mayer-Kress and Haken also consider the effect of noise on the LCE and correlations. (Figure 21.4) For $\sigma = .001$, for instance, we lose the periodic states within the chaotic regime, except for period 3, as shown in (c). For values of the knob a for which the unperturbed map has a positive LCE, the LCE of the perturbed system seems to be essentially independent of the (small) noise level σ.

For the correlation functions, the effect of noise is to produce a slow decrease in correlations even when the unperturbed system is periodic (and therefore the correlation functions are also periodic). This leads naturally to a broadband contribution to the power spectra.

Crutchfield and Huberman [21.2] considered the effect of noise on the period doubling route to chaos for the Duffing oscillator and the logistic map. For the Duffing example they considered the system

$$\ddot{x} + \alpha\dot{x} + x - 4x^3 = \Gamma\cos\mu t + f(t) \quad (21.2)$$

where the noise term f(t) has zero mean and delta correlation:

$$\langle f(t)\rangle = 0 \quad (21.3a)$$

$$\langle f(0)f(t)\rangle = 2A\delta(t) \quad (21.3b)$$

It was observed that, with increasing noise level, a "bifurcation gap" developed, corresponding to a depletion of available periodic states both in the periodic and chaotic regimes of the unperturbed system. This is consistent with the findings of Mayer-Kress and Haken. [21.1] A similar bifurcation gap was found for the logistic map with noise added as in equation (21.1).

Noise also affects the dimensions of a strange attractor. We

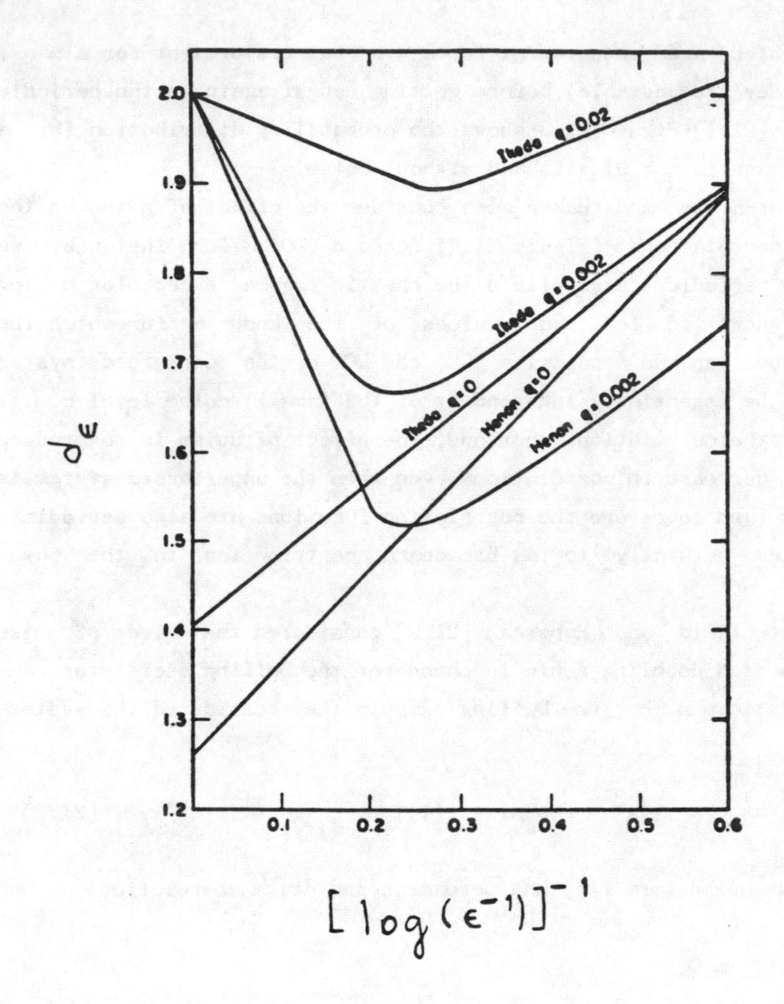

Figure 21.5 Approximate fractal dimension $d(\epsilon)$ vs. $[\log(\epsilon^{-1})]^{-1}$ computed for the Henon and Ikeda attractors with and without noise. From A. Zardecki, Reference [21.3], with permission.

mention just two examples, corresponding to the Hénon attractor and the Ikeda attractor considered later in connection with optical bistability. Figure 21.5 shows results obtained by Zardecki [21.3] for the fractal dimension $d(\epsilon) = \log N(\epsilon)/\log\epsilon$ for the Hénon and Ikeda attractors with and without noise added. The parameter q is a measure of the noise level. On a very fine scale of resolution (i.e., ϵ small), below the noise level, $d(\epsilon)$ is close to 2.0, corresponding to the fractal dimension associated with the two-dimensional Gaussian noise. In other words, at high resolution $d(\epsilon)$ characterizes the noise and not the (unperturbed) dynamical system. For coarse resolution, on the other hand, $d(\epsilon)$ is in fairly good agreement with the fractal dimensions of the attractors in the absence of noise.

In laboratory experiments, and to a lesser extent in numerical experiments on a computer, noise is of course unavoidable. We do not know what laboratory experiment presently holds the record for the most period doubling bifurcations observed before the onset of chaos; experiments exhibiting 3 or 4 or 5 period doubling bifurcations are considered as good confirmations of the theory.

22. Maxwell-Bloch Equations

The chaotic behavior discussed in the next few sections, and later in connection with the Jaynes-Cummings model, is described in terms of the so-called Maxwell-Bloch equations. For the benefit of any newcomers to quantum optics we present here a derivation of the Maxwell-Bloch equations. For the reader with some experience in this field, this section will serve only to establish our notation.

The Maxwell wave equation for the electric field in a charge-neutral, homogeneous, isotropic, and lossless medium is

$$\nabla^2 E - (1/c^2)\delta^2 E/\delta t^2 = (4\pi/c^2)\delta^2 P/\delta t^2 \qquad (22.1)$$

We consider only one linear polarization component of the field. P is the polarization (dipole moment per unit volume), which is determined by the Schrödinger equation. Assuming the field is a quasimonochromatic plane wave, we write

$$E = A(z,t)\cos[\omega t - kz + \varphi(z,t)], \quad k = \omega/c \tag{22.2}$$

and make the slowly-varying envelope and phase assumption (SVEPA) that $|\delta A/\delta z| \ll k|A|$, $|\delta^2 A/\delta z^2| \ll k^2|A|$, $|\delta A/\delta t| \ll \omega|A|$, $|\delta^2 A/\delta t^2| \ll \omega^2|A|$, and likewise for the phase φ. Then

$$\nabla^2 E - (1/c^2)\delta^2 E/\delta t^2 \cong 2k[\delta A/\delta z + (1/c)\delta A/\delta t]\sin(\omega t - kz + \varphi)$$
$$+ 2kA[\delta\varphi/\delta z + (1/c)\delta\varphi/\delta t]\cos(\omega t - kz + \varphi) \tag{22.3}$$

in the SVEPA.

The polarization is written

$$P = U(z,t)\cos(\omega t - kz + \varphi) + V(z,t)\sin(\omega t - kz + \varphi) \tag{22.4}$$

where U and V are called its in-phase (with E) and in-quadrature ($\pi/2$ out of phase) components, respectively. Applying the SVEPA to (22.4), we have

$$\delta^2 P/\delta t^2 \cong - \omega^2 U\cos(\omega t - kz + \varphi) - \omega^2 V\sin(\omega t - kz + \varphi) \tag{22.5}$$

and therefore, from (22.1) and (22.3),

$$\delta A/\delta z + (1/c)\delta A/\delta t \cong -2\pi kV \tag{22.6a}$$

$$[\delta\varphi/\delta z + (1/c)\delta\varphi/\delta t]A \cong -2\pi kU \tag{22.6b}$$

Equations for U and V are determined by the Schrödinger equation. In particular, in the case of a near-resonance between the field and some active transition of the molecules of the medium, we can restrict our attention to the two levels of the resonant transition. This is the approximation of a <u>two-level atom</u>, and for many purposes it is an excellent approximation. [22.1] In the two-level approximation an atom (or molecule) is described by the state vector

$$|\Psi(t)\rangle = c_1(t)|\psi_1\rangle + c_2(t)|\psi_2\rangle \tag{22.7}$$

where $|\psi_1\rangle$ and $|\psi_2\rangle$ are the two stationary states, corresponding to the levels of energy E_1 and E_2, respectively. The probability amplitudes $c_1(t)$ and $c_2(t)$ are determined by the time-dependent Schrödinger equation, which for the two-state system takes the form

$$i\hbar\dot{c}_1 = E_1c_1 + H_{12}c_2 \tag{22.8a}$$

$$i\hbar\dot{c}_2 = E_2c_2 + H_{21}c_1 \tag{22.8b}$$

where H is the perturbation, assumed to be off-diagonal in the two-dimensional Hilbert space of the two-level atom. The polarization P is given by

$$P = N\langle\Psi(t)|e\vec{r}|\Psi(t)\rangle_x = N\langle\Psi(t)|ex|\Psi(t)\rangle$$

$$= Nd[c_1^*(t)c_2(t) + c_1(t)c_2^*(t)] \tag{22.9}$$

where N is the number density of atoms, x refers to the direction of polarization of the electric field, and

$$d = \langle\psi_1|ex|\psi_2\rangle = e\int d^3r\psi_1^*(\vec{r})x\psi_2(\vec{r}) \equiv ex_{12} \tag{22.10}$$

is the transition dipole matrix element of the two-level atom. d has been taken to be purely real, which is always possible with a proper choice of phase for the wave function, and so this assumption cannot affect any of our physical predictions.

From (22.9) it is convenient to define $X(t) = c_1^*(t)c_2(t) + c_1(t)c_2^*(t)$, which, using the Schrödinger equation (22.8), is found to satisfy the equation

$$\dot{X}(t) = -\omega_o Y(t) - (i/\hbar)(H_{12} - H_{12}^*)Z(t) \tag{22.11}$$

where $Y(t) \equiv i[c_1^*(t)c_2(t) - c_2^*(t)c_1(t)]$, $Z(t) \equiv |c_2(t)|^2 - |c_1(t)|^2$, and $\omega_o \equiv (E_2 - E_1)/\hbar$ is the (angular) transition frequency. We have taken $E_1 = -E_2 = -\hbar\omega_o/2$, which is permissible simply by choosing appropriately the zero of energy. For two-level atoms in an electric field $E(t)$, we may take $H = -exE(t)$, and so

$$H_{12} = -ex_{12}E(t) = -dE(t) \tag{22.12}$$

Then (22.11) becomes simply

$$\dot{X}(t) = -\omega_o Y(t) \tag{22.13a}$$

under our assumption that d has been made real. Similarly we obtain

$$\dot{Y}(t) = \omega_o X(t) + (2d/\hbar)E(t)Z(t) \tag{22.13b}$$

and

$$\dot{Z}(t) = -(2d/\hbar)E(t)Y(t) \tag{22.13c}$$

Equations (22.13) are the Bloch equations for a two-level atom in an electric field $E(t)$. From (22.9) we have

$$P = NdX(t) \tag{22.14}$$

and this provides the coupling of the atom and field equations via the wave equation for E.

Using (22.2), we have the Bloch equations

$$\dot{X} = -\omega_o Y \tag{22.15a}$$

$$\dot{Y} = \omega_o X + (2d/\hbar)A\cos(\omega t - kz + \phi)Z \tag{22.15b}$$

$$\dot{Z} = -(2d/\hbar)A\cos(\omega t - kz + \phi)Y \tag{22.15c}$$

From (22.4) and (22.14) we are led to write

$$X = u(z,t)\cos(\omega t - kz + \phi) - v(z,t)\sin(\omega t - kz + \phi) \tag{22.16a}$$

and similarly

$$Y = v(z,t)\cos(\omega t - kz + \phi) + u(z,t)\sin(\omega t - kz + \phi) \tag{22.16b}$$

Using (22.16) in (22.15), and making the SVEPA, we obtain the <u>optical Bloch equations</u>

$$\partial u/\partial t = -(\Delta + \partial\phi/\partial t)v \tag{22.17a}$$

$$\partial v/\partial t = (\Delta + \partial\phi/\partial t)u + (d/\hbar)Aw \tag{22.17b}$$

$$\partial w/\partial t = -(d/\hbar)Av \tag{22.17c}$$

where $w \equiv Z$ and $\Delta \equiv \omega_o - \omega$ is the <u>detuning</u> of the field carrier frequency ω from the atomic transition frequency ω_o. Equations

(22.4), (22.14), and (22.16) identify U = Ndu and V = -Ndv, so that equations (22.6) may be written

$$\partial A/\partial z + (1/c)\partial A/\partial t = (2\pi Nd/\hbar)v \tag{22.18a}$$

$$(\partial\phi/\partial z + (1/c)\partial\phi/\partial t)A = -(2\pi Nd/\hbar)u \tag{22.18b}$$

Equations (22.17) and (22.18) are called Maxwell-Bloch equations because they couple the Maxwell equations (22.18) for the field to the optical Bloch equations (22.17) describing the resonant transition; the latter, of course, are just a way of writing the Schrödinger equation for a two-level atom. u and v are respectively the in-phase and in-quadrature components of the "Bloch vector," [22.1] and w is the population difference, i.e., the upper-state probability minus the lower-state probability. There are very many applications of the Maxwell-Bloch equations in their various forms. The reader who is not familiar with some of these applications may wish to consult, for instance, the monograph of Allen and Eberly. [22.1]

The use of the SVEPA in the derivation of the optical Bloch equations (22.17) is called the rotating-wave approximation (RWA) in quantum optics and electronics. We have already encountered this approximation in connection with the Duffing oscillator in Section 11. Later we will consider the non-RWA Maxwell-Bloch equations in connection with chaos in the Jaynes-Cummings model.

The Maxwell-Bloch equations in the form (22.17) and (22.18) are applicable when atomic and field relaxation processes are negligible, e.g., for electromagnetic pulses very short compared with any atomic relaxation times. More generally we must include the effects of such relaxation processes in the Maxwell-Bloch equations. We do this phenomenologically by writing

$$\partial u/\partial t = -(\Delta + \partial\phi/\partial t)v - \beta u \tag{22.19a}$$

$$\partial v/\partial t = (\Delta + \partial\phi/\partial t)u - \beta v + (d/\hbar)Aw \tag{22.19b}$$

$$\partial w/\partial t = -\gamma(w - w_o) - (d/\hbar)Av \tag{22.19c}$$

$$\partial A/\partial z + (1/c)\partial A/\partial t = -(1/c)\gamma_c A + (2\pi Nkd)v \tag{22.19d}$$

$$(\partial\phi/\partial z + (1/c)\partial\phi/\partial t)A = -(2\pi Nkd)u \tag{22.19e}$$

β and γ are called "transverse" and "longitudinal" relaxation rates, respectively, the terminology being a carry-over from magnetic resonance theory. Frequently β is written as $1/T_2$ and γ as $1/T_1$. β is 2π times the homogeneous linewidth (HWHM) of the transition, and γ is the rate at which the population difference relaxes to its equilibrium value w_o in the absence of an applied field. γ_c is the field attenuation rate due to (non-saturable) absorption, scattering, conductivity, etc. The subscript c denotes "cavity," for we will be concerned with the field in an optical cavity, where the loss is due mainly to transmission, diffraction, and scattering at the mirrors.

23. The Lorenz Model and the Single-Mode Laser

In the case of a single longitudinal mode of a cavity we can take the amplitude A and phase ϕ in (22.2) to be functions only of t, which means we can replace $\partial A/\partial z$ and $\partial\phi/\partial z$ by zero in the Maxwell-Bloch equations (22.19). (The amplitude and phase will not vary strongly with z if the cavity losses are not too large.) Suppose the atoms are exactly resonant with the field carrier frequency, so that $\omega_o = \omega$ and therefore $\Delta = 0$. If $w(0) = \pm 1$ (i.e., if the atom is at some time in the upper or lower level of the transition) then it follows from (22.19) that $u = \dot{\phi} = 0$ for all times. Then we have Maxwell-Bloch equations in the form

$$\dot{v} = -\beta v + (d/\hbar)Aw \tag{23.1a}$$

$$\dot{w} = -\gamma(w - w_o) - (d/\hbar)Av \tag{23.1b}$$

$$\dot{A} = -\gamma_c A + (2\pi N d\omega)v \tag{23.1c}$$

These equations may be used to model a homogeneously broadened, single-mode, unidirectional ring laser, assuming line-center ($\Delta = 0$) oscillation. The field damping rate γ_c is inversely proportional to the cavity Q factor:

$$\gamma_c \equiv \omega/2Q = (c/2)[a - (1/2L)\log R_1 R_2] \tag{23.2}$$

where a is the distributed loss coefficient, R_1 and R_2 are the mirror reflectivities, and L is the mirror separation.

Haken [23.1] showed that the system (23.1) is equivalent to the Lorenz model equations (14.1). We will briefly demonstrate this equivalence. First we define the scaled variables $X = A/A_s$, $Y = v/v_s$, and $Z = w/w_s$, where A_s, v_s, and w_s are the fixed points of (23.1), i.e., the steady-state values of A, v, and w for which the left-hand sides are zero. We also define

$$\lambda = w_o/w_s - 1 \tag{23.3a}$$

$$r = \lambda + 1 = w_o/w_s \tag{23.3b}$$

$$\sigma = \gamma_c/\beta \tag{23.3c}$$

$$b = \gamma/\beta \tag{23.3d}$$

$$\tau = \beta t \tag{23.3e}$$

and the new dependent variables

$$x = \sqrt{b\lambda}X, \; y = \sqrt{b\lambda}Y, \; z = r - Z = (w_o - w)/w_s \tag{23.4}$$

In terms of these new variables we can write (23.1) in the form

$$\dot{x} = -\sigma(x - y) \tag{23.5a}$$

$$\dot{y} = -y - xz + rx \tag{23.5b}$$

$$\dot{z} = xy - bz \tag{23.5c}$$

where now the derivatives are with respect to the dimensionless time τ. These are just the Lorenz model equations (14.1): <u>the system (23.1) for a single-mode, line-center, homogeneously broadened ring laser is exactly equivalent to the Lorenz model</u>. Before discussing the possibility of chaos in a single-mode, homogeneously broadenened laser, we will summarize some purely mathematical aspects of the Lorenz model. A detailed discussion may be found in Sparrow's monograph. [23.2]

First a few words about where the Lorenz model comes from and its historical importance for the modern developments in chaotic dynamics. Lorenz [1.1] obtained the set of equations (23.5) in connection with some partial differential equations of hydrodynamics (Navier-Stokes equations). Using a spatial-mode expansion with time-dependent coefficients, and a rather drastic truncation of the number of modes, he wound up with (23.5), which he then analyzed in great detail both theoretically and numerically. He described a number of interesting features of (23.5), including their property of very sensitive dependence on initial conditions for certain parameter regimes. (Recall our discussion of the Lorenz butterfly metaphor in Section 1.) Although these early papers did not attract much attention for a decade or so , they are now regarded as cornerstones in the literature on chaos in dissipative systems. In particular, the

Lorenz model is often cited as an example of the enormously complex behavior that can occur in a relatively simple system. (Recall that a three-dimensional phase space is the smallest phase space in which non-quasiperiodic, chaotic behavior can occur.)

Obviously (23.5) is invariant under the transformation $(x, y, z) \to (-x, -y, z)$. Furthermore the z-axis is invariant in the sense that any trajectory starting at (or passing through) a point $(0, 0, z)$ on the z-axis remains on the z-axis. Moreover all such trajectories approach the origin.

The origin (0, 0, 0) is in fact a fixed point, and it is the only (real) fixed point for $r < 1$. For $r > 1$ there are two additional fixed points:

$$x^* = y^* = \pm[b(r - 1)]^{1/2}, \; z^* = r - 1 \tag{23.6}$$

The stability of these fixed points is determined in the usual way by linearizing (23.5) about a fixed point and finding the eigenvalues of the resulting 3×3 matrix. For the linearized flow near the origin, these eigenvalues are

$$\lambda = -b, \quad -\tfrac{1}{2}(\sigma + 1) \pm \tfrac{1}{2}[(\sigma + 1)^2 + 4\sigma(r-1)]^{1/2} \tag{23.7}$$

One of these eigenvalues is > 0 whenever $r > 1$, and so (0, 0, 0) is an unstable fixed point for $r > 1$. (Actually it is more appropriately called "non-stable" for $r > 1$, since "stability" normally means that, starting near to a point, we remain near to it for <u>any</u> point near it. But we have already noted that the z-axis is invariant and that the origin attracts any trajectory starting on the z-axis. The origin is thus unstable for "most" but not all trajectories near it, and so is not unconditionally stable or unstable. Often the term "unstable" is used whenever one of the eigenvalues of the linearized flow has a positive real part, and we follow that practice here.)

The other two fixed points (23.6) are found to be unstable when

$$\sigma > b + 1 \tag{23.8a}$$

and

$$r > \sigma(\sigma + b + 3)/(\sigma - b - 1) \tag{23.8b}$$

For $\sigma = 3$ and $b = 1$, for instance, (23.8) is satisfied whenever $r > 21$. One of the eigenvalues for the linearized flow is real and negative for all r, whereas for $r > 21$ the other two eigenvalues have positive real parts. The transition at $r = 21$, where the complex eigenvalues cross the imaginary axis and take on positive real parts, is an example of a Hopf bifurcation. (Section 13)

Exercise: Verify (23.8). Show that a Hopf bifurcation occurs at $r = 21$ for $\sigma = 3$ and $b = 1$.

We now show some numerical results obtained by a fourth-order Runge-Kutta integration of (23.5). For these computations we choose $\sigma = 3$, $b = 1$, and let r be our "knob." We choose in each case the initial conditions $(x(0), y(0), z(0)) = (1, 1, 1)$. Figure 23.1 shows $y(t)$ for $r = 1/2$; $y(t)$ approaches the fixed point at the origin, consistent with the considerations above. For $r = 4$, the origin is unstable, but the fixed point $(\sqrt{3}, \sqrt{3}, 3)$ predicted by (23.1) is stable, and Figure 23.2 shows $y(t)$ approaching this fixed point for $r = 4$. For $r = 21$ $y(t)$ begins as an orderly but growing oscillation, and then apparently breaks into chaos. (Figure 23.3) Similar behavior is seen in Figure 23.4 for $r = 22$.

Figure 23.5 shows the power spectrum of $y(t)$ for $r = 22$ (Figure 23.4), obtained by applying a Fast Fourier Transform with a cosine bell window (Section 5) to $y(t)$ after initial transients have died away. Note the characteristic broadband spectrum of chaos. In Figure 23.6 we illustrate the property of very sensitive dependence on initial conditions by plotting for $r = 22$ the distance

$$d = [(y(t) - y'(t))^2]^{1/2} \tag{23.9}$$

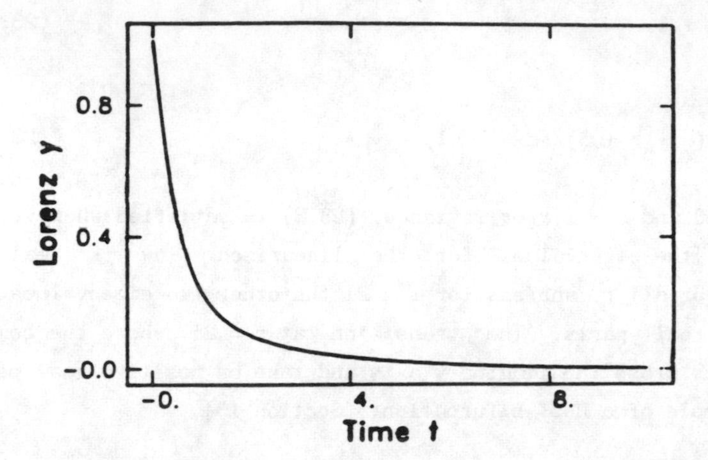

Figure 23.1 y(t) of the Lorenz model for $\sigma = 3$, b= 1, r = 1/2, x(0) = y(0) = z(0) = 1.0.

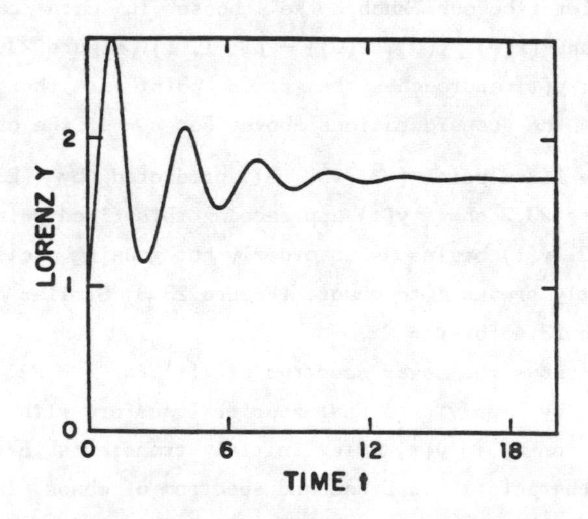

Figure 23.2 As in Figure 23.1, but with r = 4.

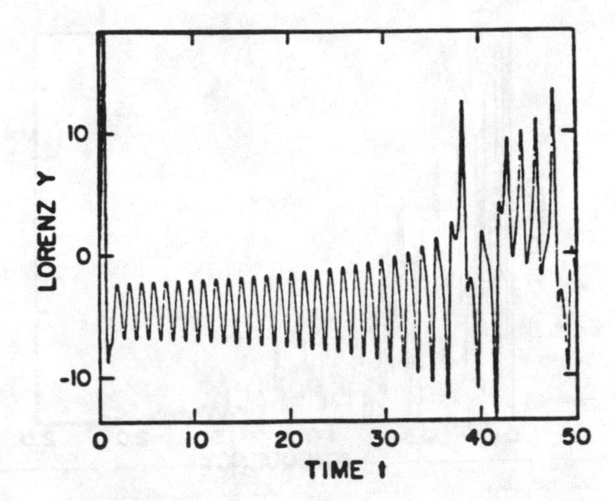

Figure 23.3 As in Figure 23.1, but with r = 21.

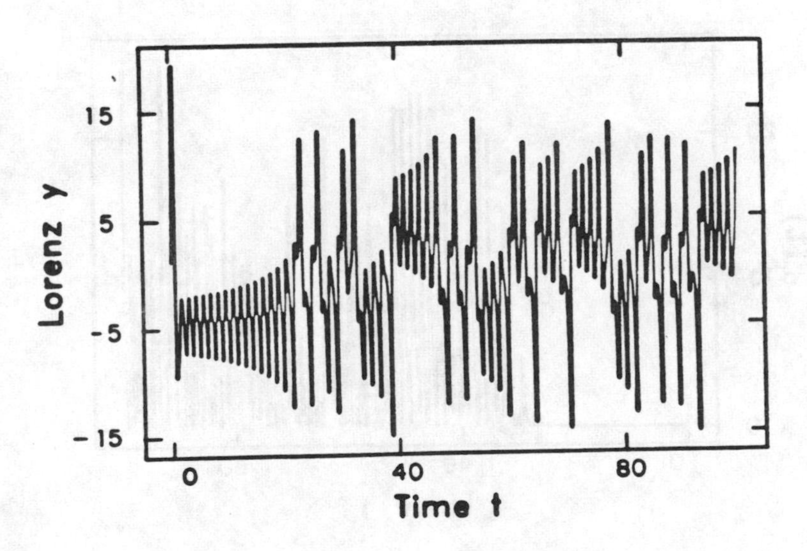

Figure 23.4 As in Figure 23.1, but with r = 22.

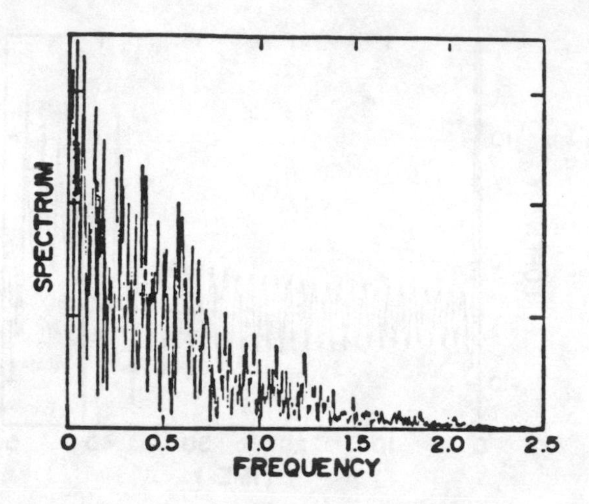

Figure 23.5 Power spectrum of $y(t)$ shown in Figure 23.4.

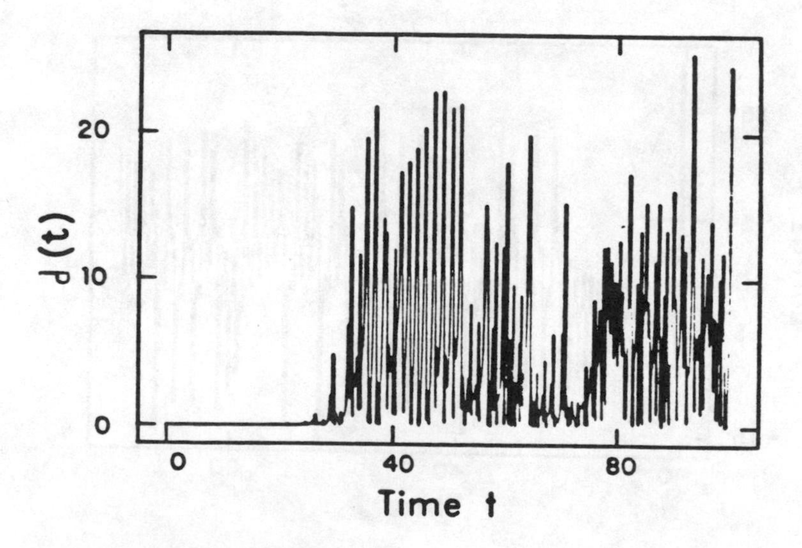

Figure 23.6 The distance (23.9), illustrating the property of very sensitive dependence on initial conditions.

where $y(t)$ and $y'(t)$ are obtained with initial conditions (1.0, 1.0, 1.0) and (1.001, 1.001, 1.001), respectively. Similar results are obtained for $x(t)$ and $z(t)$, and with much smaller differences in initial conditions.

In Figure 23.7 we show $z(t)$ corresponding to Figure 23.4. Following Lorenz, we plot in Figure 23.8 the n<u>th</u> local maximum of $z(t)$ vs. the (n-1)<u>th</u> maximum for $t > 20$, i.e., after initial transients have damped out and the trajectory has settled onto the chaotic attractor. The map obtained by this procedure looks something like the parabola of the logistic map, although the maximum is clearly much sharper than quadratic. Evidently this map has no stable n-cycle, but instead generates a chaotic sequence.

Lorenz focused on the case $\sigma = 10$ and $b = 8/3$. For these values (23.8a) is satisfied and (23.8b) becomes $r > 470/19 \cong 24.71 \equiv r_c$. For $r < 1$ all trajectories approach the origin, and two of the eigenvalues of the linearized flow about the origin are always negative, indicating a two-dimensional <u>stable manifold</u> of the origin - the set of points where a starting trajectory approaches the origin as $t \to \infty$. The one-dimensional <u>unstable manifold</u> of the origin for $r > 1$, on the other hand, consists of points approaching the origin as $t \to -\infty$. This is illustrated in Figure 23.9a. At $r = 1$ the fixed points (23.6) are born, and a trajectory starting on the unstable manifold of the origin for $r > 1$ heads for that fixed point lying in the same half space. As illustrated in Figure 23.9b, the trajectory spirals toward the fixed point (which makes it what is called a stable spiral point). This is consistent with the oscillatory approach to fixed points found numerically in Figure 23.2, for instance, and is simply a consequence of having a complex eigenvalue associated with the linearized flow. As r increases the spiral loops get larger.

When r is raised beyond $13.926... \equiv r'$, the spiral loops have become large enough that they become attracted to the fixed point in the opposite half space, as illustrated in Figure 23.10. In Figure 23.11a we show $y(t)$ vs. $x(t)$ for $r = 13.84 < r'$ ($\sigma = 10$ and $b = 8/3$),

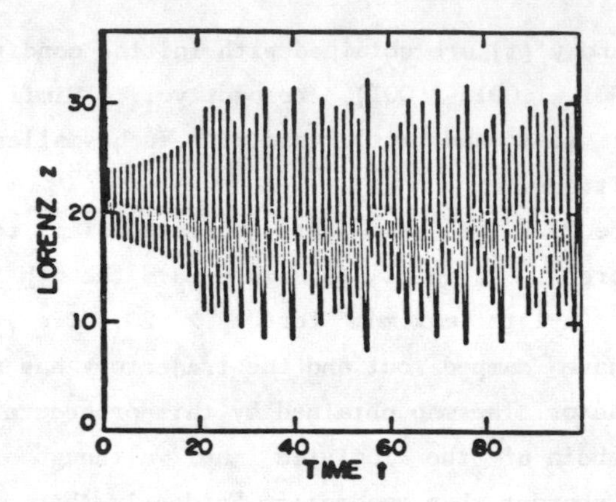

Figure 23.7 z(t) of the Lorenz model for the parameters of Figure 23.4.

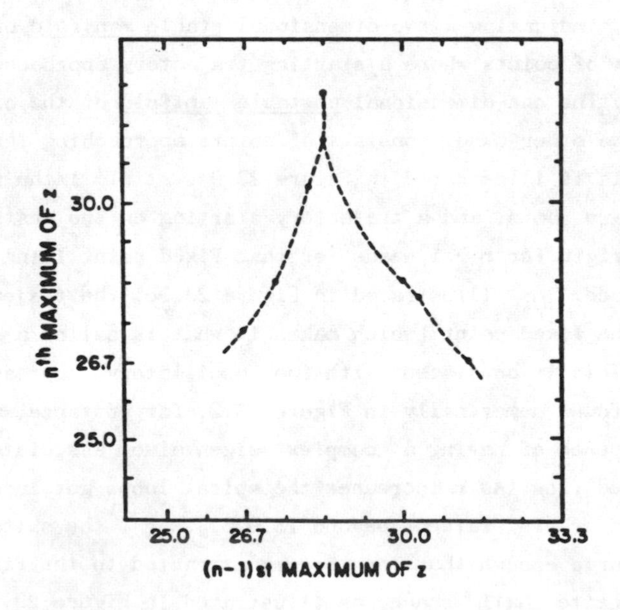

Figure 23.8 The nth maximum of z(t) (Figure 23.7) vs. the (n - 1)th maximum for t > 20.

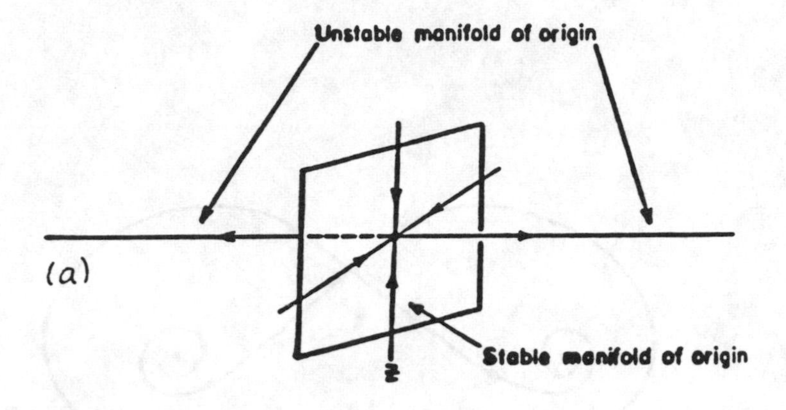

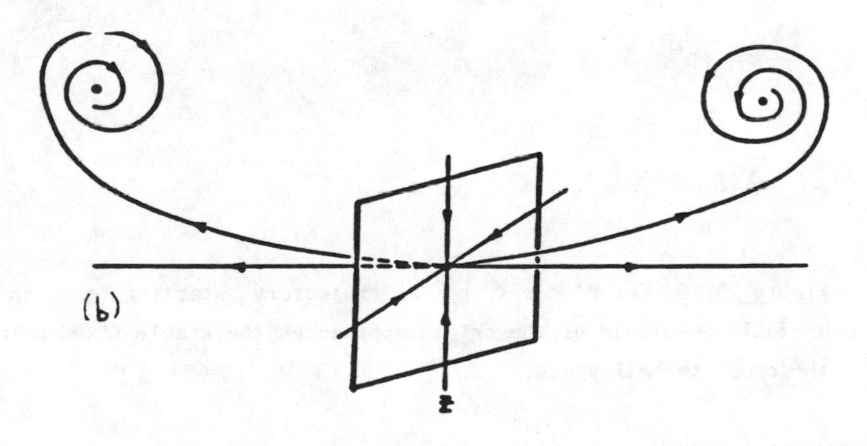

Figure 23.9 (a) Stable and unstable manifolds of the origin in the Lorenz model. (b) For $1 < r < r'$ a trajectory starting out on the unstable manifold of the origin approaches the stable fixed point in the same half space.

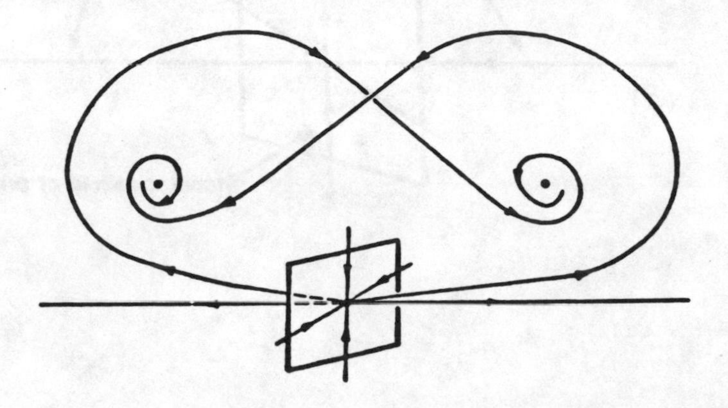

<u>Figure 23.10</u> For $r' < r < r_c$ a trajectory starting out on the unstable manifold of the origin approaches the stable fixed point in the opposite half space.

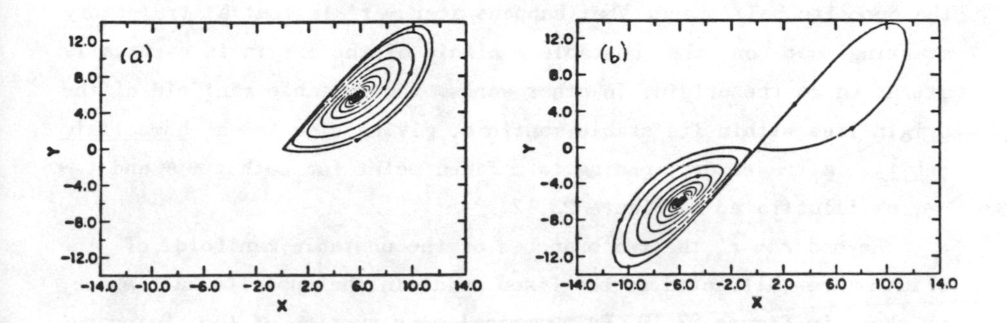

Figure 23.11 (a) y(t) vs. x(t) for r = 13.84, illustrating the bahavior shown in Figure 23.9b. (b) y(t) vs. x(t) for r = 13.96, illustrating the bahavior shown in Figure 23.10.

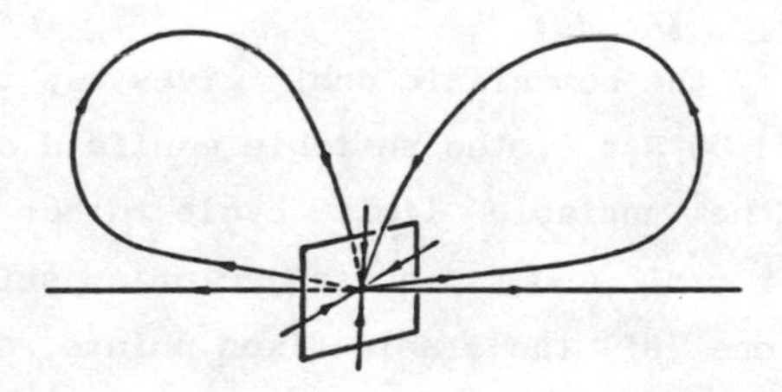

Figure 23.12 Homoclinic orbits at r = r'.

assuming an initial condition on the unstable manifold of the origin. In Figure 23.11b is shown the corresponding result for $r = 13.96 > r'$, indicating how the trajectory crosses over to the fixed point in the opposite half space. What happens at $r = r'$ is that a trajectory starting out on the unstable manifold of the origin is eventually attracted to the origin. In other words, the unstable manifold of the origin lies within its stable manifold, giving rise to a homoclinic orbit - a trajectory tending to a fixed point for both $t \to \infty$ and $t \to -\infty$, as illustrated in Figure 23.12.

Beyond $r = r'$ the two branches of the unstable manifold of the origin are attracted to the fixed point in the opposite half space, as shown in Figure 23.10. By numerical construction of the Poincare' map with surface of section $z = r - 1$, Kaplan and Yorke [23.3] have shown that there exists an infinite number of both chaotic and periodic orbits, all of which are unstable, however, because most trajectories are attracted to one of the two stable fixed points that exist for $r < r_c$. This leads to the possibility of "preturbulence," or "metastable chaos." [23.4] An example is shown in Figure 23.13 for $r = 22.2$.

At $r = r'$ the homoclinic orbit gives way to an unstable limit cycle. At $r \cong 24.06 \equiv r''$, the unstable manifold of the origin now spirals onto the unstable limit cycle rather than a stable fixed point. Between $r = r''$ and $r = r_c$ trajectories settle onto a chaotic attractor or one of the stable fixed points, depending on initial conditions. At $r = r_c$ the unstable limit cycles collapse onto the fixed points , and beyond $r = r_c$ all three fixed points are unstable. Only the stable chaotic attractor (the "standard Lorenz attractor") remains. Results like those shown in Figures 23.3 - 23.7 are typical of this chaotic regime. Figure 23.14 shows $y(t)$ vs. $x(t)$ for $r = 26$. As this picture evolves the trajectory appears to switch randomly from the neighborhood of one (unstable) fixed point to the other.

For large values of r nearly all trajectories are attracted to a stable limit cycle. [23.5] Figure 23.15 shows $y(t)$ vs. $x(t)$ for $r = 230$, 220, and 216; we see the early stages of the period doubling

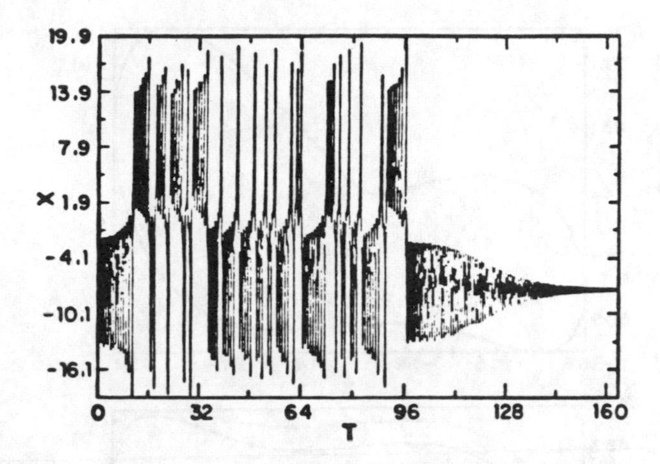

Figure 23.13 Example of "preturbulent" bahavior for r = 22.2.

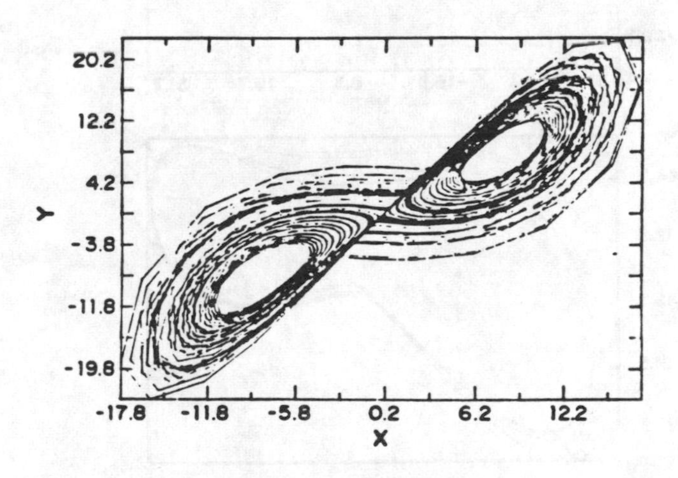

Figure 23.14 y(t) vs. x(t) for r = 26.

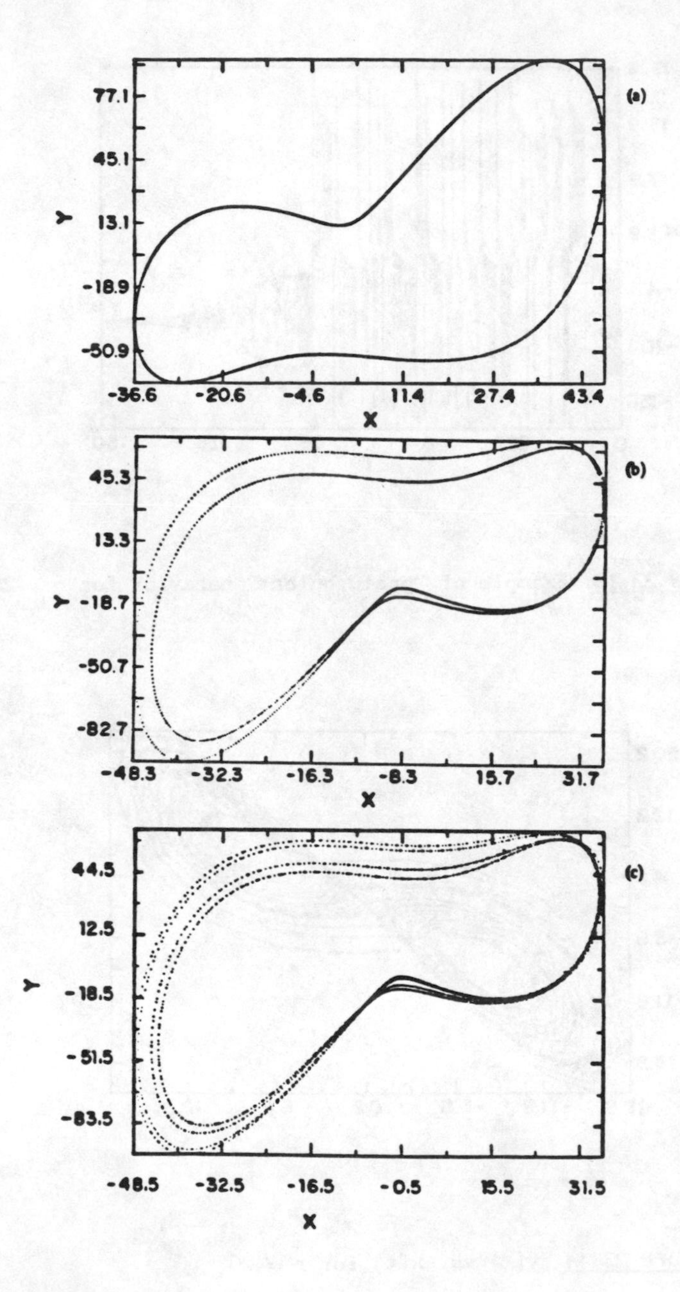

Figure 23.15 A period doubling sequence in the Lorenz model. (a) r =230, (b) r = 220, (c) r = 216.

route to chaos. Periodic windows occur within the chaotic regime, as in the case of the logistic map.

The chaotic behavior of the Lorenz model can be <u>proven</u> by computing LCE. For instance, Shimada and Nagashima [23.6] have computed a maximal LCE ≅ 1.37 for $\sigma = 16$, $b = 4$, and $r = 40$. The fractal dimension of the chaotic attractor in this case is ≅ 2.06.

24. Single-Mode Instabilities: Homogeneous Broadening

The condition $r > 1$ for the fixed point (0, 0, 0) of the Lorenz model to be unstable is equivalent, according to (23.2), to $w_0 > w_s$. Since w_o and w_s are proportional to the threshold and steady-state gain (= loss), respectively, the condition $r > 1$ for a laser is simply the condition that the laser be above threshold. In other words, for $r > 1$ we can have laser oscillation.

The fixed points (steady-state solutions) of the equations (23.1) for a single-mode, homogeneously broadened laser (SMHBL) satisfy

$$w_s = (\hbar\beta\gamma_c/2\pi N d^2\omega) \tag{24.1a}$$

$$v_s = (d/\hbar\beta)A_s w_s \tag{24.1b}$$

$$A_s^2 = (\hbar^2\gamma\beta/d^2)(w_o/w_s - 1) \tag{24.1c}$$

Defining the steady-state intensity $I_s = (c/8\pi)A_s^2$ and the saturation intensity

$$I_{SAT} = \hbar^2\gamma\beta c/8\pi d^2 \tag{24.2}$$

we have

$$w_s = \frac{w_o}{1 + I_s/I_{SAT}} \tag{24.3a}$$

$$I_s = I_{SAT}(w_o/w_s - 1) = I_{SAT}(g_o/g_t - 1) \tag{24.3b}$$

where g_o and g_t are respectively the small-signal (low-intensity) and threshold gain coefficients. Equation (24.3a) implies

$$g = \frac{g_o}{1 + I_s \ I_{SAT}} \tag{24.4}$$

for the gain coefficient, and so (24.3b) is just the condition that the gain coefficient in steady state is clamped at its threshold value, i.e., that the gain and loss are exactly balanced in steady-state oscillation. These relations, of course, are ubiquitous in the laser literature.

The conditions (23.8) for the instability of these steady-state solutions become

$$\gamma_c > \beta + \gamma \tag{24.5a}$$

$$w_o/w_s > \frac{\gamma_c(\gamma_c/\beta + \gamma/\beta + 3)}{\gamma_c - \gamma - \beta} \tag{24.5b}$$

The first condition is equivalent to

$$\gamma_c > 2\pi\delta\upsilon_o + \gamma \tag{24.6a}$$

or

$$cg_t/2 > 2\pi\delta\upsilon_o + \gamma \tag{24.6b}$$

where $\delta\upsilon_o$ is the (HWHM) homogeneous linewidth of the transition. This says that the steady-state laser oscillation is unstable if the Lorentzian width of the cavity mode is larger than the sum of the homogeneous linewidth of the transition and the population decay rate. However, in most lasers it is just the opposite: the mode width due to cavity damping is usually much smaller than the laser transition linewidth. These conditions for instability of the laser steady state therefore represent a bad-cavity instability, because the cavity loss must exceed a certain value before the instability can occur.

The condition (24.5b) requires that the laser be pumped strongly enough for any instability to arise. To find the minimum pumping level required, we calculate the value of γ_c/β that minimizes $w_o/w_s = g_o/g_t$. In the language of the Lorenz model, this is the same as finding the value of σ that minimizes r in (23.8b). We obtain $\sigma_{min} = b + 1 + [4(b + 1)(b + 2)]^{1/2}$, which gives

$$r_{min} = (w_o/w_s)_{min} = (g_o/g_t)_{min} = 5 + 3b + [8(b + 1)(b + 2)]^{1/2} \tag{24.7}$$

with $b = \gamma/\beta$. Normally γ is considerably smaller than β, except when spontaneous emission is the dominant damping mechanism in the Bloch equations. [22.1] If we take $\gamma \ll \beta$ then

$$(g_o/g_t)_{min} \cong 9 \tag{24.8}$$

In other words, the Lorenz model instability requires a laser to be operated at least nine times above threshold. This degree of pumping is not realized in most lasers.

In most lasers, furthermore, the rate-equation approximation is applicable, because $\beta \gg \gamma$. In this case we may assume that v follows w adiabatically in (23.1):

$$v(t) \cong (d/\hbar\beta)A(t)w(t) \tag{24.9}$$

and, using this approximation in (23.1b) and (23.1c), we have the rate equations

$$\dot{w} = -\gamma(w - w_o) - (d^2/\hbar\beta)A^2w \tag{24.10a}$$

$$\dot{A} = -\gamma_c A + (2\pi N d^2\omega/\hbar\beta)Aw \tag{24.10b}$$

or, in terms of intensity,

$$\dot{w} = -\gamma(w - w_o) - (8\pi d^2/\hbar^2\beta c)Iw \tag{24.11a}$$

$$\dot{I} = -2\gamma_c I + (4\pi N d^2\omega/\hbar\beta)Iw \tag{24.11b}$$

Since w is proportional to $N_2 - N_1$ (for non-degenerate levels), and I is proportional to the photon number n, these rate equations are essentially the same as those used in Section 12 in connection with the Arecchi CO_2 laser experiment. As noted there, these simple rate equations cannot exhibit chaotic behavior, and furthermore they have a stable fixed point corresponding to the steady-state laser oscillation described by equations (24.3). In other words, <u>the rate-equation approximation does not account for the "bad-cavity" or Haken-Lorenz instability</u> that occurs when the conditions (24.5) are satisfied.

The Lorenz model is one of the paradigms for the study of chaos in ODE systems, and as such has been studied in great detail. It would therefore be nice to have an unambiguous laboratory experiment for a system described by the Lorenz model. However, this is not possible in fluid systems, and in the case of a SMHBL it has been difficult to realize the instability conditions (24.5) for the so-called "second laser threshold." The difficulty stems from the large gain required for the SMHBL instability. For lasers, (24.5a) is typically not satisfied, and if it is, the condition (24.5b) requires a pumping rate difficult to achieve in most systems. In masers the first condition is usually satisfied, but not the second.

Weiss and Klische [24.1] have suggested that far infrared lasers could satisfy both conditions, and they have carried out experiments showing single-mode instabilities. Figure 24.1 shows a partial energy level diagram of the NH_3 far infrared laser, optically pumped by the 10.78 μm line of the N_2O laser. [24.2] The rotational levels are labelled as usual by the quantum numbers (J,K), and each of these is split ("inversion splitting") into two levels (a and s) due to the tunneling of the N atom through the H_3 plane. The selection rules for allowed transitions are a $\leftrightarrow$ s and $\Delta K = 0$. Laser action occurs on purely rotational transitions such as the 81.5 μm line indicated in Figure 24.1. (In the first maser a population inversion was obtained on a rotational transition of the ground vibrational state of NH_3 by using electrostatic state selection. [24.3].) Weiss and Klische [24.1] noted that in such a far-infrared laser only one velocity group is optically pumped, and therefore the lasing transition is homogeneously broadened. (Note that the inhomogeneous broadening of the pure rotational transition is small compared with that of the optically pumped vibrational transition, although their homogeneous widths are comparable.) Since spontaneous emission is very weak at these wavelengths, the transition is pressure broadened, and this broadening is very small - on the order of a few MHz or less at typical operating pressures. (Low-pressure operation is necessary to avoid rapid rotational relaxation.) This small homogeneous broadening, together with moderately high gains owing to the relatively small rotational partition function of NH_3, makes it possible to satisfy the "bad-cavity" condition (24.5a) while still obtaining laser oscillation. Thus it would appear that such a laser can in principle be made to operate in the regime of the Haken-Lorenz instability.

Experiments of Weiss and Klische [24.4] have been carried out with NH_3 and also CH_2F_2 far-infrared lasers. The pump frequency selects the lasing velocity group via the Doppler effect. Pumping near line center allows both forward- and backward-propagating traveling-wave modes in a ring cavity configuration, whereas pumping

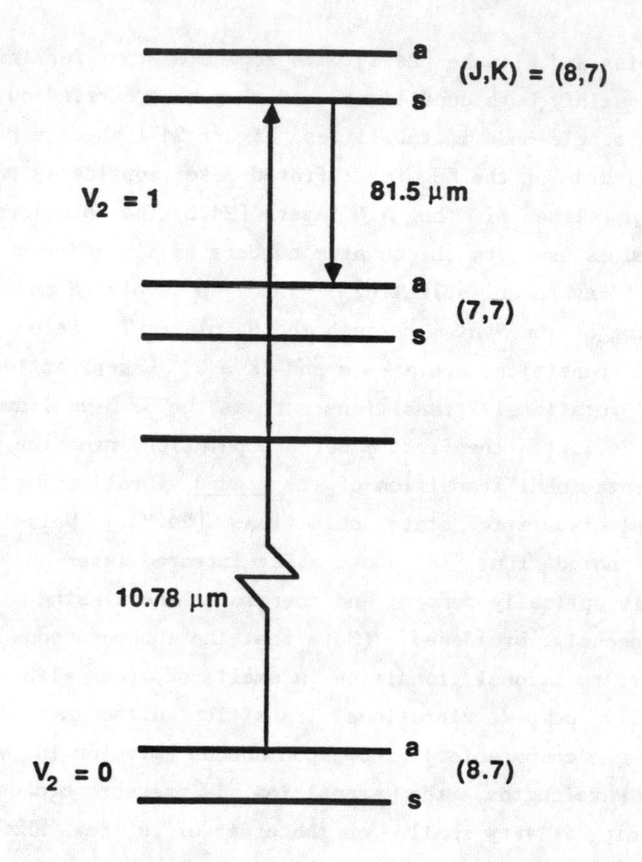

Figure 24.1 Partial energy level diagram for the NH_3 far infrared laser.

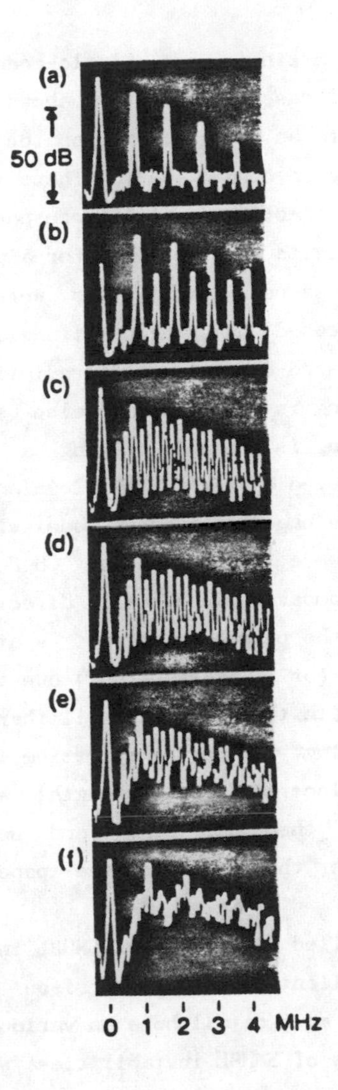

Figure 24.2 Period doubling to chaos in an NH_3 far infrared laser as the cavity is tuned closer to line center. From C.O. Weiss, et al., Reference [24.5], with permission.

off line center favors a single mode. Single-mode instabilities were found in a number of cases. Figure 24.2 shows experimental results for output power spectra as the laser resonator was tuned towards line center in the case of NH_3. [24.5] There is evidently a period doubling to chaos, the chaotic case (f) corresponding to tuning at line center. These results were obtained for a pressure of about 0.1 Torr. At higher pressures no instabilities were found, presumably because of the increased homogeneous linewidth. Note that these results cannot be compared directly to predictions of the Lorenz model, since a detuning is involved, and also because some degree of inhomogeneous broadening is probably playing a role, as discussed below. It might be noted, however, that Zeghlache and Mandel [24.6] find period doubling to chaos in a SMHBL model with detuning.

Similar results were obtained with CH_2F_2. As noted, these results cannot be compared in a really direct way with the Lorenz model. For one thing, the pump beam is not spatially uniform. The resonant Stark effect (or AC Stark effect) due to the interaction of the lasing molecules with the pump beam will therefore result in an inhomogeneous broadening, since the splitting in any given molecule (proportional to the electric field strength) will depend on its position. Furthermore the pump beam (and therefore pump rate) is being depleted. We refer the reader to the paper by Weiss for a detailed discussion.

For a more detailed discussion of SMHBL instabilities we refer the reader to the excellent review by Harrison and Biswas. [24.7] Instead of going into more detail here on various other experimental and theoretical studies of SMHBL instabilities, we next introduce a basic physical mechanism in the onset of these single-mode instabilities, namely the phenomenon of mode splitting.

25. Mode Splitting

We have been discussing instabilities in a single-mode laser. But how can laser operation on a single longitudinal mode produce unstable oscillatory behavior involving different frequencies? The answer lies in the fact that <u>different frequencies may be associated with one longitudinal mode</u>. This phenomenon, which was apparently first noted by Casperson and Yariv in 1970, [25.1] is called <u>mode splitting</u>.

Let us return to the Maxwell-Bloch equations (22.19). In steady state

$$d\phi/dz = (\Delta g/2\beta) \tag{25.1}$$

where $g = 4\pi Nkd^2\beta w_s/[\hbar(\Delta^2 + \beta^2]$ is the (steady-state) gain coefficient. Now from (22.2) we may interpret

$$k_m = k - d\phi/dz = k - (\Delta g/2\beta) \tag{25.2}$$

as the mode wave number, which is restricted to one of the values k_m (= $2\pi m/kL$) allowed by the cavity. If we use the steady-state gain clamping condition $g = 2\gamma_c/c$, we may write

$$k_m = k - \Delta\gamma_c/\beta c \tag{25.3a}$$

or

$$\omega_m - \omega = -\Delta\gamma_c/\beta = - (\gamma_c/\beta)(\omega_o - \omega) \tag{25.3b}$$

This is the familiar <u>frequency pulling</u> condition for a homogeneously broadened line, giving the laser frequency

$$\omega = (\gamma_c\omega_o + \beta\omega_m)/(\beta + \gamma_c) \tag{25.4}$$

Here $\omega_m = k_m c = 2\pi mc/L$ (m is an integer and L the cavity length) is a cavity mode frequency.

We can write (25.2) in terms of the refractive index $n(\omega)$ associated with the laser transition. From (22.14) and (22.16) we may identify Ndu as the in-phase component of the polarization P. In steady state this is $N du_s = -(\Delta/4\pi k\beta) g A_s$, which identifies $\chi = -\Delta g/4\pi k\beta$ as the susceptibility. Since $n(\omega) - 1 \cong 2\pi\chi$, we have

$$n(\omega) - 1 = -\Delta g/2k\beta = (k_m - k)/k \tag{25.5}$$

where we have used (25.2). Thus

$$[n(\omega) - 1]\omega = \omega_m - \omega \tag{25.6}$$

or, more generally,

$$(\ell/L)[n(\omega) - 1]\omega = \omega_m - \omega \tag{25.7}$$

where ℓ is the gain length. This is the form used by Casperson and Yariv. [25.1]

Another equivalent way of deriving (25.7) is to consider the total optical path length $k\ell n + k(L - \ell)$ involved in traversing the distance L. This must be equal to $2\pi m = k_m L$ in order to have a cavity mode, and (25.7) follows immediately.

Equation (25.7) shows there may be several lasing frequencies associated with one and the same cavity mode frequency ω_m. This is clear from Figure 25.1, where we plot the familiar dispersion curve for $n(\omega) - 1$:

$$n(\omega) - 1 = -[\beta g(\omega_o)/2k] \frac{\omega_o - \omega}{(\omega - \omega_o)^2 + \beta^2} \tag{25.8}$$

together with the straight lines $\omega_m - \omega$. In other words, several frequencies (ω) can lase on a single longitudinal mode (ω_m), although from the figure it is seen that the different frequencies will be appreciably displaced from the center of the gain curve at $\Delta = 0$.

Various authors have discussed the role of mode splitting in both homogeneously and inhomogeneously broadened systems, and in both single-mode and multimode instabilities. The analysis is straight-

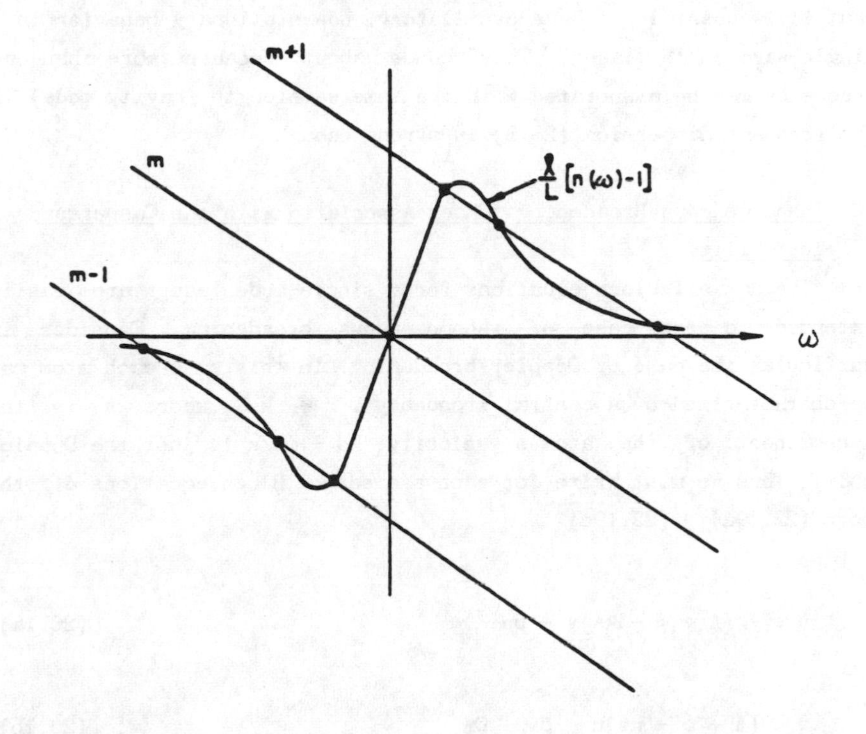

Figure 25.1 Solutions of Equation (25.7).

forward but rather involved, and so at this point we continue with our simplified theoretical framework and go on to the case of inhomogeneous broadening. The main message of this section is simply that it is possible to have oscillatory, non-stationary behavior in a single-wavelength laser. This comes about because more than one frequency may be associated with the same wavelength (cavity mode) if the resonant dispersion (25.8) is strong enough.

26. Inhomogeneous Broadening: Chaos Associated with the Casperson Instability

The Maxwell-Bloch equations for a single-mode laser are easily extended to the case of inhomogeneous broadening. Consider in particular the case of Doppler broadening. In this case each atom may be characterized by a central frequency $\omega_o - ks$, where s is the z-component of the atom's velocity; $ks = \omega s/c$ is just the Doppler shift. Thus we must write for each s a set of Bloch equations of the form (22.19a) - (22.19c):

$$\dot{u} = -(\Delta + \dot{\varphi} - ks)v - \beta u \tag{26.1a}$$

$$\dot{v} = (\Delta + \dot{\varphi} - ks)u - \beta v + \Omega w \tag{26.1b}$$

$$\dot{w} = R - \gamma w - \Omega v \tag{26.1c}$$

where $R \equiv \gamma w_o$ and $\Omega = dA/\hbar$ is the Rabi frequency. For Ω and ϕ we have

$$\dot{\Omega} = -\gamma_c\Omega + K\int_{-\infty}^{\infty} ds W(s)v(s,t) \tag{26.1d}$$

$$\dot{\varphi}\Omega = -K\int_{-\infty}^{\infty} ds W(s)u(s,t) \tag{26.1e}$$

where $K \equiv 2\pi Nd^2\omega/\hbar$ and $W(s)$ is the normalized (one-dimensional) Maxwell-Boltzmann velocity distribution associated with motion along the z-axis. The last two equations are obtained from (22.19d) and (22.19e) by adding up the contributions from all the atomic velocity groups. Again we assume the variations of Ω and ϕ with z may be ignored, which for most purposes is an excellent approximation for "good" cavities (reflectivities > ≈ 50 %), and is often a reasonably good approximation for "bad" cavities as well.

It turns out that the fixed points of (26.1) become unstable for much smaller values of the threshold parameter g_o/g_t than in the case $W(s) = \delta(s)$ considered in the preceding sections. That is, an instability is more easily found in a single-mode inhomogeneously broadened laser than in a homogeneously broadened laser. This instability was first discovered experimentally by Casperson. [26.1]

Casperson's experiments were performed with low-pressure, electric-discharge He-Xe lasers at 3.51 μm. These devices typically have large small-signal gains ($g_o \approx 1$ cm^{-1}) owing to the narrow Doppler linewidth ($\delta\nu_D \approx 100$ MHz). The combination of high gain and narrow linewidth leads to particularly strong resonant ("anomalous") dispersion effects like frequency pulling and mode splitting. [25.1] High gain and narrow linewidth also enhance the resonant dispersion in the case of homogeneous broadening, as is clear from equation (25.5), where $n(\omega) - 1$ is seen to be proportional to g/β. (Recall that the HWHM homogeneous linewidth $\delta\nu_o$ is $\beta/2\pi$.) As already noted, however, the case of inhomogeneous broadening can lead to instabilities at much lower pumping levels. We will discuss this point in more detail later. For now let us simply focus our attention on numerical solutions of the Maxwell-Bloch equations for the case of inhomogeneous (Doppler) broadening, as well as the early experiments that led to the formulation of the Casperson instability in single-mode, inhomogeneously broadened lasers (SMIBL).

Figure 26.1 shows experimental results reported by Casperson in 1978. [26.1] The plots show the output intensity of a He-Xe laser at discharge currents of 40 mA (a) and 50 mA (b). Note that the output

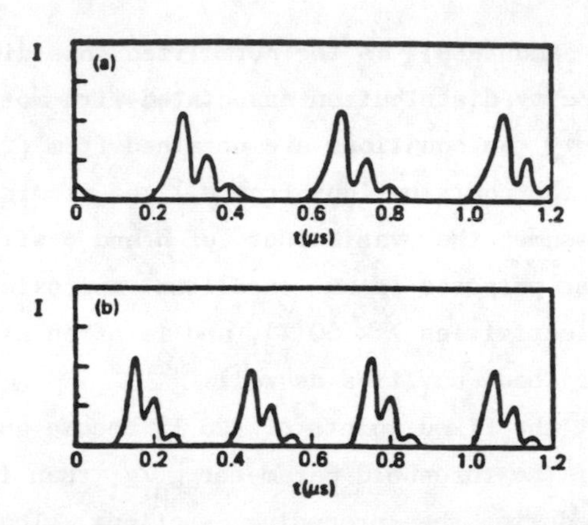

Figure 26.1 Experimental results for self-pulsing in a He-Xe laser with discharge current (a) 40 mA and (b) 50 mA. From L.W. Casperson, Reference [26.1], with permission.

L. W. Casperson, *IEEE J. Quantum Electron* QE-14 (1978) 756.

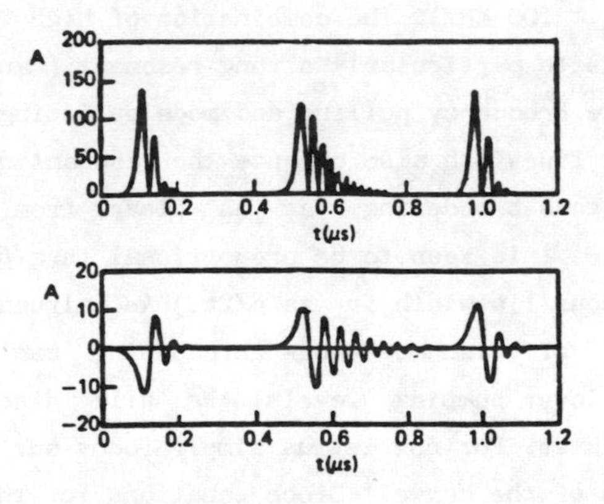

Figure 26.2 Theoretical plots of the self-pulsing in a 3.51 μm He-Xe laser. From L.W. Casperson, Reference [26.1], with permission.

L. W. Casperson, *IEEE J. Quantum Electron* QE-14 (1978) 756.

is oscillatory even though the pumping and loss parameters are not. For this reason the kind of behavior shown in Figure 26.1 is called a self-pulsing instability. ("Instability" refers to the fact that the cw, steady-state behavior normally expected of a continuously pumped system is not found; the steady state is unstable.) At a current of 70 mA an apparently chaotic output, with a broadband spectrum, was observed. [26.2] Since in addition a period doubling has evidently occurred in going from (a) to (b), there is already some indication here of deterministic chaos.

In order to understand what was going on in these experiments, Casperson solved numerically the Maxwell-Bloch equations for a SMIBL, generalized from (26.1) to allow more realistic upper- and lower-level decays, and found that these numerical simulations accounted for self-pulsing. Figure 26.2 shows results of a numerical experiment for parameters corresponding approximately to He-Xe at a pressure of .005 Torr. Casperson noted that this self-pulsing instability could not be explained within a rate-equation analysis. In particular, the oscillations in Figure 26.2 are fast compared with the dipole coherence time ($\beta^{-1} \approx 10^{-6}$ sec). Furthermore the pulsing is slow compared with a cavity transit time 2L/c, and the analysis is single-mode, and so the pulsations of Figure 26.2 have nothing to do with mode-locked pulsing involving many longitudinal modes.

For the numerical results shown in Figure 26.2, $\gamma_c = 1 \times 10^9$ sec^{-1}, $\beta = 2.5 \times 10^7$ sec^{-1}, and the lower-level decay rate (γ) was taken to be 2×10^7 sec^{-1}. Thus the bad-cavity condition $\gamma_c > \beta + \gamma$ is satisfied. The Casperson instability for a SMIBL is in fact a bad-cavity instability but, unlike the case of a SMHBL, the pumping level for instability need not be anything like nine times above threshold.

Equations (26.1) are easily generalized to allow for distinct upper- and lower-level decay rates, denoted γ_2 and γ_1, respectively. We also allow for pumping rates R_2 and R_1 to the upper and lower levels from lower-lying levels:

$$\dot{u} = -(\Delta + \dot{\varphi} - ks)v - \beta u \tag{26.2a}$$

$$\dot{v} = (\Delta + \dot{\varphi} - ks)u - \beta v + \Omega(z_2 - z_1) \tag{26.2b}$$

$$\dot{z}_2 = R_2 - \gamma_2 z_2 - \tfrac{1}{2}\Omega v \tag{26.2c}$$

$$\dot{z}_1 = R_1 - \gamma_1 z_1 + \tfrac{1}{2}\Omega v \tag{26.2d}$$

$$\dot{\Omega} = -\gamma_c \Omega + K \int_{-\infty}^{\infty} ds W(s) v(s,t) \tag{26.2e}$$

$$\dot{\varphi} = -(K/\Omega) \int_{-\infty}^{\infty} ds W(s) u(s,t) \tag{26.2f}$$

Here z_2 and z_1 are the occupation probabilities for the upper and lower levels of the laser transition.

We have carried out extensive numerical experiments on the system (26.2), and in the remainder of this section we describe typical results of these simulations. The "bottom line" is that for different parameter ranges in the bad-cavity regime we have found period doubling, two-frequency, and intermittency routes to chaos. [26.3, 26.4] As discussed in the following section, these routes to chaos in a SMIBL have also been seen experimentally. Indeed these experimental observations provided the stimulus for our numerical studies.

The population decay rates are associated predominantly with spontaneous emission, and are taken to be $\gamma_1 = 2.3 \times 10^7$ sec^{-1} and $\gamma_2 = 7.5 \times 10^5$ sec^{-1}. [26.1] From γ_2 we derive a transition dipole moment d = 5.9 Debye. β is held fixed at 61 MHz, consistent with a mix of 175 mTorr Xe and 0.7 Torr He. [26.5] The (FWHM) Doppler width $\delta\upsilon_D$, which has an isotopic contribution, is taken to be 110 MHz. [26.1, 26.5] The number density N was set at 5.8×10^{15} cm^{-3}, corresponding to an ideal gas at 175 mTorr. The small-signal gain at line center is given by the formula

$$g_o(\omega_o) = (4\pi N d^2\omega/\hbar c)(R_2/\gamma_2\delta\upsilon_D)\sqrt{\log 2/\pi} \qquad (26.3)$$

for $R_1 = 0$, and the threshold parameter r is defined as

$$r = (g_o \ell/2\gamma_c L) = (\ell/L)g_o/g_t \qquad (26.4)$$

For $r = 2.3$, $\ell = 10$ cm, and $L = 16.5$ cm, corresponding approximately to one of the experiments of Gioggia and Abraham , [26.6] equations (26.3) and (26.4) imply $R_2 = 0.35$ $sec^{-1} = 5.7 \times 10^{-9}\beta$.

In our numerical experiments we have employed both predictor-corrector and (fourth-order) Runge-Kutta integration schemes for solving (26.2). In order to accurately "resolve" the Maxwell-Boltzmann distribution, we found it necessary to work with at least 50 velocity (s) groups, so that (26.2) is replaced by at least 3(50) + 2 equations. We have included as many as 200 velocity groups in order to check on the accuracy of our "resolution." In practice, then, the system (26.2) is replaced by ≈ 200 ODE. The numerical integrations were checked using different integrators, but also by using different step sizes in the Runge-Kutta method, and different error tolerances in the predictor-corrector method.

We find, in the bad-cavity regime $\gamma > \beta + \gamma_c$, that there are pulsations in the electric field. Consider the following parameter values: $\Delta = 0$, $R_1 = 0$, $R_2 = 8.5 \times 10^{-9}\beta$, $\gamma_c = 5.4\beta$, $\gamma_1 = 0.38\beta$, $\gamma_2 = .012\beta$, $\beta = 61$ MHz, $K = 6.4 \times 10^{23}$ sec^{-2}, and $\delta\upsilon_D = 110$ MHz. Figure 26.3a shows the computed intracavity intensity as a function of time, after the decay of initial transients (relaxation oscillations).

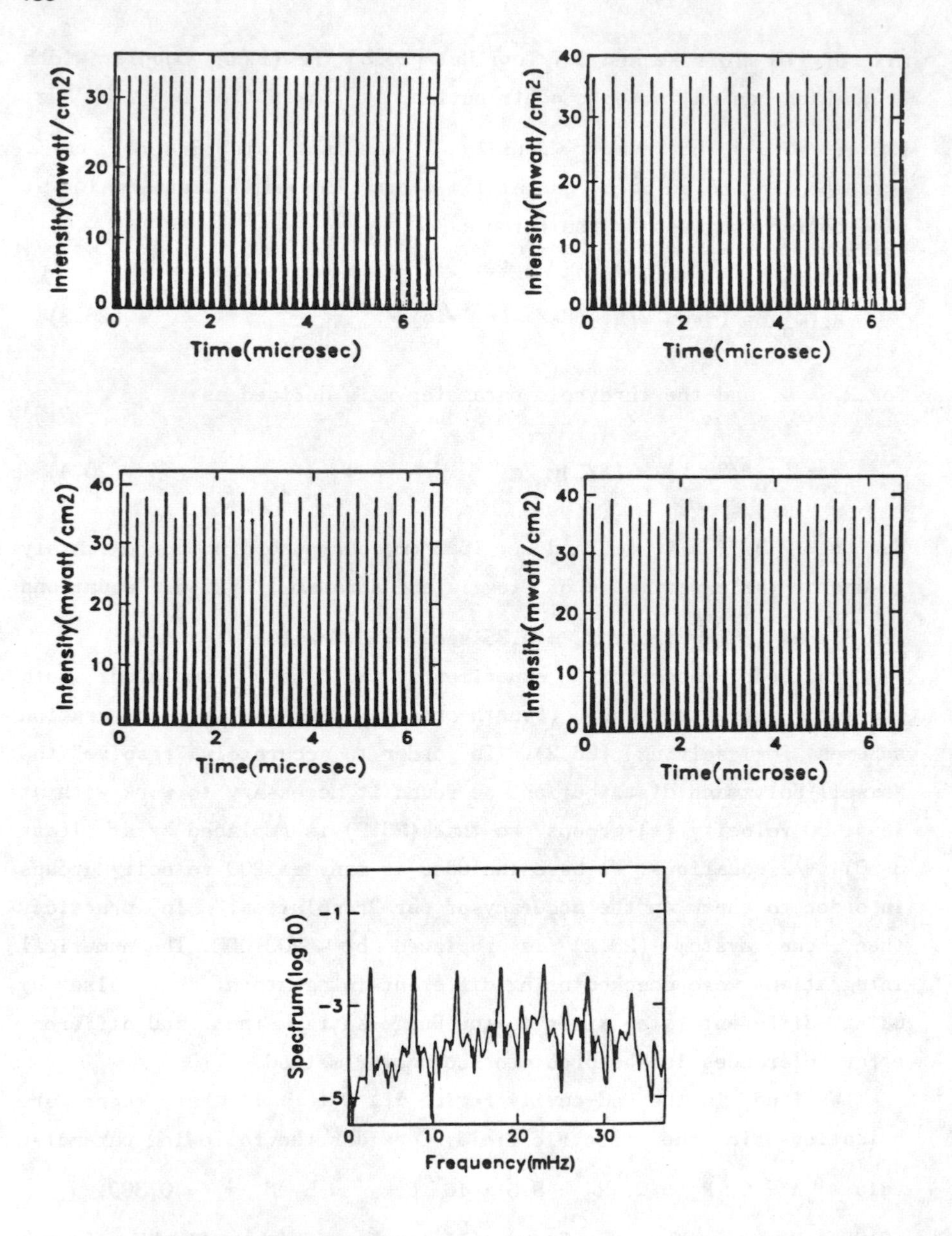

Figure 26.3 Period doubling to chaos as the pump rate R_2 is increased. (See text) Part (e) shows the power spectrum in the chaotic regime for $R_2 = 9.6 \times 10^{-9}\beta$.

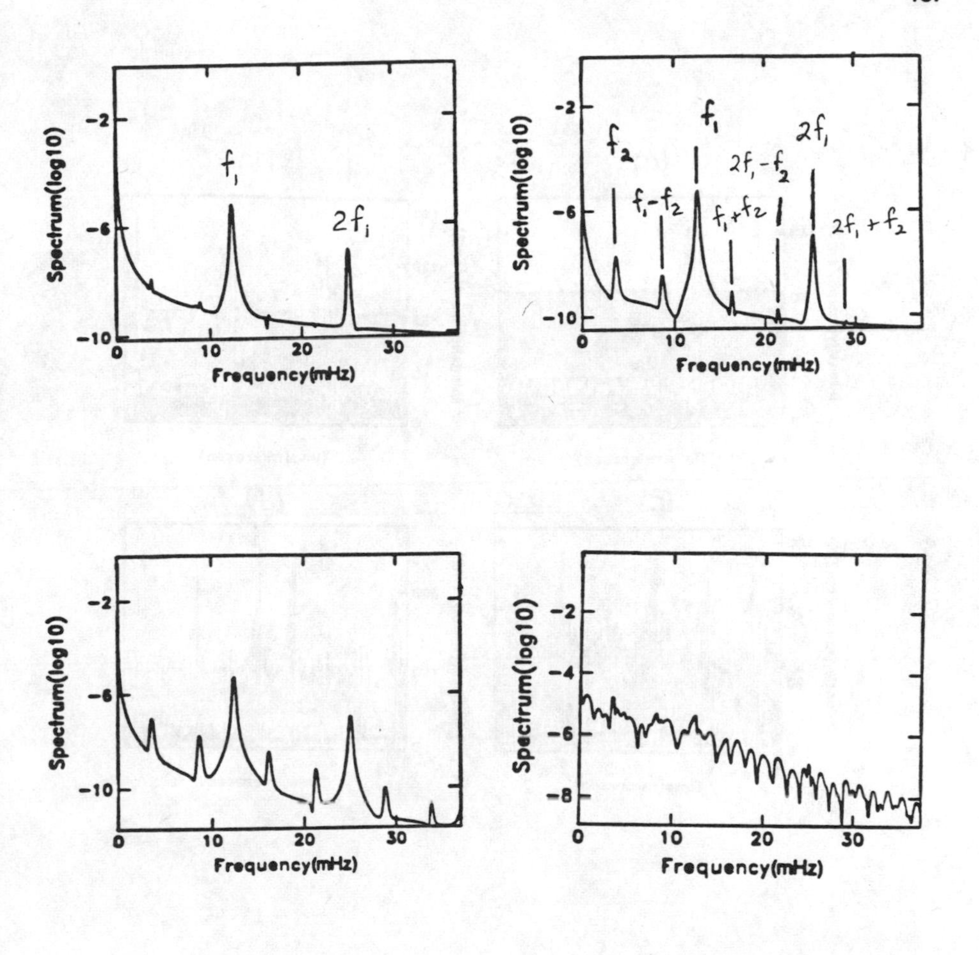

Figure 26.4 Two-frequency route to chaos as the detuning is varied. (See text)

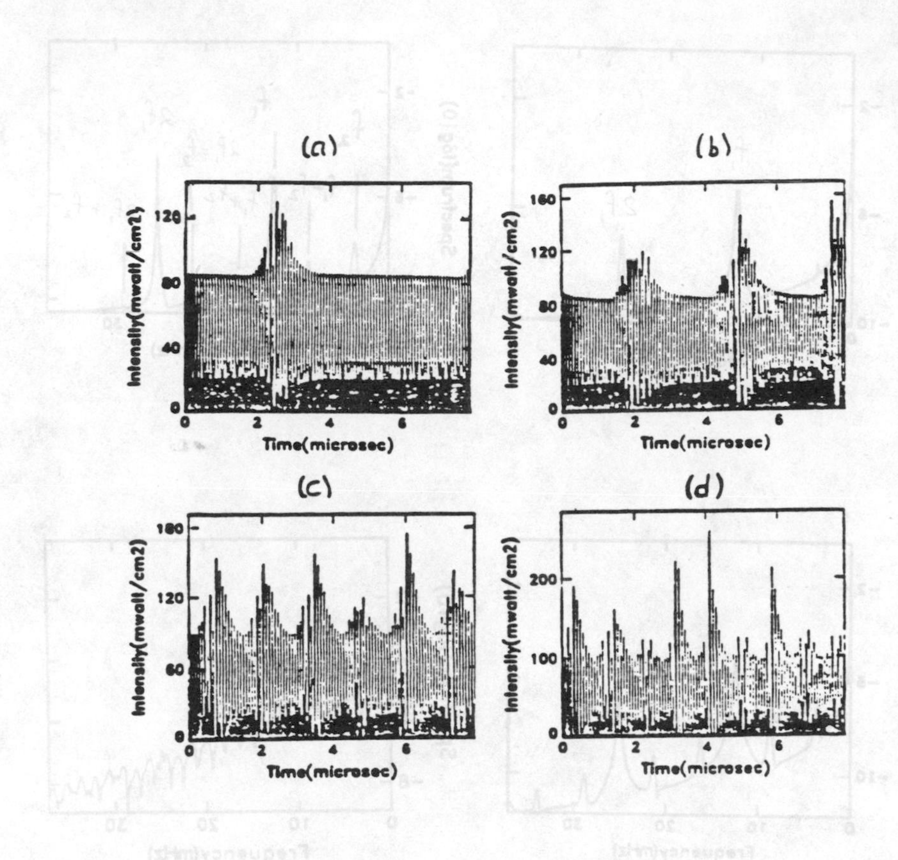

Figure 26.5 Development of chaos via intermittency. (See text)

Raising R_2 to $9.0 \times 10^{-9}\beta$, we obtain the result shown in Figure 26.3b. Note that a period doubling bifurcation has occurred. Figures 26.3c and 26.3d show the intracavity intensity for $R_2 = 9.3$ and $9.4 \times 10^{-9}\beta$, respectively, revealing more period doublings. Slight further increases in R_2 produce more period doublings and chaos as the period doubles ad infinitum. Figure 26.3e, for instance, shows the power spectrum of $\Omega(t)$ for $R_2 = 9.6 \times 10^{-9}\beta$. We also find expected periodic windows with increasing R_2 in the chaotic regime. Owing to the expense of these computations (each takes about 10 minutes of CPU time on the CRAY computer used at Los Alamos), we have not attempted to accurately check the Feigenbaum universality theory for the parameter $\delta = 4.6692...$. However, our results seem qualitatively consistent with the universality theory for period doubling.

As an example of the two-frequency route to chaos, consider the case $R_2 = 5.6 \times 10^{-9}\beta$, $\Delta \neq 0$, and all the other parameters the same as in Figure 26.3. Figure 26.4 shows power spectra of the field for $\Delta = 3.725\beta$ (a), 3.735β (b), 3.74β (c), and 3.75β (d). At first there is a single basic frequency and its harmonics, but as Δ is increased two-frequency motion develops (b), after which the broadband component of the spectrum rises and the motion is evidently chaotic (d).

Intermittency has also been found in various parameter regimes in our numerical experiments. Figure 26.5 is an example. In this case we chose $\beta = 38$ MHz, $\gamma_1 = 1.8\beta$, $\gamma_2 = .057\beta$, $\gamma_c = 9\beta$, $K = 3.6 \times 10^{21}$ sec^{-2}, $\delta\upsilon_D = 110$ MHz, and $R_1 = 0$, and varied the pumping rate R_2. Figures 26.5a - 26.5d are for $R_2 = 1.2875$, 1.29, 1.33, and $1.40 \times 10^{-5}\beta$, respectively.

We have also observed things like "metastable chaos," i.e., a long chaotic period followed by an abrupt transition to quasiperiodicity.

Casperson [26.7] has performed a linear stability analysis for the Maxwell-Bloch equations with Doppler broadening, and has obtained the bad-cavity instability condition $\gamma_c > \beta$ under some reasonable approximations. He interpreted the instability in terms of mode splitting, as discussed later. A simplified analysis of Mandel [26.8] suggests that the single-mode instability is associated with a Hopf bifurcation, as might be surmised from the numerical simulations.

Our numerical experiments show that the three best-known routes to chaos for dissipative systems occur in the regime of the Casperson instability. The period doubling route has been found also in a similar numerical study. [26.9] In retrospect it seems remarkable that so few numerical studies were made in the past on equations of the Maxwell-Bloch type relevant to lasers (as opposed to pulse propagation studies, for instance), although we have already mentioned in Section 1 the early work of Buley and Cummings [1.2] that in fact uncovered a "chaotic" aspect of the SMHBL.

27. Inhomogeneous Broadening: Experiments

The period doubling, two-frequency, and intermittency scenarios for the transition to chaos have all been realized in SMIBL experiments.

Gioggia and Abraham [26.6] reported these three routes to chaos in a bad-cavity He-Xe SMIBL. Different routes were observed in different ranges of detuning, and each route could be followed as the detuning was varied. Figure 27.1 shows their results for the period doubling and two-frequency routes to chaos, as investigated with the intensity power spectrum. These results are for a standing-wave SMIBL, although chaotic emission has been similarly observed in a ring SMIBL by employing Faraday rotation isolators for unidirectional operation. [27.1, 27.2]

For details of these and other experiments we refer the reader to the cited literature. Although a detailed, one-to-one correspondence cannot be made between the numerical simulations done thus far and the laboratory experiments, [26.4] it seems fair to say that the essential features of the experiments are well accounted for by the standard SMIBL theory in the form of (26.2). In any event, the

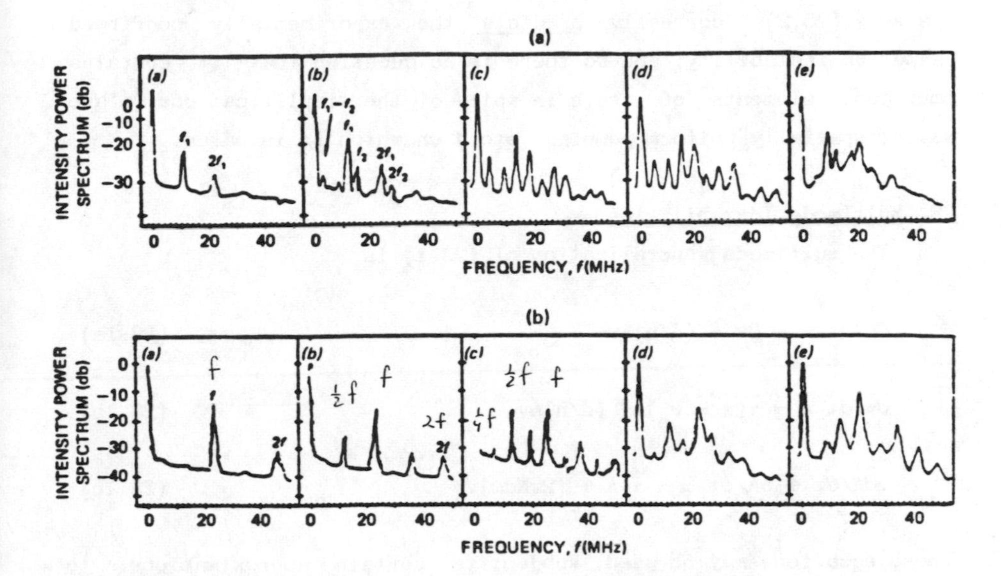

Figure 27.1 Two-frequency (a) and period doubling (b) routes to chaos in a 3.51 μm He-Xe laser. From R.S. Gioggia and N.B. Abraham, Reference [27.1], with permission.

transitions to chaos found in the system (26.2) are certainly consistent with the behavior found experimentally. Furthermore the system (26.2) correctly <u>predicts</u> the experimentally confirmed Casperson instability, and so there is no question that it contains important elements of truth in spite of the simplifications (plane waves, spatially uniform pumping, etc.) on which it is based.

28. Multimode Instabilities

The multimode generalization of (23.1) is

$$\partial v/\partial t = - \beta v + (d/\hbar)Aw \tag{28.1a}$$

$$\partial w/\partial t = - \gamma(w - w_o) - (d/\hbar)Av \tag{28.1b}$$

$$c\partial A/\partial z + \delta A/\delta t = - \gamma_c A + (2\pi Nd\omega)v \tag{28.1c}$$

These equations may be used, subject to certain approximations, to describe a multimode homogeneously broadened laser (MMHBL) in the case of a unidirectional ring cavity. The principal approximations are (a) the RWA and SVEPA, (b) plane waves, and (c) the two-level treatment of the laser transition, with a single relaxation rate (γ) for the population difference w between nondegenerate levels. In particular, (28.1c) allows for an arbitrary number of (longitudinal) modes, since we are not making any mode expansion. Implicit in the system (28.1) is the assumption that there is one mode exactly resonant with the atomic transition frequency ω_o. In general we should also include u and $\partial\phi/\partial t$, $\partial\phi/\partial z$ in these equations. However, these may be set to zero in the (single-mode) steady state, and they may be shown to remain zero at the onset of the MMHBL instability considered below.

Equations (28.1) must be solved with the boundary conditions $v(z + L, t) = v(z,t)$, $w(z + L, t) = w(z,t)$, and $A(z + L, t) = A(z,t)$. This problem was first investigated by Risken and Nummedal [28.1] and Graham and Haken. [28.2] They found an instability that does not require the bad-cavity condition $\gamma_c > \beta + \gamma$ necessary for the Lorenz-type SMHBL instability. This so-called Risken-Nummedal instability corresponds to the onset of multimode, pulsating oscillation, with the lasing modes locked together in phase, as found in numerical solutions of the Maxwell-Bloch equations. [28.1] In other words, the instability corresponds to the onset of a sort of spontaneous mode locking. The Risken-Nummedal instability, although not a bad-cavity instability, does require that

$$r > 5 + 3b + [8(b + 1)(b + 2)]^{1/2} \tag{28.2}$$

in the notation (23.3) ($r = w_o/w_s$, $b = \gamma/\beta$). This instability condition is the same as (24.7). In other words, the Risken-Nummedal instability, like the (single-mode) Lorenz-Haken instability, requires a pumping at least nine times above the threshold for laser oscillation. In addition to (28.2), the Risken-Nummedal instability requires that the mode wave numbers $\alpha = (2\pi c/L)m$, $m = 0, \pm1, \pm2, \ldots$, have magnitudes between $\alpha_\pm$, where

$$\alpha_\pm = \sqrt{(\gamma/2)(3\lambda\beta - \gamma \pm R)}\,[1 - 2\gamma_c/\{\beta(\lambda - 2) - \gamma \pm R\}] \tag{28.3a}$$

with

$$R = [\lambda^2\beta^2 - 2(4\beta + 3\gamma)\lambda\beta + \gamma^2]^{1/2} \tag{28.3b}$$

and again $\lambda = r - 1 = w_o/w_s - 1$. These wave numbers enter into the theory when the perturbations δv, δw, and δA are written as $\exp[i(\alpha x/c + \beta t)]$, and the instability corresponds to $\mathrm{Re}\beta > 0$.

The Risken-Nummedal instability is sometimes said to represent a "second laser threshold." When the pumping is raised beyond the ("first") threshold for laser oscillation, the single-mode oscillation is stable until the second threshold condition (28.2) is met, at which point the oscillation becomes multimode. Beyond this second laser threshold Risken and Nummedal found pulse-train solutions of (28.1) in which v, w, and A are functions only of $t - z/v_p$, where v_p is some velocity determined by the cavity boundary conditions. These solutions satisfy the ODE

$$\dot{v} = -\beta v + (d/\hbar)Aw \tag{28.4a}$$

$$\dot{w} = -\gamma(w - w_o) - (d/\hbar)Av \tag{28.4b}$$

$$\epsilon\dot{A} = -\gamma_c A + (2\pi N d\omega)v \tag{28.4c}$$

where $\epsilon \equiv 1 - c/v_p$ and the derivatives are with respect to the single independent variable $t - z/v_p$. It is easy to show that these equations are isomorphic to the Lorenz model with $r = \lambda + 1 = w_o/w_s$, $\sigma' = \gamma_c/\beta\epsilon$, and $b = \gamma/\beta$. However, the numerical studies of Risken and Nummedal in cases with $r < r_c$ revealed apparently stable periodic solutions, whereas in the corresponding Lorenz model there are no stable limit cycles for $r < r_c$. This difference evidently arises from the periodic boundary conditions imposed on solutions of (28.4) in the case of a laser cavity. [28.3]

Graham has obtained (28.2) in a way that connects it directly to the Lorenz-model instability conditions. [28.3] Since v_p and therefore σ' in the MMHBL model are not specified in the equations, it is interesting to find that value σ_{min} of σ' (= $\gamma_c/\beta\epsilon$) that minimizes r_c. The resulting expression for r_{min} has already been given (equation (24.7)), and coincides with (28.2). Thus both the SMHBL and MMHBL instabilities may be described by analogy with the Lorenz model.

Equations (28.1) may be simplified using a "thin-sheet gain approximation." [28.4, 28.5] First let us re-write (28.1) as

$$\partial v/\partial\tau = -\beta v + (d/\hbar)Aw \tag{28.5a}$$

$$\partial w/\partial\tau = -\gamma(w - w_o) - (d/\hbar)Av \tag{28.5b}$$

$$c\partial A/\partial Z = -\gamma_c A + (2\pi Nd\omega)v \tag{28.5c}$$

where now the independent variables are $\tau = t - z/c$ and $Z = z$. In the thin-sheet gain approximation we assume the entire gain medium is effectively concentrated in a thin sheet at $z = z_o$: $v(z,t) = L\delta(z - z_o)v(t)$, $w(z,t) = L\delta(z - z_o)w(t)$. Then

$$c\partial A/\partial Z = -\gamma_c A + (2\pi NLd\omega)\delta(Z - Z_o)v(\tau + Z/c) \tag{28.6}$$

and after integration we have

$$A(t + T) = e^{-\gamma_c T}[A(t) + (2\pi NLdk)v(t)] \tag{28.7}$$

where $T = L/c$ is the field transit time in the ring cavity and $A(t)$ is the field amplitude just outside the gain sheet. Combining this result with (28.5a) and (28.5b), we have the delay-differential system

$$dv/dt = -\beta v + (d/\hbar)Aw \tag{28.8a}$$

$$dw/dt = -\gamma(w - w_o) - (d/\hbar)Av \tag{28.8b}$$

$$A(t + T) = e^{-\gamma_c T}[A(t) + (2\pi NLdk)v(t)] \tag{28.8c}$$

These equations are obtained assuming $A(Z_o + L,\tau) = A(Z_o,\tau)$, and so they have the cavity mode condition built in. In the limit $T \to 0$ - which is the single-mode (short-cavity) limit - (28.8c) becomes

$$dA/dt = -\gamma_c A + (2\pi Nd\omega)v \tag{28.9}$$

and we are back to the SMHBL model (23.1).

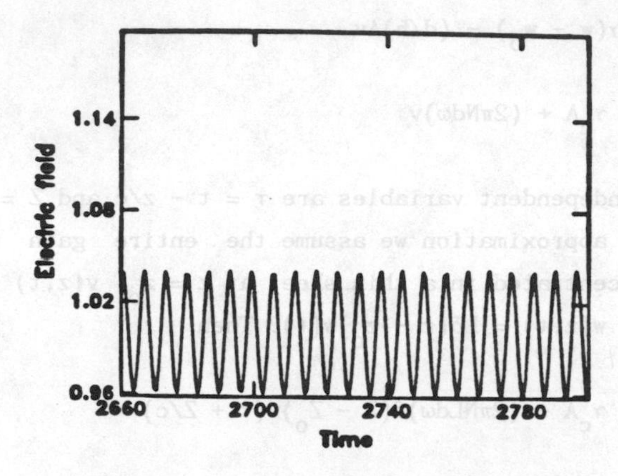

Figure 28.1 Solution of equations (28.10) for the field for s = 7, showing self-pulsing due to spontaneous mode locking. (See text)

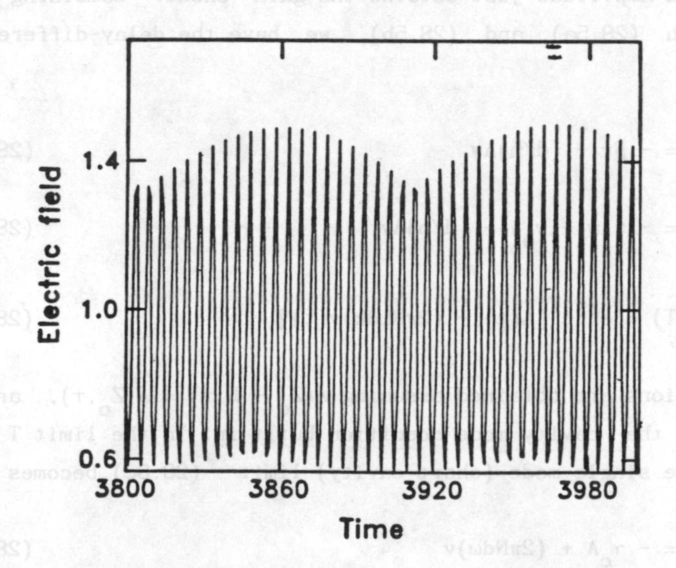

Figure 28.2 Solution of equations (28.10) for the field when s is raised to 10. (See text)

Defining $V = v/v_s$, $W = w/w_s$, and $X = A/A_s$, where v_s, w_s, and A_s are the steady-state solutions of (28.8), we may write (28.8) in a form convenient for numerical computations:

$$dV/d\tau = -V + XW \tag{28.10a}$$

$$dW/d\tau = -bW + b(\lambda + 1) - b\lambda XV \tag{28.10b}$$

$$X(\tau + s) = \alpha X(\tau) + (1 - \alpha)V(\tau) \tag{28.10c}$$

where $s = \beta T$, $\alpha = \exp(-\gamma_c T)$, and b, λ, and τ are defined by (23.3). Note that $\gamma_c T$ is the ratio of the cavity linewidth to the mode spacing, whereas s is the gain linewidth divided by the mode spacing. Figure 28.1 shows $X(\tau)$ obtained by numerical integration of (28.10) with $b = 0.1$, $\lambda = 10$, $\gamma_c T = 0.7$, and $s = 7$. When λ is lowered to 9 we obtain steady-state behavior $(X = X_s)$. As s is increased (i.e., as the number of modes under the gain curve is increased), it is easier to obtain oscillatory behavior, and the oscillations tend to be more pronounced. Figure 28.2 shows what happens when s is raised to 10 and the other parameters of Figure 28.1 remain the same.

The thin-sheet gain approximation in this context is obviously an oversimplification, and so we will not pursue it further here. The whole subject of multimode lasing is well known to be quite complicated, and it is not easy to arrive at intuitive explanations of instabilities. However, some progress has been made, and we devote the next section to a more or less physical discussion of single-mode and multimode self-pulsing instabilities.

29. Physical Explanations of Self-Pulsing Instabilities

We have already discussed mode splitiing in the context of the SMHBL instability (Section 25): several frequencies can be associated with one and the same longitudinal mode when the resonant dispersion is strong enough. What we discussed earlier may be considered as a "spontaneous" mode splitting, and occurs mainly in the wings of the gain curve.

There can also be <u>induced mode splitting</u> associated with the distortion of the dispersion curve by spectral hole burning. Figure 29.1 shows what can happen when this hole burning due to a line-center mode in a SMIBL is taken into account. Now a mode splitting close to line center can occur, and the allowed frequencies can be expected to lase; note that they are displaced from the dip burned into the center of the gain curve, and so can have gains larger than that of the saturating, line-center mode. Note also that the induced mode splitting effect is strongest at line center. Self-pulsing can result from the phase locking of the allowed frequencies. Furthermore, in addition to the mode splitting of the center mode, the sidebands can also split as they themselves burn spectral holes in the gain curve. [29.1]

This picture helps to explain why the single-mode instability is much easier to realize in the case of inhomogeneous broadening. In the case of homogeneous broadening <u>and strong pumping</u> we can think of the laser transition as undergoing a Rabi splitting which, to a weak probe field of frequency different from the line-center mode, is somewhat like a spectral hole burning. In particular, if the "probe" arises from spontaneous emission by the laser medium, there can be induced mode splitting in which the "pump" and "probe" frequencies have the same wavelength. Thus both the SMHBL and SMIBL instabilities may be thought of in terms of mode splitting, although in the former case much stronger pumping rates are required in order to realize a situation analogous to that shown in Figure 29.1.

Hendow and Sargent [29.2] have used the "weak sideband" approach for both single-mode and multimode instabilities. This approach may be summarized as follows. The field is assumed to consist of a central mode and one or two (or, in principle, more) equally spaced side modes. Single-mode oscillation is stable if the side modes do not have net positive gain. If the side modes have net gain, however, the single-mode oscillation is unstable and the side modes can also lase, taking energy from the initial central mode, and they are

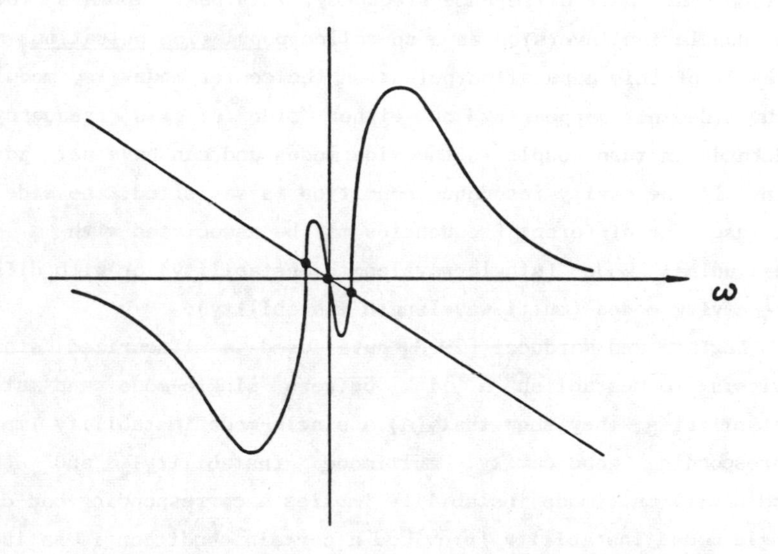

Figure 29.1 Effect of hole burning on the solutions shown in Figure 25.1.

phase-locked with the central mode.

The interference of the central mode with a side mode produces a beat note at their difference frequency. This beat appears then in the population inversion as a so-called population pulsation, and as a result of this population pulsation the center mode is modulated, with sidebands appearing on either side of its frequency. The sidebands in turn couple to the side modes and can have net positive gain. If the cavity resonance condition is satisfied, the side modes can lase. The different frequencies may be associated with a single longitudinal mode (single-wavelength instability) or with different bare-cavity modes (multi-wavelength instability).

Lugiato and Narducci [29.3] have used a linearized stability analysis to establish a link between single-mode and multimode instabilities. They show that (a) a single-mode instability implies a corresponding good-cavity, multimode instability, and (b) a good-cavity multimode instability implies a corresponding bad-cavity, single-mode instability (provided a certain condition is satisfied). Consider, for instance, the bad-cavity, single-mode instability. If the cavity is made longer (or the width of the gain curve is increased), additional cavity modes are brought closer to resonance and come into play as sideband frequencies which can grow at the expense of the initial central mode; the physical picture in terms of the weak sideband approach explains the onset of the instability. Note, however, that an increase of the cavity length (or a widening of the gain curve through an increase in β) can change a "bad" cavity to a "good" one. The details of this connection between single-mode and multimode instabilities are treated in Reference [29.3].

These arguments derived from the weak sideband approach or linear stability analysis are certainly interesting and useful, but we tend to agree with Casperson that "There is no single easy physical explanation for why the pulsations occur. It is safest to simply observe that whenever one has a physical system which is governed by several nonlinear equations, instabilities and pulsations are a likelihood." [26.2]

It might be noted that the case of a multimode inhomogeneously broadened laser also admits an instability at a low level of pumping, as shown by a study using the thin-sheet gain approximation of the preceding section. This case is cumbersome to treat in a more realistic manner, although some progress has been made.

30. Transverse Mode Effects

Thus far we have restricted our considerations to uniform plane waves. However, transverse field variations can also play an important role in laser instabilities and chaos.

As early as 1964 Fleck and Kidder [30.1] presented a coupled-mode theory for the undamped spiking behavior of ruby and other solid-state lasers. They proposed that a moderate amount of spatial inhomogeneity of the pump rate, leading to coupling of different transverse modes, could give rise to undamped spiking behavior. Their theory assumed a homogeneous linewidth much greater than either the population decay rate or the inhomogeneous linewidth, but they retained the "beat notes" between modes. Considering the highly truncated case of two coupled transverse modes, Fleck and Kidder showed numerical results for cases of undamped spiking. We have recently re-examined the two-mode equations and have found that, for parameters near those assumed by Fleck and Kidder for ruby, the "undamped spiking" is in fact chaotic, and there is a period doubling to chaos. [26.3, 30.2]

The irregular and irreproducible spiking of ruby and other solid-state lasers is often attributed to "random" factors like optical inhomogeneity of the gain crystal. Experimentally, the effect has of course generally been regarded as a nuisance rather than an interesting effect in its own right. The available evidence indicates that undamped spiking is most likely to occur when there are several transverse modes of approximately equal loss. Obviously our results suggest that, if the spiking can be attributed to pumping nonuniformity, then what is actually being observed is chaotic lasing.

Related conclusions have been reached by Hollinger, et al., [30.4] and they provide some experimental evidence using a solid-state laser. In their theoretical work they have approximated the time development of the field by coupling the Kirchhoff diffraction integral to a rate equation for the population inversion; this approach avoids the use of any mode expansion. Using parameters appropriate to a Nd:glass laser, they find that lasing on a single longitudinal mode can be chaotic if several transverse modes are oscillating. As the pump rate increases the output goes from steady-state to quasiperiodic to chaotic, and they argue that their observations are consistent with the Ruelle-Takens scenario for the onset of chaos.

Biswas and Harrison [30.5] have observed chaotic behavior associated with transverse mode structure in a cw CO_2 laser. By adjusting an intracavity aperture they could select TEM_{00} or higher-order transverse mode structure. With cavity tuning as a "knob" they found that the transverse beam distribution expanded and lost any discernible mode pattern, implying a multiplicity of transverse modes. Periodic oscillation due to the beating of two transverse modes gave way to quasiperiodic oscillation involving three modes, and on further tuning the pulsations became irregular and apparently chaotic.

31. Discussion

One point that deserves emphasis is that the Haken-Lorenz, Risken-Nummedal, and Casperson instabilities for SMHBLs, MMHBLs, and SMIBLs cannot be accounted for when the familiar rate-equation approximation is made. In fact in the case of homogeneous broadening the simplest form of rate-equation approximation, which neglects spatial hole burning and intermode beats, implies that only a single longitudinal mode can lase in steady state. The physical argument for this is well known: if the laser transition is homogeneously broadened, the gain on a mode detuned by Δ from line center may be assumed to saturate with intensity according to the formula

$$g(\Delta) = \frac{g_o \beta^2}{\Delta^2 + \beta^2 + \beta^2 I/I_{SAT}} \tag{31.1}$$

In steady-state oscillation the gain must just balance the loss, and so $g(\Delta) = g_{th}$, where g_{th} is the threshold gain for laser oscillation. (This "gain clamping" condition, of course, is only an approximation, because the actual steady-state condition is that the integrated round-trip gain must balance the loss. The approximation is well justified in most cases, where the output coupling is not too large and the variations of I with z are small. [31.1]) But if $g(0) = g_{th}$ then $g(\Delta \neq 0) < g_{th}$, and so modes off line center cannot lase; more generally the mode closest to line center will be the only one to lase.

Of course in many homogeneously broadened lasers it is indeed observed that lasing occurs on a single longitudinal mode. This is generally true in high-pressure gas lasers, where effects of spatial hole burning are washed out by molecular motion. In such systems, furthermore, the homogeneous linewidth is large enough to make the rate-equation approximation quite accurate.

The Risken-Nummedal MMHBL instability shows that <u>the familiar "gain clamping" condition is not generally valid:</u> homogeneously broadened lasers <u>can</u> lase on more than one longitudinal mode if the pumping is strong enough that the Rabi splitting of the transition can strongly couple (phase-lock) neighboring modes. Furthermore even the SMHBL can depart from the steady-state stable fixed point predicted by the rate-equation approximation if the pumping is strong enough, because mode splitting permits more than one frequency to be associated with the same wavelength (longitudinal mode). And instabilities are easy to observe in bad-cavity SMIBLs because the mode splitting is strong even near line center.

32. More Laser Instabilities

The experimental data shown in the preceding sections provide convincing evidence of deterministic chaos: irregular time dependence, broadband power spectra, and, most significantly, well-characterized routes to chaos. We should also mention that Albano, et al. [32.1] have employed the embedding techniques reviewed in Sections 18 and 19 and found strong evidence from analysis of digitized time series that the chaos in the 3.51 μm He-Xe SMIBL is indeed deterministic.

There has been so much published on laser instabilities and chaos in the past few years that we could easily fill the remainder of this volume with a review of all that has been going on. However, this is not our intention and it is hardly necessary, as Harrison and Biswas [24.7] have given a review specifically devoted to laser instabilities and chaos, and Abraham, Lugiato, and Narducci [32.2] have provided a useful review of work up to about the end of 1984. Our goal here has not been to provide anything like an exhaustive review of the field, but rather to survey it within the context of the introduction to chaos given in Sections 1 - 21. From the fact that the field is replete with nonlinearities, it should be evident by now that instabilities and chaos in this field are "all over the place." In the remainder of this section we will conclude our survey of laser instabilities and chaos by briefly discussing some of the other experimental work we have come across.

We have already mentioned the CO_2 laser experiment of Arecchi, et al. [12.1] which made use of an "external" control knob - in this case the modulation frequency of the cavity loss. (Section 12) More recently Midavaine, et al. [32.3] have elasto-optically modulated the cavity length of a CO_2 laser. They observed period doubling to chaos as the driver voltage is raised, and periodic windows within the chaotic regime. Figure 32.1 shows an oscilloscope trace of a bifurcation diagram they obtained.

Klische, et al. [32.4] have modulated the pump (at a rate ≈ the relaxation oscillation frequency) of a solid-state laser (NdP_5O_{14})

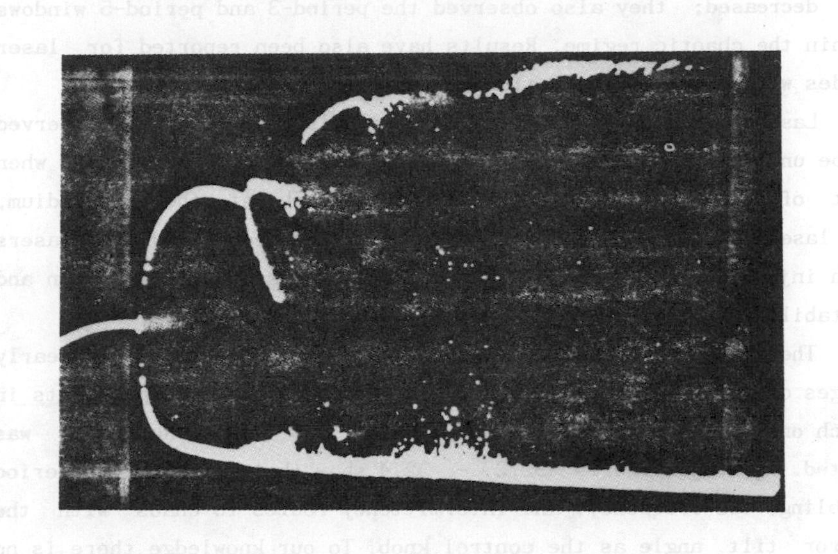

Figure 32.1 **Bifurcation diagram for a CO_2 laser with modulated cavity length, showing output vs. voltage applied to modulator (0 - 20 V). From T. Midavaine, *et al*., Reference [32.3], with permission.**

and observed a period doubling to chaos as the modulation frequency was decreased; they also observed the period-3 and period-5 windows within the chaotic regime. Results have also been reported for laser diodes with modulated injection currents. [32.5]

Lasers with externally injected signals have also been observed to be unstable and chaotic. (It has long been recognized that when part of the output of a laser is injected back into the gain medium, the laser output can become "chaotic.") It has been found that lasers with injected signals can exhibit coexisting basins of attraction and bistability. [32,6, 32,7]

The three "universal" routes to chaos were observed in the early stages of this field by Weiss and his collaborators in experiments in which one of the mirrors of a high-gain 3.39 μm He-Ne laser was tilted. [32.8] Figures 32.2 - 32.4 show their results for period doubling, two-frequency, and intermittency routes to chaos with the mirror tilt angle as the control knob. To our knowledge there is no theory explaining their observations.

Winful, Chen, and Liu [32.9] have obtained data for a diode laser in which two parameters - the modulation amplitude and frequency - are varied. They observed the closure of certain resonance regions (Arnold horns) discussed in a more general context by Aronson, et al. [32.10] Consider the example of the forced Brusselator equations [32.10, 32.11]:

$$\dot{x} = A + x^2y - Bx - x + a\cos\omega t \tag{32.1a}$$

$$\dot{y} = Bx - x^2y \tag{32.1b}$$

For $a = 0$ the (unforced) Brusselator has a limit cycle with natural frequency $\omega_o = 0.3750375$. For $a > 0$ it is convenient to consider the stroboscopic map obtained by sampling solutions of (32.1) at times $t = NT = N(2\pi/\omega)$, $N = 1, 2, 3, \ldots$; a periodic point of this map implies an entrainment of the oscillator by the periodic driving force. The oscillation frequency locks to a harmonic or subharmonic

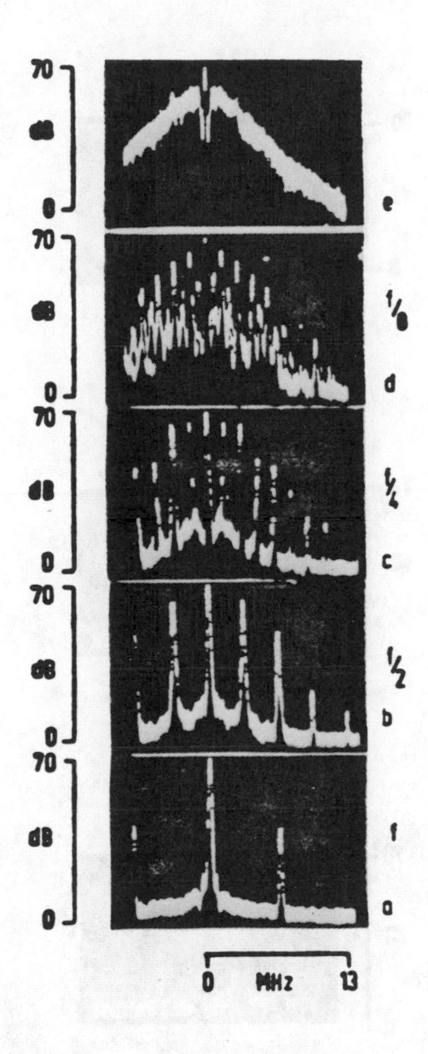

Figure 32.2 Period doubling route to chaos in a He-Ne laser as one of the mirrors is tilted. From C.O. Weiss, et al., Reference [32.8], with permission.

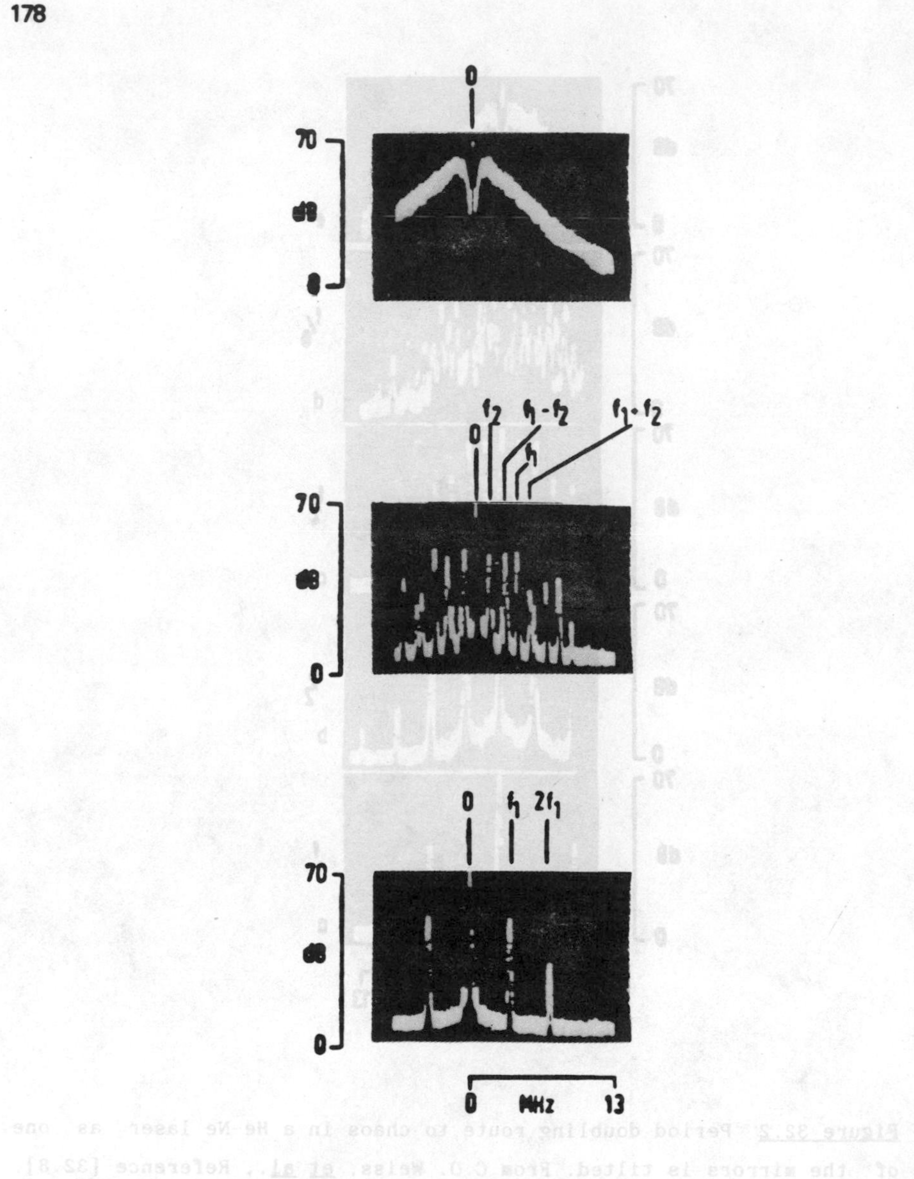

Figure 32.3 Two-frequency route to chaos in a He-Ne laser with mirror tilt. From C.O. Weiss, et al., Reference [32.8], with permission.

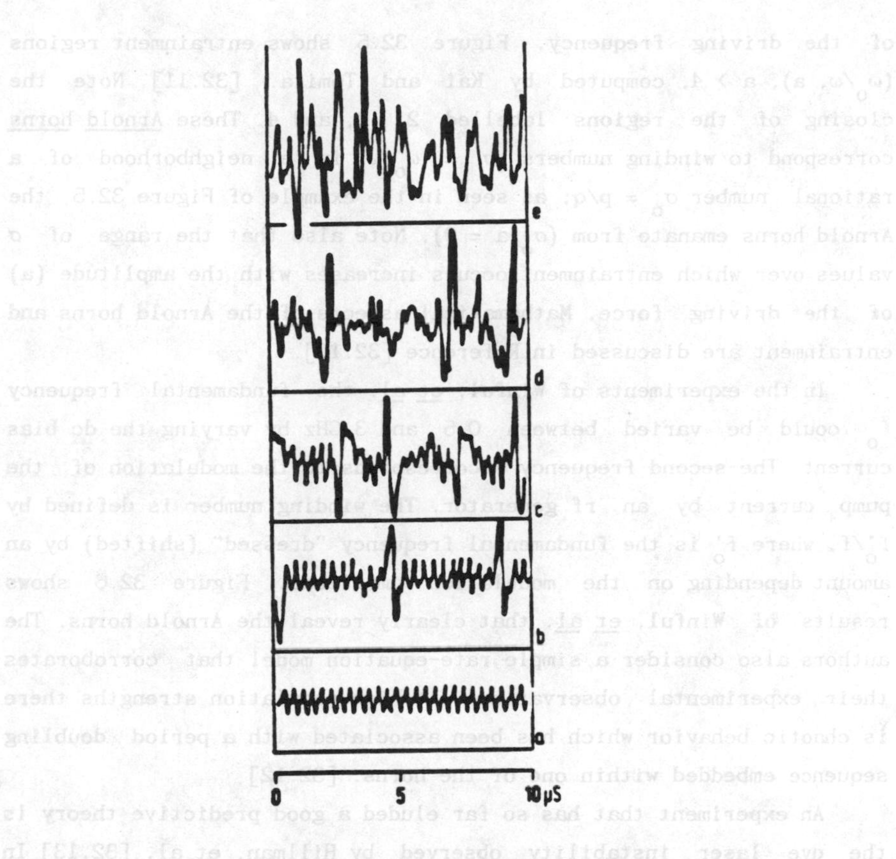

Figure 32.4 Intermittency route to chaos in a He-Ne laser with mirror tilt. From C.O. Weiss, *et al.*, Reference [32.8], with permission.

of the driving frequency. Figure 32.5 shows entrainment regions $(\omega_o/\omega, a)$, $a > 4$, computed by Kai and Tomita. [32.11] Note the closing of the regions labelled 2, 3, and 4. These Arnold horns correspond to winding numbers $\sigma = \omega_o/\omega$ in a neighborhood of a rational number $\sigma_o = p/q$; as seen in the example of Figure 32.5, the Arnold horns emanate from $(\sigma_o, a = 0)$. Note also that the range of σ values over which entrainment occurs increases with the amplitude (a) of the driving force. Mathematical aspects of the Arnold horns and entrainment are discussed in Reference [32.10].

In the experiments of Winful, et al. the fundamental frequency f_o could be varied between 0.5 and 3 GHz by varying the dc bias current. The second frequency f corresponds to the modulation of the pump current by an rf generator. The winding number is defined by f_o'/f, where f_o' is the fundamental frequency "dressed" (shifted) by an amount depending on the modulation amplitude. Figure 32.6 shows results of Winful, et al. that clearly reveal the Arnold horns. The authors also consider a simple rate-equation model that corroborates their experimental observations. At high modulation strengths there is chaotic behavior which has been associated with a period doubling sequence embedded within one of the horns. [32.12]

An experiment that has so far eluded a good predictive theory is the dye laser instability observed by Hillman, et al. [32.13] In these experiments the dye and the cavity are such that several thousand longitudinal modes lie within the homogeneous gain linewidth. The principal observations are as follows. For low pumping levels the output power grows linearly with the (argon laser) pump power. This is the region between A and B in Figure 32.7. At a point B the output power jumps to a higher value, and begins to follow a new, approximately linear growth with pump intensity, the region between C and D in Figure 32.7. (Note that this occurs at a pumping level something like 30% above threshold.) Then another jump occurs at the point F. Two hysteresis loops are evident in the figure.

Figure 32.8 shows the spectral output of the dye laser as a function of power. Evidently the laser suddenly switches to

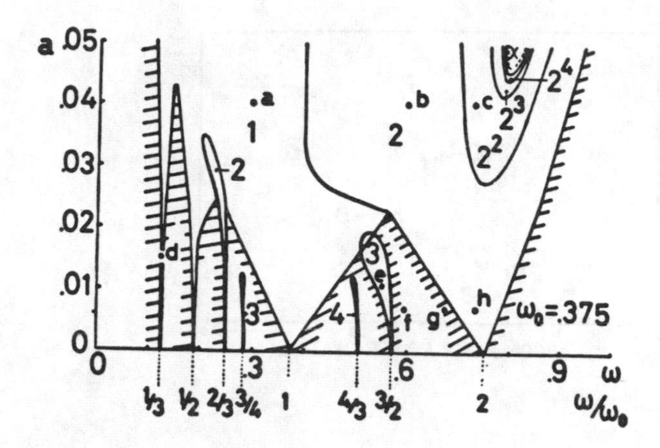

Figure 32.5 Entrainment regions for the forced Brusselator system (32.1). From T. Kai and K. Tomita, Reference [32.11], with permission.

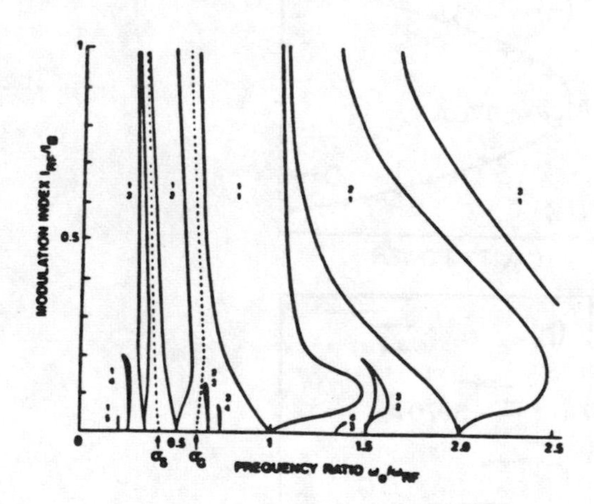

Figure 32.6 Experimental results of Winful, et al. on the entrainment regions for a diode laser with varying modulation amplitude and frequency, showing Arnol'd horns. From Reference [32.9], with permission.

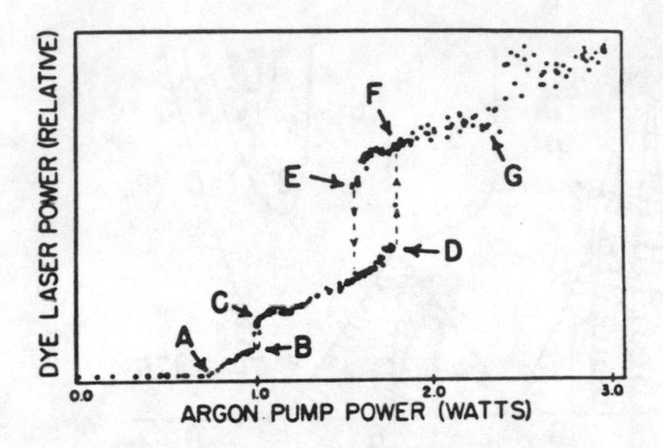

Figure 32.7 Variation of dye laser output power with pumping power observed by Hillman, et al. From Reference [32.13], with permission.

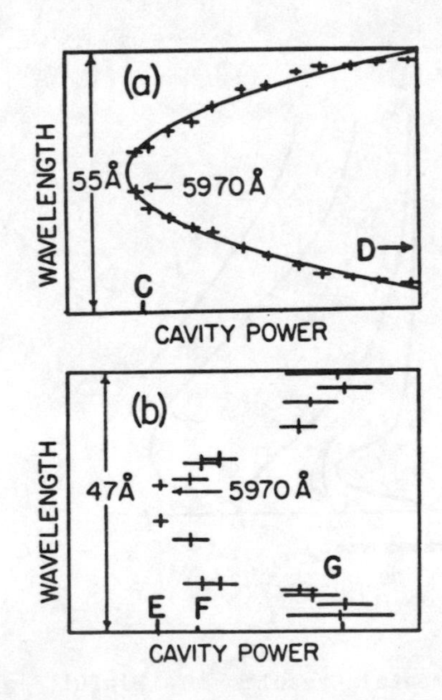

Figure 32.8 Spectral output of a dye laser as a function of pump power observed by Hillman, et al. From Reference [32.13], with permission.

bichromatic oscillation at the point C, as shown in part (a) of the figure, which corresponds to the C-D branch of Figure 32.7. Part (b) shows the spectral characteristics for the E-G branch of Figure 32.7.

We have made preliminary attempts, without much success, to explain these observations assuming (a) uniform scalar plane waves, (b) unidirectional operation, (c) purely homogeneous broadening, (d) temporally and spatially uniform pumping, and (e) various multilevel models of the dye laser transitions, including both singlet and triplet states.

33. Optical Bistability

A system is said to be bistable if it has two possible output states for one and the same input state. Optical bistability generally refers to an optical system with two possible outputs for the same input intensity. Optically bistable devices are of interest in connection with optical computers. Gibbs [33.1] has written a monograph on optical bistability, and Abraham and Smith [33.2] have published a useful review article. The concept of optical bistability goes back to early papers of Szöke, et al. [33.3] and McCall. [33.4]

Suppose a laser beam is injected into a cavity containing N two-level atoms per unit volume. The Maxwell-Bloch equations (23.1) become

$$\dot{v} = -\beta v + (d/\hbar)(A + A_o)w \tag{33.1a}$$

$$\dot{w} = -\gamma(w - w_o) - (d/\hbar)(A + A_o)v \tag{33.1b}$$

$$\dot{A} = -\gamma_c A + (2\pi N d\omega)v \tag{33.1c}$$

where A_o is the (constant) amplitude of the injected field, assumed (like the intracavity field) to be exactly resonant with the two-level atoms. The steady-state solution of (33.1) gives

$$A_o = A_T + (c\alpha_o/2\gamma_c)\,\frac{A_{SAT}^2 A_T}{A_{SAT}^2 + A_T^2} \tag{33.2}$$

for the steady-state value of the total field amplitude $A_T = A_o + A$. Here $\alpha_o = 4\pi Nd^2\omega/\hbar\beta c$ is the absorption coefficient at line center (we assume an absorbing medium, so that $w_o = -1$), and $A_{SAT}^2 = \hbar^2\beta\gamma/d^2$. Note that the saturation intensity defined by (24.2) is just $I_{SAT} = cA_{SAT}^2/8\pi$. In terms of $X = A_T/A_{SAT}$ and $X_o = A_o/A_{SAT}$ we have simply

$$X_o = X + aX/(1 + X^2) \tag{33.3}$$

where $a \equiv c\alpha_o/2\gamma_c$.

Figure 33.1 is a plot of X vs. X_o satisfying (33.3) for a = 25. This indicates that, when a field is injected into a cavity containing an absorbing medium, the total field in the cavity might be a multivalued function of the injected field. To investigate this possibility, we now consider in more detail the injection of a field into a cavity containing an absorption cell.

Consider the system shown in Figure 33.2. Mirrors 1 and 2 have reflectivity R = 1 - T, whereas mirrors 3 and 4 are assumed to be perfect reflectors. The cell inside the cavity contains N atoms per unit volume. E_I denotes the amplitude of the monochromatic input field, and the transmitted and intracavity fields are denoted by E_T and E, respectively. The cell length is L, the total ring circuit length is L_{tot}, and we denote the difference by $\ell = L_{tot} - L$.

The system shown in Figure 33.2 is described by the Maxwell-Bloch equations with the appropriate boundary conditions. The major differences between this device and a laser are that the medium is absorbing ($w_o = -1$) rather than amplifying, and there is an injected field E_I. Generalizing equations (28.1) to include a detuning Δ and the field phase ϕ ($E = Ae^{i\phi}$), we write

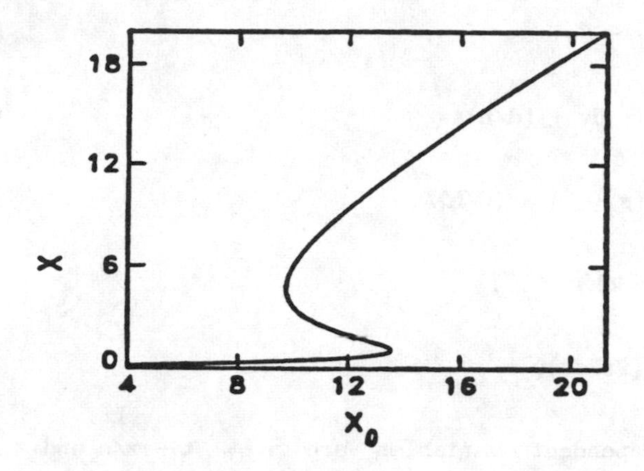

Figure 33.1 Plot of X vs. X_o satisfying equation (33.3) for a = 25.

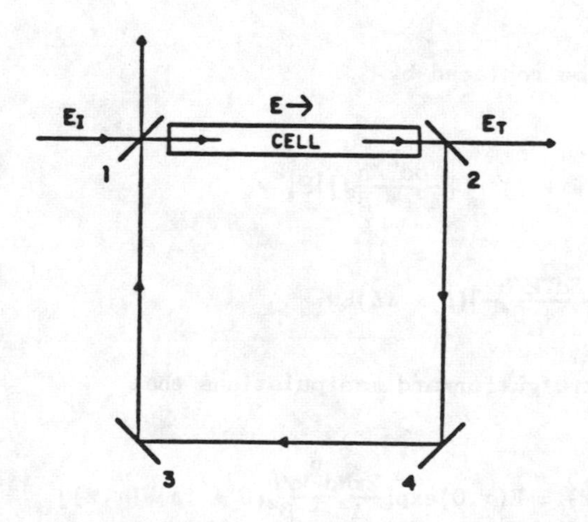

Figure 33.2 Ring cavity configuration for an optically bistable device.

$$\partial u/\partial\tau = -\Delta v - \beta u \tag{33.4a}$$

$$\partial v/\partial\tau = \Delta u - \beta v + (d/\hbar)Aw \tag{33.4b}$$

$$\partial w/\partial\tau = -\gamma(w + 1) - (d/\hbar)Av \tag{33.4c}$$

$$\partial A/\partial z = (2\pi Ndk)v \tag{33.4d}$$

$$A\partial\phi/\partial z = -(2\pi Ndk)u \tag{33.4e}$$

where the independent variables are $\tau = t - z/c$ and z. Loss is assumed to occur only at the mirrors, so that $\gamma_c = 0$, and the absorber is assumed to be homogeneously broadened.

Assume β is large enough that the rate-equation approximation is justified:

$$u \cong -(\Delta/\beta)v \tag{33.5a}$$

$$v \cong (\Delta/\beta)u + (d/\hbar\beta)Aw \tag{33.5b}$$

Then (33.4) may be replaced by

$$\partial w/\partial\tau = -\gamma(w + 1) - [\frac{\beta d^2/\hbar}{\Delta^2 + \beta^2}]|E|^2 w \tag{33.6a}$$

$$\partial E/\partial z = [\frac{2\pi Nd^2 k/\hbar}{\Delta^2 + \beta^2}](\beta + i\Delta)Ew \tag{33.6b}$$

It follows by straightforward manipulations that

$$E(\tau + z/c, z) = E(\tau,0)\exp[\frac{2\pi Nd^2 k/\hbar}{\Delta^2 + \beta^2}(\beta + i\Delta)W(\tau,z)] \tag{33.7a}$$

$$\partial W(\tau,z)/\partial\tau = -\gamma[W(\tau,z) + z] - (1/4\pi N\hbar k)|E(\tau,0)|^2$$

$$\times \{\exp[\frac{4\pi Nd^2 k\beta/\hbar}{\Delta^2 + \beta^2} W(\tau,z)] - 1\} \tag{33.7b}$$

where

$$W(\tau,z) \equiv \int_0^z dz' w(\tau + z'/c, z') \tag{33.7c}$$

The cavity boundary conditions may be written in terms of the fields at the ends of the absorption cell [33.5]:

$$E(t,0) = \sqrt{T} E_I(t) + R\exp(ikL_{tot})E(t - \ell/c, L) \tag{33.8a}$$

$$E_T(t) = \sqrt{T} E(t,L)\exp(ikL) \tag{33.8b}$$

Equations (33.7) and (33.8) are the bases of the analysis by Ikeda. [33.5] He defines the population difference in such a way that $w = 1/2$ ($-1/2$) for an atom in the upper (lower) state. The connection with Ikeda's work is therefore facilitated by replacing W by 2W, so that W below will be the same as Ikeda's. Defining $\Theta = 2\pi Nd^2k/\hbar$, $\Delta\omega = \Delta$, $\gamma_\perp = \beta$, $\gamma_\parallel = \gamma$, and $\mu = d/\hbar$, we can write (33.7a) and (33.7b) as equations (4) and (5) of Ikeda's paper [33.5]:

$$E(\tau + z/c, z) = E(\tau,0)\exp[2\Theta W(\tau,z)(i\Delta\omega + \gamma_\perp)/(\Delta\omega^2 + \gamma_\perp^2)] \tag{33.9a}$$

$$\partial W(\tau,z)/\partial\tau = -\gamma_\parallel(W + z/2) - (\mu^2/2\Theta)|E(\tau,0)|^2$$

$$\times \{\exp[4\Theta\gamma_\perp W/(\Delta\omega^2 + \gamma_\perp^2] - 1\} \tag{33.9b}$$

Following Ikeda, define the dimensionless variables

$$x = \gamma_\parallel t, \qquad \kappa = \gamma_\parallel L_{tot}/c, \qquad \Lambda = \Delta\omega/\gamma_\perp$$

$$\phi(t) = (1/L)W(t - L_{tot}/c, L)$$

$$\epsilon(t,z) = (\mu/2)E(t,z)[\gamma_{\perp}\gamma_{\parallel}(1 + \Delta^2)]^{-1/2}$$

$$\epsilon_T(t) = (\mu/2)E_T(t - \ell/c)[\gamma_{\perp}\gamma_{\parallel}(1 + \Delta^2)]^{-1/2}$$

$$\epsilon_I(t) = (\mu/2)E_I(t)[\gamma_{\perp}\gamma_{\parallel}(1 + \Delta^2)]^{-1/2}$$

in terms of which (33.8) and (33.9) may be combined to give [33.5]

$$\epsilon(x,0) = \sqrt{T}\epsilon_I(x) + R\epsilon(x-\kappa, 0)\exp[\alpha L\phi(x)] \times \exp\{i[\alpha L\Delta(\phi(x) + \tfrac{1}{2}) - \delta_o]\} \quad (33.10a)$$

$$d\phi/dx = - [\phi(x) + \tfrac{1}{2}] - 2|\epsilon(x-\kappa, 0)|^2(1/\alpha L)\{\exp[2\alpha L\phi(x)] - 1\} \quad (33.10b)$$

$$\epsilon_T(x) = \sqrt{T}\epsilon(x-\kappa, 0)\exp[\alpha L\phi(x)]\exp\{i[\alpha L\Delta(\phi(x) + \tfrac{1}{2}) - (\delta_o + k\ell)]\} \quad (33.10c)$$

where

$$\alpha = 2\theta\gamma_{\perp}/(\Delta\omega^2 + \gamma_{\perp}^2) \quad (33.11a)$$

$$\delta_o = -k[(1 - \frac{2\pi N\mu^2\Delta\omega}{\Delta\omega^2 + \gamma_{\perp}^2})L + \ell] + 2\pi M \quad (33.11b)$$

and $2\pi M$ is introduced as the integral multiple of 2π closest to the first term on the right-hand side of (33.11b).

The steady-state limit of (33.10) is obtained by setting $d\phi/dx = 0$ and $\epsilon(x,0)$ = constant. If $\delta_o = \Delta = 0$, we obtain after simple algebra the steady-state relation

$$\log[1 + T(\epsilon_I/\bar{\epsilon}_T - 1)] + (\bar{\epsilon}_T^2/2T)\{[1 + T(\epsilon_I/\bar{\epsilon}_T - 1)]^2 - 1\} = \alpha L/2 \tag{33.12}$$

where $\bar{\epsilon}_T$ denotes the steady-state value of ϵ_T. This is the relation for "absorptive" optical bistability derived by Bonifacio and Lugiato. [33.6] They showed that, for R greater than some critical value depending on L, there are two allowed values of $\bar{\epsilon}_T$ for the same ϵ_I, i.e., <u>there is optical bistability</u>. (This is called "absorptive" bistability because $\Delta = 0$ implies there is no resonant dispersion.) If αL, $\alpha\Delta L$, $|\delta_o|$, and $|\bar{\epsilon}_T|^2/T$ are all small compared with 1, we have the case of "dispersive" bistability observed in experiments of Gibbs, <u>et al</u>. [33.7]

For a detailed treatment of optical bistability in the system of absorbing atoms in a ring cavity, we refer the reader again to the review of Abraham and Smith. [33.2] We now turn our attention to chaos in optically bistable systems.

34. Chaos in Optical Bistability

If the population difference has a relaxation time much shorter than the ring transit time L_{tot}/c, so that $\kappa \gg 1$, we may set $d\phi/dx \cong 0$ in (33.10), and the delay differential equations in this limit reduce to a discrete mapping:

$$\epsilon_{0,n} = \sqrt{T}\epsilon_I + R\epsilon_{0,n-1}\exp(\alpha L\phi_n)\exp\{i[\alpha L\Delta(\phi_n + \tfrac{1}{2}) - \delta_o]\} \tag{34.1a}$$

$$\epsilon_{T,n} = \sqrt{T}\epsilon_{0,n-1}\exp(\alpha L\phi_n)\exp\{i[\alpha L\Delta(\phi_n + \tfrac{1}{2}) - (\delta_o + k\ell)]\} \tag{34.1b}$$

$$\phi_n + \tfrac{1}{2} = (2/\alpha L)|\epsilon_{0,n-1}|^2[1 - \exp(\alpha L\phi_n)] \qquad (34.1c)$$

where

$$\epsilon_{0,n} \equiv \epsilon(x_0 + n\kappa, 0) \qquad (34.2a)$$

$$\epsilon_{T,n} \equiv \epsilon_T(x + n\kappa) \qquad (34.2b)$$

Ikeda [33.5] iterated the map (34.1) for the example $\alpha\Delta L = 6$, $\alpha L = 4$, $\delta_o = 0$, and $R = 0.95$. Within a range of values of $|\epsilon_I|$, <u>the sequence $\{\epsilon_{T,n}\}$ of transmitted field strengths is chaotic, and follows a period doubling route to chaos</u>.

Consider now a medium of two-level atoms far removed from resonance with the input field. If the absorption coefficient of the medium is nonsaturable and the two-level atoms, being far from resonance, make no significant contribution to it, we may replace $\alpha L\phi$ in (33.10) by $-\alpha L/2$. Defining e and θ by

$$\epsilon(x,0) = [2\Delta(1 - e^{-\alpha L})]^{-1/2}e(x) \qquad (34.3a)$$

$$\phi(x) + \tfrac{1}{2} = (1/\alpha\Delta L)\theta(x) \qquad (34.3b)$$

we then obtain from (33.10) the delay differential system

$$e(t) = A + Be(t - t_R)\exp[i(\theta(t) - \delta_o)] \qquad (34.4a)$$

$$\gamma_{\parallel}^{-1}\dot{\theta}(t) = -\theta(t) + |e(t - t_R)|^2 \qquad (34.4b)$$

where

$$A = [2T\Delta(1 - e^{-\alpha L})]^{1/2}\epsilon_I \tag{34.5a}$$

$$B = Re^{-\alpha L/2} \tag{34.5b}$$

$$t_R = L_{tot}/c \tag{34.5c}$$

Equations (34.4) for the case of "dispersive" optical bistability are identical to those of Ikeda, et al. [34.1] for the case in which the cell contains a medium with a nonlinear refractive index. In their case the longitudinal relaxation rate $\gamma_{\parallel}$ of the optical Bloch equations is replaced by the relaxation rate used in the Debye theory of dielectric relaxation, and A is defined differently. They show that the equations (34.4) admit chaotic behavior, and Nakatsuka, et al. [34.2] have observed this chaotic behavior experimentally, as discussed below.

When the cavity transit time is much larger than the material relaxation time (i.e., $\gamma_{\parallel}t_R >> 1$), we may use an adiabatic approximation to $\theta(t)$, setting the left-hand side of (34.4b) to zero so that $\theta(t) \cong |e(t - t_R)|^2$. Then from (34.4a) it follows that

$$e(t) = A + Be(t - t_R)\exp\{i[|e(t - t_R)|^2 - \delta_o]\} \tag{34.6}$$

In this approximation, therefore the differential-difference system (34.4) is replaced by the map

$$e_n = A + Be_{n-1}\exp\{i[|e_{n-1}|^2 - \delta_o]\} \tag{34.7}$$

studied by Ikeda, et al. [34.1] This takes the form of a real two-dimensional mapping that has a period doubling route to chaos. The results of a detailed study of this map have been reported by Carmichael, et al. [34.3]

The fixed points $\bar{e}$ of the map (34.7) may be found numerically.

In particular, the amplitudes $|\bar{e}|$ are determined by the equation

$$A = |\bar{e}|[1 + B^2 - 2B\cos(|\bar{e}|^2 - \delta_o)]^{1/2} \tag{34.8}$$

assuming for simplicity that A is real. Figure 34.1 shows $|\bar{e}|$ as a function of A for $\delta_o = 0$ and $B = 0.5$, and also indicates which branches of the solution are stable. Proceeding along the lower stable branch from A = 0, stability is lost at $A \cong 1.24775$, and a period doubling bifurcation occurs. At A = 1.5, for instance, the iteration of (34.7) starting from $e_0 = 0$ yields $|e_n| = 1.807953$, 0.6226517, 1.807953, 0.6226517, ... after initial transients die out. At $A \cong 1.511525$ a 4-cycle is found, and small increases in A lead to more period doublings and chaos. Carmichael, et al. [34.3] have verified that the rate of period doubling is governed by the Feigenbaum constant δ.

Figure 34.2 shows the chaotic attractor in the complex plane as computed by Ikeda, et al. [34.1] for the case B = 0.4, A = 4.39. The enlargement shown in Figure 34.2b shows a self-similar structure typical of a strange attractor. (Section 17)

If $B \ll 1$ but $A^2B \approx 1$, we can write the following approximate version of (34.4):

$$\gamma_{\parallel}^{-1}\dot{\theta}(t) = -\theta(t) + A^2\{1 + 2B\cos[\theta(t - t_R) - \delta_o]\} \tag{34.9}$$

By numerical integration of this equation, and Fourier transforming the time series so obtained, Ikeda, et al. [34.1] find a transition to chaos as A is varied; the power spectrum of $\theta(t)$ changes from a discrete spectrum of sharp spikes to a broadband spectrum.

The experiments of Nakatsuka, et al. [34.2] on chaos in an "all-optical" bistable system support these predictions. Their experimental arrangement is shown in Figure 34.3. A single-mode optical fiber with a quadratic nonlinear index is used as the

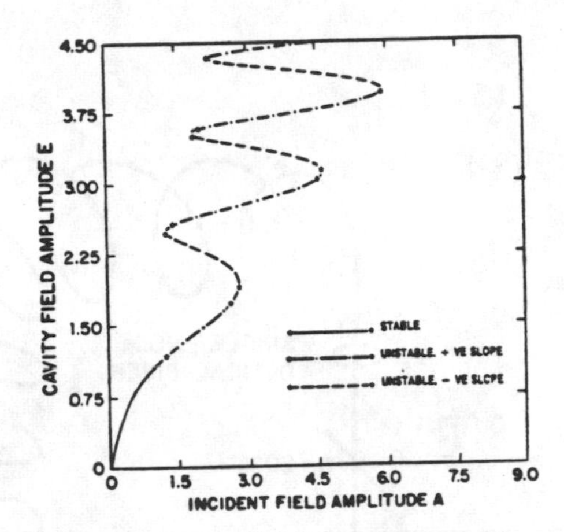

Figure 34.1 $|\bar{e}|$ as a function of A for $\delta_o = 0$ and B = 0.5, indicating which branches of the solution are stable. From H.J. Carmichael, et al., Reference [34.3], with permission.

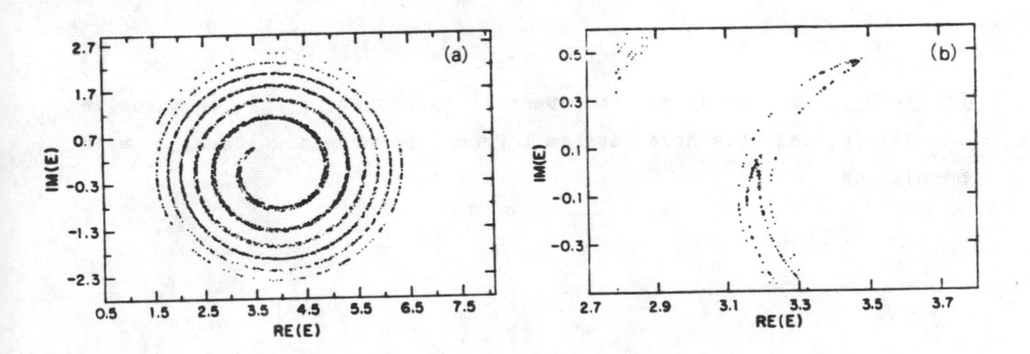

Figure 34.2 Chaotic Ikeda attractor in the complex e plane for B = 0.4 and A = 4.39 (a), and an enlargement (b) showing the self-similar structure characteristic of a strange attractor.

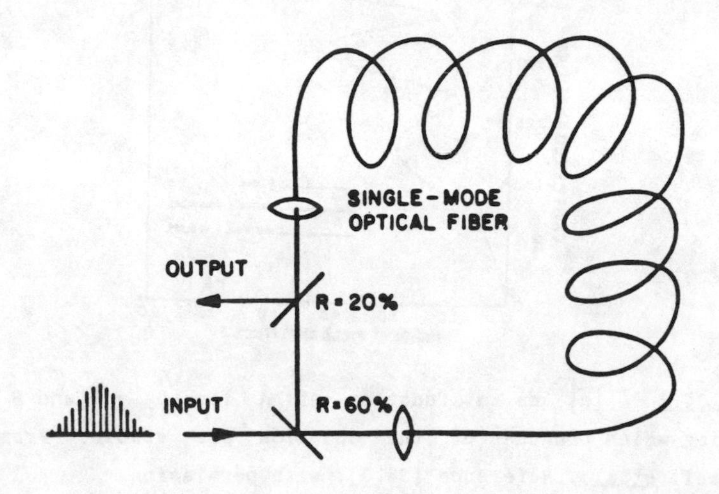

Figure 34.3 Schematic of experiment of Nakatsuka, et al. on chaos in an all-optical bistable system. From Reference [34.2], with permission.

nonlinear "cell" in the ring. The high power level necessary for the observation of chaos with this system was obtained from the second harmonic of a mode-locked YAG laser (pulse separation ≈ 7.6 nsec). As the bifurcations to chaos stem from the interference of the input and cavity fields, it was necessary to match the ring transit time to the period of the mode-locked pulses. The cavity was such that $B \cong 0.4 - 0.5$, and the parameter A could be increased sufficiently, by raising the input power level, to realize a chaotic regime of the system. When the peak power of the input pulse train was raised from 50 to 160 W, a period doubling of the output train was observed. At a power level of 300 W the output was chaotic. The results were consistent with the theoretical model.

Chaos in a "hybrid" optically bistable device, in which a delay time was introduced electronically using a delayed feedback line, was observed in 1980 by Gibbs, et al. [34.4] Their hybrid device is well described by equation (34.9) of Ikeda, et al., [34.1] although of course the origin of the delay time t_R is different.

Chaos in an all-optical ring resonator system employing an ammonia cell and injected CO_2 laser pulses has been reported by Harrison, et al. [34.5] As in the experiments of Gibbs, et al. there is evidence of a period doubling route to chaos. Their approach enjoys considerable flexibility for the variation of parameters, and furthermore enjoys the theoretical neatness associated with a homogeneously broadened two-level system.

The development of chaos in a ring resonator pumped by a sequence of sech pulses has been studied numerically by Blow and Doran. [34.6] Their analysis is based on a nonlinear, Schrödinger-like wave equation in which the dispersive nonlinearity is due to the Kerr effect. Chaos via period doubling was observed with increasing pump intensity, followed by an inverse Feigenbaum sequence.

PART TWO

HAMILTONIAN SYSTEMS

A. A violent order is disorder; and
B. A great disorder is order. These
Two things are one.

Wallace Stevens, Connoisseur of Chaos (1942)

35. Classical Hamiltonian Systems

A Hamiltonian system is defined by the existence of a function $H(q_1, q_2, q_3, \ldots, q_N; p_1, p_2, p_3, \ldots, p_N)$ of generalized coordinates q_i and momenta p_i such that the equations of motion take the form

$$\dot{q}_i = \partial H / \partial p_i \tag{35.1a}$$

$$\dot{p}_i = - \partial H / \partial q_i \tag{35.1b}$$

q_i and p_i are said to be conjugate variables, and any set of q's and p's whose time evolution is governed by (35.1) is called a set of canonical variables.

A Hamiltonian system of N degrees of freedom has a 2N-dimensional phase space of q's and p's. There is at least one constant of the motion on every trajectory in this phase space, namely the Hamiltonian function. In Hamiltonian systems phase-space volumes are conserved and there are no attractors; thus strange attractors cannot occur in Hamiltonian systems. But chaos can and does occur in Hamiltonian systems, and it may be identified in the same way as in dissipative systems: there is at least one positive Lyapunov exponent, implying very sensitive dependence on initial conditions and a broadband power spectrum of the type shown in Figure 1.2.

Hamiltonian systems are more fundamental than dissipative systems. When we say a system is dissipative it is often implied that there are certain degrees of freedom left out, except insofar as they act as an approximately irreversible decay channel for what we have chosen to call the "system." The friction (dissipation) itself is typically not treated dynamically but rather is modelled by some damping coefficient in the equations of motion for the system. But to be completely rigorous we should include everything in a Hamiltonian and write 2N Hamilton equations of motion of the form (35.1) for a closed system of N degrees of freedom.

When N is very large it is usually impractical to solve these equations, and we resort to techniques that collectively are called statistical mechanics. Here the primary goal is to predict average values of things that can be measured. We will see that there are interesting implications of chaotic behavior for statistical mechanics and in particular for the ergodic hypothesis. For the remainder of this section, however, we will simply review a few well known (and important) aspects of Hamiltonian systems.

As noted above, phase-space volumes are preserved in Hamiltonian systems. This result (Liouville's theorem) may be obtained by considering as in Section 17 the divergence of the "velocity" vector $(\dot{\vec{q}},\dot{\vec{p}})$ in phase space:

$$\sum_i \left(\frac{\partial \dot{q}_i}{\partial q_i} + \frac{\partial \dot{p}_i}{\partial p_i}\right) = \sum_i \left[\frac{\partial}{\partial q_i}\left(\frac{\partial H}{\partial p_i}\right) - \frac{\partial}{\partial p_i}\left(\frac{\partial H}{\partial q_i}\right)\right] = 0 \tag{35.2}$$

Liouville's theorem holds regardless of whether the Hamiltonian is a function of time. Unless otherwise noted we will focus our attention first on conservative (or autonomous) Hamiltonian systems, in which H is independent of time. Later, in connection with the theory of atoms or molecules in applied, time-varying fields, we will be concerned also with time-dependent Hamiltonians.

It may also be shown that a flow that preserves phase-space volume, and for which the velocity vector is sufficiently well behaved, is Hamiltonian.

Stable fixed points of Hamiltonian systems must be centers. (Section 2) That is, the eigenvalues associated with the linearized flow near a stable fixed point must be purely imaginary, so that trajectories in the neighborhood of a stable fixed point are closed loops. (Of course these loops are not limit cycles, since they are not isolated, as discussed in Section 2.) Centers are also called elliptic fixed points in the literature.

We have already noted that a conservative Hamiltonian flow in a 2N-dimensional phase space has at least one constant of the motion on each trajectory, namely the Hamiltonian function itself. Each trajectory is therefore confined to a surface of dimension $\leq$ 2N-1 in the 2N-dimensional phase space. For the case N = 1 with Hamiltonian H = $p^2/2m + V(q)$, these surfaces are defined by

$$H(q,p) = p^2/2m + V(q) = E \tag{35.3}$$

For the simple harmonic oscillator with $V(q) = \frac{1}{2}kq^2$ $(k > 0)$, for instance, these curves are just closed loops in the plane. There is one fixed point, namely the center $(q^*,p^*) = (0,0)$. In general periodic motion is of two types: if $\dot{q}$ always has the same sign we have rotation; if $\dot{q}$ changes sign we have libration, as in the simple harmonic oscillator. For N = 1 any closed loop in phase space corresponds, of course, to libration.

The concept of a separatrix in phase space is useful in charting out the types of motion (e.g., rotations, librations, unbounded motions) that can be exhibited by a system. Consider the example H = $p^2/2m - \frac{1}{2}kq^2$ $(k > 0)$, i.e., the case of a linear repulsive force in one dimension. In this case the (unstable) fixed point at the origin is a hyperbolic point, as is easily shown. In other words, the two eigenvalues associated with the linearized flow near the origin are real and of opposite sign. The constant-energy surfaces defined by (35.3) satisfy

$$(p + \sqrt{km}q)(p - \sqrt{km}q) = 2mE \tag{35.4}$$

Thus we have the straight lines $p = \pm\sqrt{km}\, q$ as well as the hyperbolas that join these straight lines asymptotically. (Figure 35.1) The straight lines $p = \pm \sqrt{km}\, q$ in this example form the separatrices: they divide the phase space into regions of different kinds of motion, as is clear from Figure 35.1. In general a separatrix is a curve in phase space that passes through a hyperbolic fixed point.

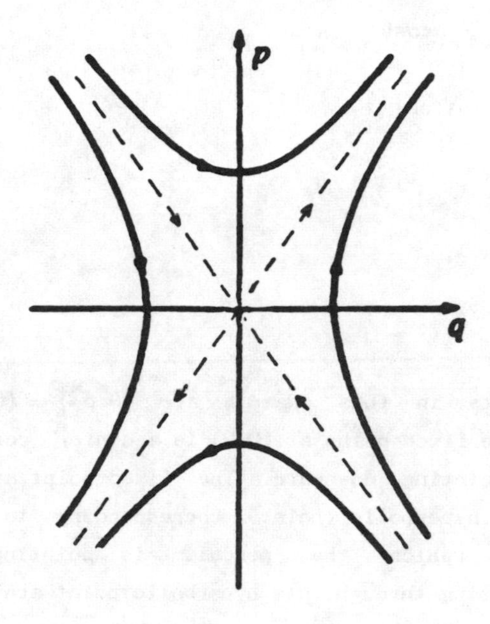

Figure 35.1 Phase curves for the linearly repulsive potential, showing the separatrices (dashed lines) defined by (35.4).

A simple but useful example is the pendulum with Hamiltonian

$$H(\theta,p) = \tfrac{1}{2}p^2 - \alpha\cos\theta \qquad (35.5)$$

and equations of motion

$$\dot{\theta} = p \qquad (35.6a)$$

$$\dot{p} = -\alpha\sin\theta \qquad (35.6b)$$

The fixed points in this example are $(\theta^*,p^*) = (0,0)$ and $(\pi,0)$. (Figure 35.2) The fixed point at (0,0) is a center, corresponding to the pendulum pointing downward. The fixed point at $(\pi,0)$, on the other hand, is a hyperbolic point, corresponding to the unstable equilibrium in which the pendulum is pointing upward. The separatrices passing through this hyperbolic point are defined by $p = \pm\sqrt{2E + 2\alpha\cos\theta}$ and, since $E = \alpha$ at $(\pi,0)$, $E = \alpha$ on the separatrices. Thus the separatrices are defined by

$$p = \pm\sqrt{2\alpha(1 + \cos\theta)} = \pm 2\sqrt{\alpha}\,\sin(\theta/2) \qquad (35.7)$$

The separatrices are indicated by the dashed lines in Figure 35.2. As in Figure 35.1, they divide the phase space into regions of qualitatively different behavior.

Exercise: Describe the pendulum motion in the different regions of phase space defined by the separatrices shown in Figure 35.2.

The canonical transformation theory of Hamiltonian systems is covered in detail in such books as Goldstein [35.1] and ter Haar. [35.2] Nevertheless it is probably worthwhile to review here a few pertinent points, especially in connection with action-angle

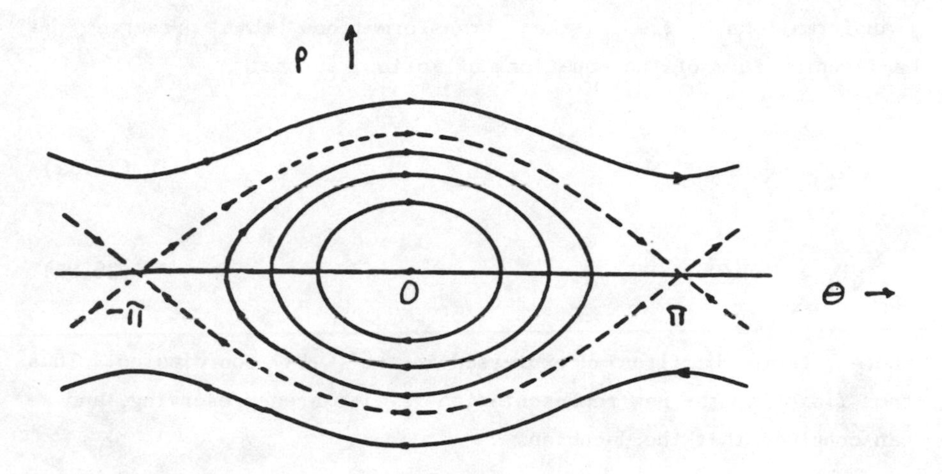

Figure 35.2 Phase curves for the pendulum with Hamiltonian (35.5), showing the separatrices as dashed curves.

variables.

Transformations from variables (q,p) to (Q,P) may be used to simplify a problem. Of special interest are the canonical transformations, i.e., the transformations that preserve the Hamiltonian form of the equations of motion, so that

$$\dot{Q}_i = \partial\bar{H}/\partial P_i \tag{35.8a}$$

$$\dot{P}_i = -\,\partial\bar{H}/\partial Q_i \tag{35.8b}$$

where $\bar{H}$ is the Hamiltonian expressed in the (Q,P) coordinates. Thus the flow in the new representation is also area-preserving, and we can conclude that the Jacobian

$$\frac{\partial(Q,P)}{\partial(q,p)} = \frac{\partial Q}{\partial q}\frac{\partial P}{\partial p} - \frac{\partial Q}{\partial p}\frac{\partial P}{\partial q} = 1 \tag{35.9}$$

(For simplicity we will consider explicitly the case N = 1 and assume that the coordinates q,p and Q,P are Cartesian.)

Consider the area S enclosed by a closed curve C in the q,p phase space. According to Stokes' theorem we can write this area as

$$\int_S dq\,dp = \oint_C p(q)\,dq \tag{35.10}$$

Similarly in the Q,P representation we have

$$\int_S dQ\,dP = \oint_C P(Q)\,dQ \tag{35.11}$$

and therefore

$$\oint_C (p\,dq - P\,dQ) = 0 \tag{35.12}$$

In writing (35.10) - (35.12) we are viewing the transformation $(q,p) \to (Q,P)$ in the "passive" sense of a coordinate transformation: (q,p) and (Q,P) are viewed as different ways of labelling the same phase point. This allows us to refer to the same S and C in writing these equations in the different coordinate systems. Since

$$\oint_C d(QP) = \oint_C (QdP + PdQ) = 0 \tag{35.13}$$

we can also write (35.12) as

$$\oint_C (pdq + QdP) = 0 \tag{35.14}$$

Since the closed curve C is arbitrary, it follows from (35.14) that $pdq + QdP$ must be the exact differential of some function $F_2(q,P)$:

$$pdq + QdP = dF_2(q,P) \tag{35.15a}$$

or

$$p = \partial F_2(q,P)/\partial q \tag{35.15b}$$

$$Q = \partial F_2(q,P)/\partial P \tag{35.15c}$$

which determines (Q,P) in terms of (q,p). $F_2(q,P)$ is called the <u>generating function</u> of the transformation $(q,p) \to (Q,P)$. An example is $F_2 = qP$, which gives $Q = q$ and $P = p$, the identity "transformation."

The generating function may also be a function of (p,P) or (p,Q) or (q,Q). From (35.12), for instance, we can conclude that $pdq - PdQ$ is the exact differential of some function $F_1(q,Q)$, and

$$p = \partial F_1(q,Q)/\partial q \qquad (35.16a)$$

$$P = -\ \partial F_1(q,Q)/\partial Q \qquad (35.16b)$$

The four types of canonical transformation are discussed in the standard texts, typically from the Lagrangian viewpoint. We have followed the notation of Goldstein [35.1] for the generating functions F_1 and F_2.

36. Integrability and Action-Angle Variables

A great simplification is achieved if the generating function is chosen in such a way that $\bar{H}$ depends only on Q or P. Thus if $\bar{H} = \bar{H}(P)$ then

$$\dot{P} = -\ \partial\bar{H}/\partial Q = 0 \qquad (36.1a)$$

$$\dot{Q} = \partial\bar{H}/\partial P = \text{constant} \equiv \beta \qquad (36.1b)$$

or

$$P = \text{constant} \qquad (36.2a)$$

$$Q = \beta t + \delta \qquad (36.2b)$$

In the case of N degrees of freedom (36.2) generalizes to P_i = constant = γ_i and $Q_i = \beta_i t + \delta_i$. The β_i and δ_i are the 2N constants of integration.

The problem of reducing the Hamiltonian to the form $\bar{H}(P)$, for instance, amounts to finding the right generating function S. If S is of the type $F_2(q,P)$, then from (35.15) we have

$$p = \partial S/\partial q \qquad (36.3a)$$

$$Q = \partial S/\partial P \tag{36.3b}$$

Since $H(q,p) = H(q,\partial S/\partial q) = \bar{H}(P) = \text{constant} \equiv E(P)$, S must satisfy the Hamilton-Jacobi equation

$$H(q,\partial S/\partial q) = E(P) \tag{36.4a}$$

or more generally

$$H(q_1,q_2,\ldots,q_N;\ \partial S/\partial q_1,\partial S/\partial q_2,\ldots,\partial S/\partial q_N) = E \tag{36.4b}$$

Once the solution S of the Hamilton-Jacobi equation is found, the solution of the problem in terms of Q's and P's is trivial (equation (36.2)), and then we can perform the inverse transformation $(Q,P) \to (q,p)$ to find the original (q,p) variables as functions of time.

Example: $H = p^2/2m + \frac{1}{2}kq^2$. The Hamilton-Jacobi equation is

$(1/2m)(\partial S/\partial q)^2 + \frac{1}{2}kq^2 = P$ if we choose $E(P) = \bar{H}(P) = P$ for simplicity. Then $S(q,P) = \int_{q_o}^{q} dq\sqrt{2m(P - \frac{1}{2}kq^2)}$ and

$\dot{Q} = \partial\bar{H}/\partial P = 1$ so $Q = t - t_o$

$Q = \delta S/\delta P = \sqrt{m/2}\int_{q_o}^{q} dq[P - \frac{1}{2}kq^2]^{-1/2} = t - t_o$

The last equation implies

$q = q_o + \sqrt{2P/k}\,\sin[\sqrt{k/m}(t-t_o)]$

Other examples, like n-dimensional harmonic oscillators and the Kepler problem, are worked out in the standard texts.

In general the solution of the Hamilton-Jacobi partial differential equation is a far from trivial task. In some problems, however, coordinates may be chosen in such a way that the Hamilton-Jacobi equation is separable, with one independent equation for each of the N degrees of freedom. In this case the system is said to be separable, and it is also integrable.

Example: $H = p_1^2/2m + p_2^2/2m + \frac{1}{2}k_1q_1^2 + \frac{1}{2}k_2q_2^2$

The Hamilton-Jacobi equation is

$(\partial S/\partial q_1)^2 + (\partial S/\partial q_2)^2 + k_1mq_1^2 + k_2mq_2^2 = 2mE$. Write

$S = S_1(q_1,P_1) + S_2(q_2,P_2)$. Then

$(\partial S_i/\partial q_i)^2 + k_imq_i^2 = \alpha_i$, where the separation constants α_i satisfy $\alpha_1 + \alpha_2 = E$.

The separation constants for the Hamilton-Jacobi equation of an integrable system are called isolating integrals; they are invariants of the motion. A system with N degrees of freedom is integrable if and only if N such independent isolating integrals exist. The isolating integrals must be "in involution," i.e., the Poisson brackets of all pairs vanish. This guarantees that they form a complete set of new momenta.

(Note: Integrability and separability are not generally the same. For instance, the integrable system of a particle in a central force field is separable in spherical coordinates but not in Cartesian coordinates.)

To introduce action-angle variables we first consider a conservative Hamiltonian system with a single degree of freedom. Since the Hamiltonian is a constant of the motion, all systems with N = 1 are integrable. Furthermore the bounded motion of such a system is periodic (recall the Poincare'-Bendixson theorem mentioned in Section 2), and this periodic motion is either libration or rotation. Let S be the generating function for the transformation $(q,p) \rightarrow (\theta,J)$, and a solution of the Hamilton-Jacobi equation. Then the transformed Hamiltonian is a function only of J, and from (36.3) we have

$$p = \partial S/\partial q \tag{36.5a}$$

$$\theta = \partial S/\partial J \tag{36.5b}$$

Thus far all we have done is to write (θ, J) instead of (Q, P). Now we use the fact that the motion is periodic. In particular

$$\partial\theta/\partial q = \partial/\partial q(\partial S/\partial J) = \partial/\partial J(\partial S/\partial q) = \partial p/\partial J \tag{36.6}$$

and so, letting $\oint$ denote an integration over one period of the motion, we have

$$\frac{\partial}{\partial J}\oint p dq = \oint \frac{\partial\theta}{\partial q} dq = \oint d\theta = 1 \tag{36.7}$$

where we have chosen θ such that the constant $\oint d\theta$ is unity. From (36.7) it follows that we may take

$$J = \oint p dq \tag{36.8}$$

J and θ are the <u>action-angle variables</u>; they have dimensions of angular momentum and angle, respectively. From (36.1) we have

$$J = \text{constant} \tag{36.9a}$$

$$\dot{\theta} = \partial\bar{H}/\partial J = \text{constant} \equiv \nu, \ \theta = \nu(t - t_0) \tag{36.9b}$$

From (36.7) and (36.9) it follows that $\nu T = 1$, where T is the period of the motion. In other words, $\nu = \partial\bar{H}/\partial J = \partial E/\partial J$ is the frequency. And <u>once we have the Hamiltonian as a function of the action variable J, we can calculate the frequency without having to solve the equations of motion</u>. This is a major advantage of introducing action-angle variables.

Example: $H = p^2/2m + \frac{1}{2}kq^2 = E$, $p = \sqrt{2m(E - \frac{1}{2}kq^2)}$

$$J = \oint \sqrt{2m(E - \tfrac{1}{2}kq^2)}\, dq = 2E(m/k)^{1/2} \int_0^{2\pi} d\theta \cos^2\theta$$

$$= 2\pi E(m/k)^{1/2}$$

$$\nu = \partial E/\partial J = (1/2\pi)(k/m)^{1/2}$$

For separable systems of $N > 1$ degrees of freedom we may write the solution of the Hamilton-Jacobi equation in the form

$$S = \sum_i S_i(q_i;\ \alpha_1, \alpha_2, \ldots, \alpha_N) \tag{36.10}$$

where now we denote the P_i by α_i to follow a convention and to emphasize that they are constants of the motion. The Hamilton-Jacobi equation for S is separable into N ordinary differential equations

$$H_i(q_i, \partial S_i/\partial q_i) = \alpha_i \tag{36.11}$$

where

$$H = \sum_i H_i = \sum_i \alpha_i = E \tag{36.12}$$

Equations (36.11) may be solved simply by quadrature, i.e., by solving for $\partial S_i/\partial q_i$ and integrating over q_i. The action-angle variables (θ_i, J_i) for integrable systems form a set of canonical variables defined by

$$\dot{J}_i = - \partial\bar{H}/\partial\theta_i = 0 \tag{36.13a}$$

$$\dot{\theta}_i = \partial\bar{H}/\partial J_i \equiv \nu_i(\vec{J}) \tag{36.13b}$$

or

$$\vec{J} = \text{constant} \tag{36.14a}$$

$$\vec{\theta} = \vec{\nu}(\vec{J})t + \vec{\delta} \tag{36.14c}$$

in the notation $\vec{J} \equiv (J_1, J_2, \ldots, J_N)$. These are obvious generalizations of (36.9).

The concept of integrability is very important for our purposes because, as discussed in the next section, <u>integrability implies quasiperiodicity</u>. That is, integrable systems exhibit regular motion, not chaos. We will end this section with a brief digression on the importance of action-angle variables in connection with adiabatic invariance, the old quantum theory, and "semiclassical quantization."

First let us note that by writing

$$\psi(x) = \exp[iS(x)/\hbar] \tag{36.15}$$

in the time-independent Schrödinger equation

$$-(\hbar^2/2m)\partial^2\psi/\partial x^2 + V(x)\psi(x) = E\psi(x) \tag{36.16}$$

for one-dimensional motion, we obtain for S the equation

$$(1/2m)(\partial S/\partial x)^2 - (i\hbar/2m)\partial^2 S/\partial x^2 + V(x) = E \tag{36.17}$$

In the limit $\hbar \to 0$ this reduces to the Hamilton-Jacobi equation. Of course (36.17) is the basis for the WKB (or "semiclassical") approximation.

The action variables are <u>adiabatic</u> <u>invariants</u> in the sense that they do not change when the parameters in the Hamiltonian are varied slowly compared with the frequency of the motion. That is, if $H = H(q, p, \lambda)$, and $T|d\lambda/dt| \ll |\lambda|$, where T is the period of the motion with λ = constant, then the average of dJ/dt over a period vanishes. A neat proof of the adiabatic invariance of the action is given by Landau and Lifshitz. [36.1]

In the days of the old quantum theory Ehrenfest emphasized that the things that are quantized should be adiabatic invariants. The slow variation of a parameter in the Hamiltonian means that no transitions can occur as the parameter varies, because all the frequencies in its Fourier spectrum will be small compared with any Bohr transition frequency. This implies that the quantum numbers characterizing the state of the system are unchanged, and therefore that the physical quantitities associated with these quantum numbers are also unchanged - they are adiabatic invariants. In particular, the action is an adiabatic invariant and therefore can be quantized. This leads to the Bohr-Wilson-Sommerfeld quantization rules. (See also Section 51.)

Example: Assume the quantization rule $J = \oint p\,dq = (n + \frac{1}{2})h$

For the harmonic oscillator $\nu = \partial E/\partial J$, $E = J\nu = (n + \frac{1}{2})h\nu$. For nonlinear systems the semiclassical quantization procedure does not in general give exact energy eigenvalues, although the approximation can be excellent.

How should semiclassical quantization be done if the system is nonseparable? This question was first raised, apparently, by Einstein in 1917. With the advent of the Schrödinger equation the question became irrelevant, for of course the Schrödinger equation assumes nothing about separability. But Einstein had, in the context of the old quantum theory, raised the first question about "quantum chaos," a subject that will concern us later.

37. Integrability, Invariant Tori, and Quasiperiodicity

When the angle variables θ_i each change by one the (q_i, p_i) return to their original values before the change. This means that for integrable systems the trajectories are confined to N-dimensional tori (called invariant tori) in phase space. (Of course the fact that there are N independent constants of the motion implies that the motion is confined to an N-dimensional surface in the 2N-dimensional phase space. The point now is that these surfaces have "toroidal topology.") For the case N = 2, for instance, we have a 2-torus with frequencies ν_1 and ν_2 ($\nu_i = \dot{\theta}_i = \partial\bar{H}/\partial J_i$). See Figure 15.1. Of course this is a reason for calling the θ_i angle variables.

The value of $\vec{J}$ tells us where the N-torus is in phase space, whereas the value of $\vec{\theta}$ tells us the location of a trajectory on the torus. Since the θ_i are periodic with period 1, any trajectory can be expressed as a discrete Fourier series:

$$\vec{q}(t) = \sum_{\vec{n}} \vec{A}_{\vec{n}}(\vec{J})e^{2\pi i\vec{n}\cdot\vec{\theta}} = \sum_{\vec{n}} \vec{A}_{\vec{n}}(\vec{J})e^{2\pi i\vec{n}\cdot(\vec{\nu}t + \vec{\delta})} \tag{37.1a}$$

$$\vec{p}(t) = \sum_{\vec{n}} \vec{B}_{\vec{n}}(\vec{J})e^{2\pi i\vec{n}\cdot\vec{\theta}} = \sum_{\vec{n}} \vec{B}_{\vec{n}}(\vec{J})e^{2\pi i\vec{n}\cdot(\vec{\nu}t + \vec{\delta})} \tag{37.1b}$$

where $\vec{n}$ represents an N-dimensional set of indices. Thus the power spectra associated with integrable motion consist of sharp spikes (Figure 1.1): trajectories of integrable systems are quasiperiodic. They cannot be chaotic.

How can we tell whether a given Hamiltonian system is integrable or nonintegrable? In general there is no way to answer this question without resorting to numerical experiments. In recent years, however, people have used the "Painleve' test." Regarding the independent variable of an ODE as a complex variable, one investigates the nature of the poles of the solution at the movable singularies. (Section 2) It appears (in the absence of a proof) that systems for which these poles are simple are integrable. This test is applicable regardless of whether the system is Hamiltonian. It might also be noted that, when a partial differential equation has soliton solutions, the associated ODE appears to pass the Painleve' test. [37.1] Partial differential equations with soliton solutions are considered "integrable," and it is sometimes said that solitons and chaos are paradigms for two opposite extremes of nonlinear behavior.

38. Ergodicity, Mixing, and Chaos

Statistical mechanics is based on the assumption of equal *a priori* probabilities. Because it deals in probabilities, statistical mechanics makes use of hypothetical ensembles of identically prepared systems, whereas what is actually measured is an average over time in the evolution of a particular system. As so - from one point of view - the foundations of statistical mechanics rest on the *ergodic hypothesis*, the assumption that these ensemble and time averages are equal. Boltzmann expressed the ergodic hypothesis as follows: over the course of time a trajectory will uniformly cover the (2N-1)-dimensional surface of constant energy in the 2N-dimensional phase space. Averages on this energy surface are then equal to time averages.

The problem is to justify the ergodic hypothesis. For a system with a single degree of freedom ($N = 1$) it is trivially true, but it is demonstrably *not* true for any of the other analytically solvable textbook problems (n-dimensional harmonic oscillators, motion in a central force field, etc.). In fact for any integrable system with $N > 1$ any trajectory is confined to an N-dimensional torus, and since $N < 2N-1$ for $N > 1$, such a system cannot be ergodic.

(We hasten to add that not everyone thinks the ergodic hypothesis is necessary for the foundations of statistical mechanics. Some people argue that statistical mechanics should not be based on long-time averages, but should be developed starting from the recognition that we can only make probabilistic statements subject to our uncertainty about the precise Hamiltonian, initial conditions, etc. E.T. Jaynes, for instance, formulates statistical mechanics in terms of information theory. Furthermore we are usually concerned with a reduced probability distribution f_n, $n \ll N$, while the N-particle system may not even be equilibrated, let alone ergodic.)

The "ergodic problem" stimulated the development of the mathematical discipline called ergodic theory, but it seems that this abstract study has not produced general proofs of ergodicity under assumptions palatable to physicists. For about half a century some physicists who worried about the ergodic problem believed that small, nonintegrable perturbations of an integrable system would give rise to ergodicity. Fermi, for instance, felt this way, and when electronic computers like the MANIAC at Los Alamos became available in the early fifties he decided to test this hypothesis "experimentally." These numerical experiments are discussed in the following section.

There is still not very much that can be said about the general validity of the ergodic hypothesis. One important result that follows from rigorous mathematical work of Sinai, however, is that a hard-sphere gas of $n \geq 2$ particles in a box is ergodic. The proof is long (over a hundred pages) and abstruse, and this apparently unavoidable circumstance may be one reason why many physicists are "turned off" by ergodic theory.

Let P be some point in phase space that evolves according to Hamilton's equations into the point P_t at time t. We can write the transformation $P \rightarrow P_t$ as

$$P_t = \phi_t P \tag{38.1}$$

Now if $f(P_t) = f(P,t)$ is some function of the q's and p's its time average is

$$\bar{f} \equiv \lim_{T\to\infty}(1/T)\int_0^T dt f(P,t) \tag{38.2}$$

On the other hand the ensemble average of f is defined as

$$\langle f \rangle \equiv (1/\Omega)\int_{\Omega} d\Gamma f(P) \tag{38.3}$$

where $d\Gamma$ is the volume element in the energetically accessible phase space of volume Ω. The ergodic hypothesis is then the assumption that

$$\langle f \rangle = \bar{f} \tag{38.4}$$

Ergodicity may be related to Birkhoff's concept of <u>metrical transitivity</u>. Suppose V is an invariant subspace of the phase space associated with the transformation (38.1), so that $\phi_t V = V$. V is called metrically transitive if it cannot be decomposed into two invariant subspaces V_1 and V_2 of nonzero measure. If we define the measure $\mu(V)$ of V to be 1, then V is metrically transitive if any invariant subset of V has measure 0 or 1. It may be shown that metrical transitivity implies ergodicity, and vice versa.

Consider any two subspaces A and B of V, $\mu(V) = 1$, and let the points of A evolve according to the transformation (38.1). The dynamical system is called <u>mixing</u> if, for any A and B,

$$\lim_{T\to\infty} \mu[\phi_t A \cap B]/\mu(B) = \mu(A) \tag{38.5}$$

where $\cap$ as usual denotes set intersection. The left side is the fraction of B that is overlapped by A_t, while the right side is the fraction of V taken up by A. Thus we can think of mixing in terms of Arnold's rum and cola [38.1]: rum and cola are poured into a glass, 10% rum and 90% cola. (Arnold and Avez recommend 20% rum and 80% cola.) After the mixture is stirred a large number of times ($n \to \infty$) every little cell of it is made up of approximately 10% rum and 90% cola; it is <u>mixed</u>. Cells in the (incompressible) liquid can be thought of as cells in phase space of some dynamical system, and the stirs as iterations of a map or a continuous flow like (38.1).

Mixing implies ergodicity, but the converse is generally not true. To see this, let A be an invariant subspace ($\phi_t A = A$) and let B = A in (38.5). Then obviously $\phi_t A \cap A = A$ and (38.4) implies $\mu(A) = \mu(A)^2$, or $\mu(A) = 0$ or 1. Thus mixing implies metrical transitivity and therefore ergodicity.

An amusing example of mixing is provided by Arnold's famous cat map, the area-preserving, two-dimensional discrete mapping

$$x_{n+1} = (x_n + y_n) \bmod 1 \tag{38.6a}$$

$$y_{n+1} = (x_n + 2y_n) \bmod 1 \tag{38.6b}$$

Figure 38.1, taken from Arnold and Avez, [38.1] shows what a "crazy mixed-up cat" [38.2] we have after just two iterations of the mixing system (38.6)!

The relevance of the concept of mixing to statistical mechanics may be illustrated by consideration of the (nondissipative) baker's transformation

$$x_{n+1} = 2x_n \,,\; y_{n+1} = y_n/2 \quad \text{for } 0 \leq x_n < 1/2 \tag{38.7a}$$

$$x_{n+1} = 2x_n - 1,\; y_{n+1} = (y_n + 1)/2 \quad \text{for } 1/2 \leq x_n < 1 \tag{38.7b}$$

This transformation is reminiscent of a baker working dough by stretching, cutting, and stacking. (Figure 38.2) As such it is no surprise that the baker's transformation is mixing. (A dissipative version of a baker's transformation is discussed in Section 17.) Now consider some normalized distribution function $f_n(x,y)$ on the unit square, defined after n iterations of the map. The "Liouville equation" for $f_n(x,y)$ is simply $f_n(x,y) = f_{n-1}(x/2,2y)$ for $0 \leq y < 1/2$, $f_n(x,y) = f_{n-1}(\frac{x+1}{2},2y-1)$ for $1/2 \leq y < 1$. Let us perform an average of $f_n(x,y)$ over y, which for a physical system would correspond to averaging over phase-space coordinates that are not "interesting":

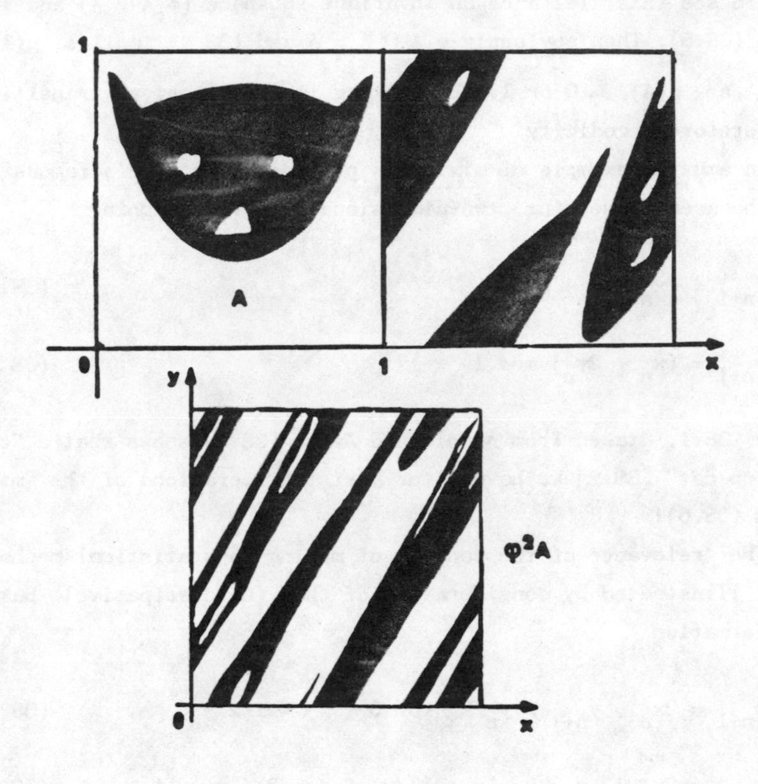

Figure 38.1 The effect of the mixing transformation (38.6) on Arnold's poor cat.

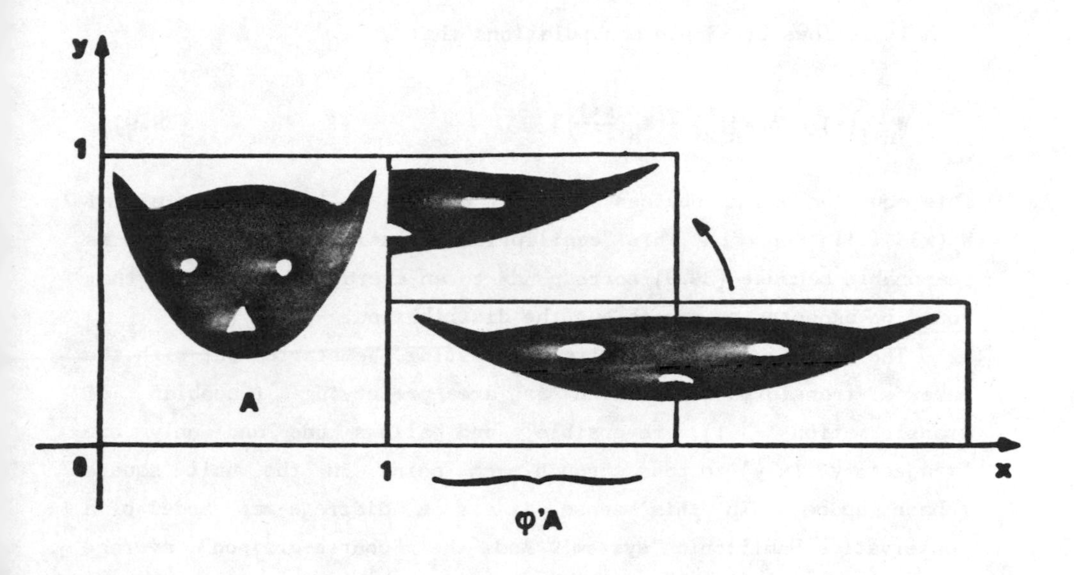

Figure 38.2 Illustrating the stretching, cutting, and stacking characteristics of the baker's transformation (38.7).

$$W_n(x) \equiv \int_0^1 dy f_n(x,y) \tag{38.8}$$

Then it follows by simple manipulations that

$$W_{n+1}(x) = \tfrac{1}{2}[W_n(\tfrac{x}{2}) + W_n(\tfrac{x+1}{2})] \tag{38.9}$$

This equation has an obvious solution: $W_n(x) = 1$. In fact any initial $W_o(x)$ will approach this "equilibrium" solution as $n \to \infty$, which is reasonable because (38.9) corresponds to an averaging procedure that would be expected to smooth out the distribution.

The result (38.9) is quite interesting. We started out with the baker's transformation, which is area-preserving (Jacobian of transformation = 1), reversible, and allows one and only one "trajectory" (x,y) to pass through each point in the unit square (phase space). In this sense it is a discrete-map model of a conservative Hamiltonian system. And the "coarse-grained" average (38.8) of the distribution function satisfies an irreversible rate equation. So what we have is a model for the irreversible approach to equilibrium (of coarse-grained averages) of a Hamiltonian system. [38.2]

We have noted that mixing implies ergodicity. Next up in the hierarchy of "disorder" are the so-called <u>K-systems</u> (K for Kolmogorov), which are systems having a positive Kolmogorov entropy. We introduced the K-entropy in Section 20; here we give an alternate definition that is typically used for conservative systems. [38.3] Once again we partition phase space into a set $\{A_j(0)\}$ of small cells of finite measure at t = 0. The backward evolution of the system by a unit time step transforms this set to $\{A_j(-1)\}$. The intersection $B(-1) \equiv \{A_i(0) \cap A_j(-1)\}$ of this new set with the first set will, of course, usually have smaller measure than $\{A_j(0)\}$. Backward evolution by another unit time step provides us with a new set $\{A_j(-2)\}$, and we consider the intersection B(-2) of this set with the previous intersection $B(-1)$: $B(-2) \equiv \{A_i(0) \cap A_j(-1) \cap A_k(-2)\}$. Continuing in this manner, we say that the system has positive K-entropy if the average measure of each element of B decreases exponentially as $t \to -\infty$. When this occurs the average exponential rate

$$h\{A_j(0)\} = -\lim_{t\to\infty}(1/t)\sum_i \mu[B_i(-t)]\log\mu[B_i(-t)] > 0 \qquad (38.10)$$

where μ denotes measure, and the Kolmogorov entropy is defined as the maximum of h over all initial, measurable partitions of phase space. The K-entropy is often defined in such a way (in regions of connected chaos, where the Lyapunov exponents are independent of initial conditions) that it equals the sum of all positive Lyapunov exponents. That is, K-systems may be characterized by the property we have emphasized for chaotic systems: very sensitive dependence on initial conditions, meaning exponential separation of initially close trajectories. This is intuitively reasonable from the definition above.

Kolmogorov and Sinai proved that K-systems are mixing. We might expect this intuitively from the extreme sensitivity to initial conditions.

Consider again the Arnold cat map (38.6). Writing $(dx_{n+1}, dy_{n+1}) = (dx_n + dy_n, dx_n + 2dy_n)$, and rotating axes, we may write the differential transformation in the diagonal form $(d\epsilon_{n+1}, d\eta_{n+1}) = (\lambda d\epsilon_n, \lambda^{-1}d\eta_n)$, where $\lambda = \frac{1}{2}(3 + \sqrt{5})$. Thus, although the area is preserved ($\lambda\lambda^{-1} = 1$), there is stretching in one direction (ϵ) and contraction in the other (η). The stretching results in exponential separation with n of initially close points, at the rate $\log\lambda = 0.962\ldots$, which is the (positive) Lyapunov exponent and, in this case of only one positive exponent, also the K-entropy.

From the chain K-system → mixing → ergodicity we can begin to sense the possible relevance of chaotic dynamics to fundamental problems of statistical mechanics (, for instance). However, we very often are confronted with near-integrable systems (like the

Hénon-Heiles model considered later), which have positive Lyapunov exponents in some regions of phase space but not others, and so cannot be characterized so neatly as K-systems. _In general we have intermingled regions of chaotic and regular behavior_.

There are systems whose time evolution is even more "stochastic" than K-systems, in the sense that they are K-systems but not vice versa. These are called _C-systems_ (or _Anosov systems_). In a C-system part of the tangent space is associated with stretching for all initial conditions, and there is also a part disjoint from this that is associated with contraction for all initial conditions. The Arnold cat map is an example of a C-system: there is stretching for all initial conditions in one direction (ϵ) and contraction for all initial conditions in the other (η). C-systems are K-systems, and therefore they are mixing and ergodic.

In summary, C-systems are K-systems, K-systems are mixing, and mixing systems are ergodic. As already mentioned, we are usually faced with near-integrable systems in which the phase space is divided into intermingled regions of chaotic and orderly behavior. Such systems do not belong to any of the above categories, but they may be _locally_ equivalent to C-systems in chaotic regions of phase space.

39. The Fermi-Pasta-Ulam Problem

If there are any constants of the motion in addition to the Hamiltonian, trajectories in the 2N-dimensional phase space are confined to regions of dimension less than the dimension 2N-1 of surfaces of constant energy. In other words, _if there are any constants of the motion in addition to the Hamiltonian, the system cannot be ergodic_. Note that nonergodicity does not necessarily imply (complete) integrability, although of course integrability implies that a system is not ergodic.

We mentioned in the preceding section the belief of some people before around 1950 that small, nonintegrable perturbations might give rise to ergodicity in otherwise integrable systems. For those who regard the ergodic hypothesis as essential to statistical mechanics, this would mean that small nonlinearities might, as a practical matter, justify the application of statistical-mechanical methods to integrable systems.

Example: The classical, linear theory of the specific heats associated with crystal lattice vibrations works well at high temperatures. But the assumption of harmonic vibrations means the theory deals with an integrable system. In particular, the energy of each normal mode of vibration is a constant of the motion, and if all the energy is initially restricted to a single normal mode of vibration, it remains so restricted because all the normal modes are independent. But the ergodic hypothesis obviously requires energy to be shared among all the modes. Does a small anharmonic coupling (so small that its contribution to energy and specific heat is negligible) give rise to energy sharing and ergodicity?

Fermi, Pasta, and Ulam studied the ergodic problem "experimentally" for particular models. [39.1] Consider a linear chain of N+1 identical oscillators with nearest-neighbor couplings:

$$H = (1/2m)\sum_{j=1}^{N-1} p_j^2 + \tfrac{1}{2}k\sum_{j=1}^{N-1}(q_{j+1} - q_j)^2 \qquad (39.1)$$

The $j = 0$ and $j = N$ oscillators are assumed to be held fixed at the ends of the chain. (Figure 39.1) We can transform (39.1) to

$$H = (1/2m)\sum_j (P_j^2 + m\omega_j^2 Q_j^2) \qquad (39.2)$$

where the normal mode coordinates (Q_j, P_j) are defined by

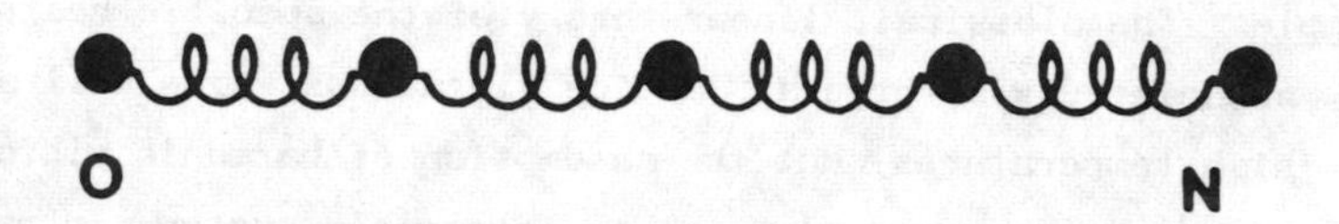

Figure 39.1 The Fermi-Pasta-Ulam model. The masses are connected by nonlinear springs, and the two masses at the ends are held fixed.

$$q_k = (2/N)^{1/2} \sum_{j=1}^{N-1} Q_j \sin(\pi kj/N) \tag{39.3a}$$

$$P_j = m\dot{Q}_j \tag{39.3b}$$

and the mode frequencies are

$$\omega_j = 2(k/m)^{1/2} \sin(\pi j/2N) \tag{39.3c}$$

Clearly this system is integrable. Consider now the addition of nonlinear terms:

$$H = (1/2m)\sum_j p_j^2 + \tfrac{1}{2}k \sum_j (q_{j+1} - q_j)^2 + (\alpha/3)\sum_j (q_{j+1} - q_j)^3 \tag{39.4a}$$

$$m\ddot{q}_j = k(q_{j+1} - 2q_j + q_{j-1}) + \alpha[(q_{j+1} - q_j)^2 - (q_j - q_{j-1})^2] \tag{39.4b}$$

In terms of normal mode coordinates we can write

$$\ddot{Q}_j = -\omega_j^2 Q_j + \alpha F_j \tag{39.5}$$

where the forces F_j are nonlinear in the Q's. Fermi, Pasta, and Ulam [39.1] considered (39.4) as well as models with quartic nonlinearity and piecewise linearity. They carried out numerical experiments to see how the energies

$$E_j = (m/2)(\dot{Q}_j^2 + \omega_j^2 Q_j^2) \tag{39.6}$$

vary with time when all the energy is initially concentrated in one or a few modes. Ergodicity in such a model would mean that the energy is eventually shared on average among all the modes.

What was found is just the opposite: the energy stayed within only a few modes, as shown in the example of Figure 39.2, which is Figure 1 of the Fermi-Pasta-Ulam paper. Fermi was surprised by this violation of the ergodic hypothesis, and considered the results an interesting "little discovery."

And so, most melancholy for the ergodic hypothesis, a small nonintegrable perturbation of an integrable system will not in general produce ergodicity and energy equipartition. At about the same time as Fermi, Pasta, and Ulam, Kolmogorov [39.2] independently stated a theorem to the effect that, if the nonintegrable perturbation is sufficiently small, most of the trajectories will remain confined to an N-dimensional surface in phase space. This celebrated KAM theorem, which is discussed in the following section, takes some of the surprise out of the Fermi-Pasta-Ulam results. At the same time, however, we will see that the KAM theorem does not of itself preclude the possibility of ergodicity - or even chaos - in near-integrable systems.

We close with two points of historical interest. First, Boltzmann originally stated the ergodic hypothesis to mean that a trajectory eventually covers all points on the energy surface. However, strictly speaking this turns out to be mathematically impossible: a trajectory is a continuous, nonintersecting, one-dimensional set of points in phase space, and it is known that one cannot make a continuous, one-to-one correspondence of such a one-dimensional space with a space of dimension $\geq$ 2. Ergodicity in the Boltzmann sense is thus modified sometimes to mean that a trajectory passes arbitrarily close to every point on the energy surface.

The Fermi-Pasta-Ulam problem also played an important role in the origin of the soliton concept, in the sense that attempts to understand the numerical results led to the discovery (by Zabusky and Kruskal [39.3]) of soliton behavior. In particular, Kruskal had been

STUDIES OF NON LINEAR PROBLEMS

E. Fermi, J. Pasta, and S. Ulam

Document LA-1940 (May 1955).

Abstract.

A one-dimensional dynamical system of 64 particles with forces between neighbors containing nonlinear terms has been studied on the Los Alamos computer Maniac I. The nonlinear terms considered are quadratic, cubic, and broken linear types. The results are analyzed into Fourier components and plotted as a function of time.

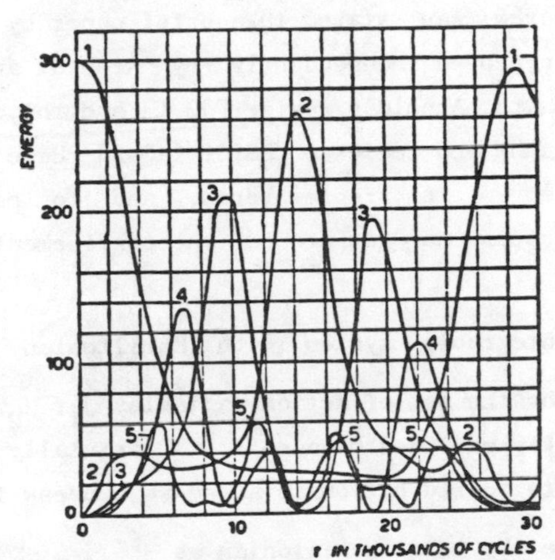

Fig. 1. – The quantity plotted is the energy (kinetic plus potential in each of the first five modes). The units for energy are arbitrary. $N = 32$; $\alpha = 1/4$; $\delta t^2 = 1/8$. The initial form of the string was a single sine wave. The higher modes never exceeded in energy 20 of our units. About 30,000 computation cycles were calculated.

Figure 39.2 Figure 1 of the Fermi-Pasta-Ulam paper. Instead of equipartition of energy, the energy remains confined to just a few modes.

interested in why there is a recurrence of energy in the initially excited mode in the Fermi-Pasta-Ulam model; [39.4] the energy appears in numerical experiments to recur nearly periodically. The model can be converted to a nonlinear partial differential equation when the oscillators are taken to be distributed continuously on the chain. One obtains the so-called Boussinesq equation, which is structurally akin to the Korteweg-deVries equation. [39.5]

40. The KAM Theorem

The KAM theorem was stated in a brief paper by Kolmogorov in 1954, and was later proved independently by Arnold and Moser. We refer the reader to Arnold's review [40.1] and the discussions in more recent monographs and reviews. [38.3, 40.2] Here we wish to introduce the theorem in simple terms, and to comment on how near-integrable systems can "get around" the confinement to invariant tori.

Consider an integrable system with Hamiltonian $H_o(\vec{J})$, where again $\vec{J}$ stands for the set of action variables $J_1, J_2, \ldots, J_N$, $N \geq 2$. (The case $N = 1$ is excluded because it is trivially integrable.) If a nonintegrable perturbation is added we express it in terms of the $(\vec{\theta},\vec{J})$ and write the total Hamiltonian as

$$H = H_o(\vec{J}) + \epsilon H_1(\vec{\theta},\vec{J}) \tag{40.1}$$

where ϵ is a parameter characterizing the strength of the perturbation. When $\epsilon = 0$ we have J_i = constant and $\theta_i = \nu_i t + \delta_i$ (equation 36.14)), with $\nu_i = \partial H_o/\partial J_i$, and the θ_i are angles on an N-torus. When $\epsilon \neq 0$ the (θ_i, J_i) are no longer action-angle variables, but they remain bona fide canonical variables.

If ϵ is not small then the system is of course nonintegrable. The interesting question is whether the system (40.1) is integrable if the perturbation is small. In particular, what happens to the invariant tori of the integrable Hamiltonian H_o? This is the question addressed by the KAM theorem.

Suppose the perturbation is sufficiently smooth and small. Suppose also that the frequencies $\nu_i(\vec{J})$ associated with the unperturbed Hamiltonian are linearly independent, i.e., for any set $(n_1, n_2, \ldots, n_N)$ of integers that are not all zero,

$$\sum_{j=1}^{N} n_j \nu_j(\vec{J}) \neq 0 \qquad (40.2)$$

(The frequencies in this case may also be said to be incommensurate.) Under these conditions, the KAM theorem says that most of the N-tori ("KAM tori") of the unperturbed system are not destroyed, only distorted slightly. We say <u>most</u> tori because the nondestruction doesn't apply for tori with commensurate frequencies, or those for which the frequencies are close enough to being commensurate.

The proof of the KAM theorem is mathematically sophisticated and well beyond our scope in these lectures. The basic idea of the proof is to seek a canonical transformation to new action-angle variables, so that the transformed Hamiltonian depends only on the new action variables. This leads to a complicated equation for the generating function S that is solved in terms of a perturbation expansion in ϵ. Here one incurs the infamous "small denominators" difficulty of perturbation theories, which in the proof is circumvented by an accelerated convergence procedure. We will not pursue these matters here, since we will be mainly interested in how the KAM tori can be <u>destroyed</u> rather than preserved. That is, we want to know how systems can beat the apparent constraints of the KAM theorem and even evolve chaotically.

The key to the destruction of KAM tori is resonance and, in particular, the notion of "resonance overlap." For a simple example (or reminder) of the importance of resonance let us return to the Fermi-Pasta-Ulam model with two degrees of freedom (N + 1 = 4 particles, two held fixed). The normal mode equations (39.5) take the form

$$\ddot{Q}_1 = -\omega_1^2 Q_1 - (\alpha\sqrt{2})Q_1 Q_2 \tag{40.3a}$$

$$\ddot{Q}_2 = -\omega_2^2 Q_2 - (\alpha/\sqrt{2})(Q_1^2 - 3Q_2^2) \tag{40.3b}$$

Assuming a perturbation expansion in α, it is found that the solutions for Q_1 and Q_2 involve various denominators that vanish when $n_1\omega_1 + n_2\omega_2 \approx \alpha$, where n_1 and n_2 are small integers. (We assume now that ω_1 and ω_2 are variable rather than being fixed by (39.3c).) It is precisely in such resonant cases that there can be substantial sharing of energy between the two oscillators. [40.3] That is, resonances can apparently destroy the 2-tori associated with the unperturbed, integrable Hamiltonian.

In summary, the KAM theorem says that for small nonintegrable perturbations, most of the tori will be preserved. But the theorem leaves open the possibility that resonance phenomena might after all destroy the tori associated with the unperturbed system.

41. Overlapping Resonances

The lowest-dimensional case of interest is N = 2. To get some hint of the enormous complexity possible in this "simplest" of cases, we follow the discussion of Walker and Ford. [41.1]

Suppose the Hamiltonian is of the form

$$H = H_o(J_1,J_2) + f(J_1,J_2)\cos(m\theta_1+n\theta_2) \tag{41.1}$$

For $f = 0$ the motion of the system is confined to 2-tori, positions on which are specified by the angle variables θ_1,θ_2. If we can find a canonical transformation to new action-angle variables $(\theta_1',\theta_2' ; J_1',J_2')$, so that the transformed Hamiltonian $\bar{H} = \bar{H}(J_1',J_2')$, then the motion of the perturbed system is also confined to 2-tori. Now the generating function

$$F_2(\theta_1,\theta_2\ ;\ J_1',J_2') = \theta_1 J_1' + \theta_2 J_2' + B(J_1',J_2')\sin(m\theta_1+n\theta_2) \quad (41.2)$$

transforms (41.1) to

$$\bar{H} = H_o(J_1',J_2') + \{[m\omega_1(J_1',J_2') + n\omega_2(J_1',J_2')]B(J_1',J_2') + f(J_1',J_2')\}$$

$$\times \cos(m\theta_1'+n\theta_2') \quad (41.3a)$$

with

$$\omega_i(J_1',J_2') \equiv \partial H_o(J_1',J_2')/\partial J_i' \quad (41.3b)$$

if we retain only lowest-order terms, assuming the generating function (41.2) is close to that for the identity transformation for which $B(J_1',J_2') \equiv 0$. The Hamiltonian (41.3) is a function only of the actions if $B(J_1',J_2')$ is chosen as

$$B(J_1',J_2') = \frac{-f(J_1',J_2')}{m\omega_1(J_1',J_2') + n\omega_2(J_1',J_2')} \quad (41.4)$$

With this choice we can expect the perturbed tori to be close to those for the unperturbed system, since the transformation has been chosen to be nearly the identity transformation.

This argument breaks down, however, if (41.4) is not small, for then the generating function (41.2) is not close to that for the identity transformation. In particular, if there are frequencies ω_i for the system (41.1) satisfying the resonance condition

$$|m\omega_1(J_1,J_2) + n\omega_2(J_1,J_2)| \ll |f(J_1,J_2)| \quad (41.5)$$

then the angle-dependent perturbation in (41.1) might be expected to greatly distort the unperturbed tori. In other words, we can expect a resonance such as $m\omega_1 + n\omega_2 \approx 0$ to substantially distort the unperturbed tori.

In general we can suppose the perturbation to have a Fourier expansion like

$$V = \sum_m \sum_n f_{mn}(J_1, J_2)\cos(m\theta_1 + n\theta_2) \tag{41.6}$$

and if there is a resonance associated with $m\theta_1+n\theta_2$, we might expect the tori distorted by this resonance to be strongly affected also by terms $m'\theta_1' + n'\theta_2'$ such that m'/n' is sufficiently close to m/n. Because of the (J_1, J_2)-dependence of ω_1 and ω_2, and because J_1 and J_2 can be thought of loosely as "radii" for the unperturbed tori, a relation like (41.5) determines the distortion of a <u>zone</u> of unperturbed tori. If such resonance zones <u>overlap</u>, so that in the region of overlap the unperturbed tori are distorted by a large number of terms like $\cos(m\theta_1+n\theta_2)$, we can perhaps anticipate the complete <u>destruction</u> of the unperturbed tori. And this possibility becomes stronger as the perturbation is increased, for this makes (41.4) larger and takes the transformation further and further from the identity.

To see what happens when there are multiple resonances and resonance overlap we continue to follow Walker and Ford. Suppose first that the Hamiltonian is given by (41.1), $\omega_i = \partial H_o/\partial J_i$, and that (41.5) can be satisfied. There are two constants of the motion, namely H and $I = nJ_1 + mJ_2$; the constancy of the latter may be verified by taking its Poisson bracket with H. Since there are two constants of the motion in involution, the system (41.1) is still integrable. Walker and Ford consider the unperturbed Hamiltonian

$$H_o = J_1 + J_2 - J_1^2 - 3J_1J_2 + J_2^2 \tag{41.7}$$

where the actions J_1, J_2 are related to the Cartesian coordinates $(\vec{q}, \vec{p})$ by the formulas

$$q_i = \sqrt{2J_i}\cos\theta_i \tag{41.8a}$$

$$p_i = -\sqrt{2J_i}\sin\theta_i \tag{41.8b}$$

which means that $J_i = \frac{1}{2}(p_i^2 + q_i^2)$. Therefore the unperturbed motion in the (q_i, p_i) planes is confined to concentric circles. The condition that

$$\omega_1 = 1 - 2J_1 - 3J_2 \tag{41.9a}$$

and

$$\omega_2 = 1 - 3J_1 + 2J_2 \tag{41.9b}$$

be positive implies $0 \lesssim E \lesssim 3/13$.

In the case of a "2-2 resonance" with

$$H = H_o(J_1, J_2) + \alpha J_1 J_2 \cos(2\theta_1 - 2\theta_2) \tag{41.10}$$

there is the additional constant of motion

$$I = J_1 + J_2 \tag{41.11}$$

This additional constant of the motion makes it possible to study the system in considerable detail using only simple analytical methods. From (41.7), (41.10), and (41.11) one obtains

$$(3 + \alpha\cos 2\theta_1)J_1^2 - (5I + I\cos 2\theta_1)J_1 + I + I^2 = E \tag{41.12}$$

where θ_2 has been set equal to $3\pi/2$. Equations (41.12) and (41.8) may be used to determine algebraically the (q, p) curves in the plane surface of section defined by $q_2 = 0$, $p_2 = \dot{q}_2 \geq 0$. A typical result is shown in Figure 41.1. Note that the concentric circles associated

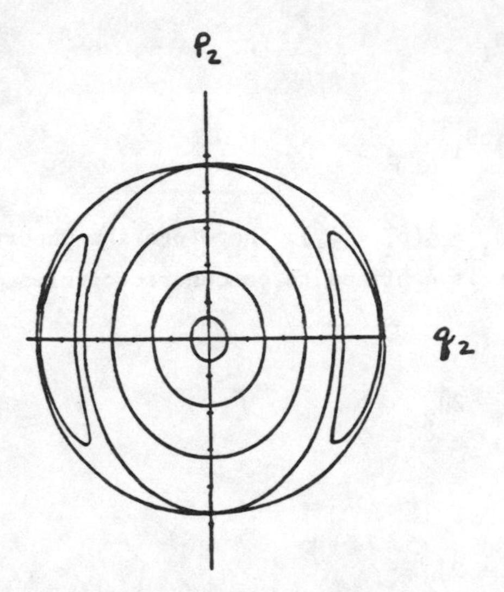

Figure 41.1 A typical surface of section obtained algebraically from equations (41.8) and (41.12), showing the distortion of the concentric circles of the unperturbed motion. From G.H. Walker and J. Ford, Reference [41.1], with permission.

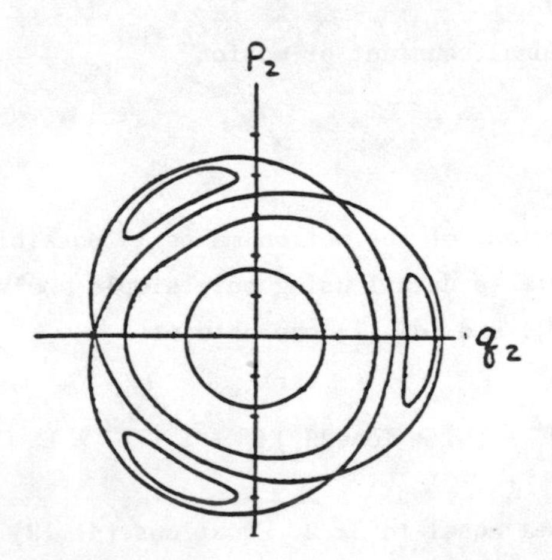

Figure 41.2 Distortion of the concentric circles of the unperturbed motion in the case of a 3-2 resonance. From G.H. Walker and J. Ford, Reference [41.1], with permission.

with the unperturbed motion are only slightly distorted except in the "2-2 resonance zone" enclosed by the self-intersecting separatrix, where the closed curves are crescent-shaped.

We can relate the more strongly distorted (crescent-shaped) curves to the resonance condition (41.5), which in this case becomes $2\omega_1 \cong 2\omega_2$ or, from (41.9),

$$J_1 \cong 5J_2 \tag{41.13}$$

Consider the fixed points at the center of each crescent. These points correspond to stable <u>periodic</u> trajectories, with J_1, J_2 constant and $\dot{\theta}_1 = \dot{\theta}_2$. From the equations of motion

$$\dot{J}_1 = -\partial H/\partial\theta_1 = -2\alpha J_1 J_2 \sin(2\theta_1 - 2\theta_2) \tag{41.14a}$$

$$\dot{J}_2 = -\partial H/\partial\theta_2 = 2\alpha J_1 J_2 \sin(2\theta_1 - 2\theta_2) \tag{41.14b}$$

$$\dot{\theta}_1 = \partial H/\partial J_1 = 1 - 2J_1 - 3J_2 + \alpha J_2 \cos(2\theta_1 - 2\theta_2) \tag{41.14c}$$

$$\dot{\theta}_2 = \partial H/\partial J_2 = 1 - 3J_1 + 2J_2 + \alpha J_2 \cos(2\theta_1 - 2\theta_2) \tag{41.14d}$$

we then infer that $2\theta_1 - 2\theta_2 = \pi$ or 3π and

$$J_1/J_2 = \frac{5+\alpha}{1+\alpha} \tag{41.15}$$

The solutions $2\theta_1 - 2\theta_2 = \pi$ or 3π are chosen instead of 0 or 2π in order to investigate the points at the centers of the crescents in Figure 41.1, for which $p_1 = 0$. (Recall (41.8) and the fact that θ_2 is held fixed at $3\pi/2$ in Figure 41.1.) For small α (41.15) is equivalent to (41.13). This confirms that tori are distorted most strongly in the 2-2 resonance zone.

The choices $\theta_1 - \theta_2 = 0$ or π, which also give $\dot{J}_1 = \dot{J}_2 = 0$, correspond to the two points at the self-intersections of the separatrix in Figure 41.1. These two points represent <u>unstable</u> periodic motion. For these points we obtain (41.15) with the sign of α changed.

If $2\omega_1 = 2\omega_2$ then (41.9) and (41.7) imply the following values of J_1 and J_2 on the <u>unperturbed</u> 2-2 torus;

$$J_1 = 5[1 - (1 - 13E/3)^{1/2}]/13 \qquad (41.16a)$$

$$J_2 = [1 - (1 - 13E/3)^{1/2}]/13 \qquad (41.16b)$$

Thus the unperturbed 2-2 torus, and the perturbed 2-2 resonance zone, exist for all allowed energies ($0 \leq E \leq 3/13$). As E is increased the resonance zone of strongly distorted tori moves out from the origin.

A 3-2 resonance associated with a Hamiltonian like [41.1]

$$H = H_o(J_1, J_2) + \alpha J_1^{3/2} J_2 \cos(3\theta_1 - 2\theta_2) \qquad (41.17)$$

may be similarly studied algebraically due to the additional constant of motion $I = 2J_1 + 3J_2$. [41.1] Figure 41.2 shows how the preceding results are modified. In this case the points at the center of each of the crescents shown correspond to a <u>single</u> periodic solution, so that the 3-2 resonance is said to consist of a <u>chain of three islands</u>. (In Figure 41.1 the points at the center of each crescent represent two distinct periodic orbits, not a chain of two islands.) For a 2-3 resonance similar results are found, except that the three-island chain now appears in the J_2 plane. [41.1]

In general an m-n resonance with $m \neq n$ gives rise to a chain of m islands in the J_1 plane and a chain of n islands in the J_2 plane. The resonance distorts the unperturbed tori, creating new stable and unstable periodic orbits in pairs. A resonance zone, enclosed by a separatrix passing through the unstable periodic points, appears beyond some threshold energy E.

What happens if <u>two</u> resonances are possible? Walker and Ford consider the Hamiltonian

$$H = H_o(J_1,J_2) + \alpha J_1 J_2 \cos(2\theta_1 - 2\theta_2) + \alpha J_1 J_2^{3/2} \cos(2\theta_1 - 3\theta_2) \quad (41.18)$$

in which the unperturbed tori can now be distorted by both 2-2 and 2-3 resonances. In this case the surfaces of section had to be studied by numerical integration of the equations of motion (with α = 0.02).

Using an argument similar to that used in arriving at (41.16), it may be shown that the unperturbed 2-3 torus does not exist for E $\lesssim$ 0.16. Figure 41.3 shows numerical results [41.1] for (q_2,p_2) curves of the Hamiltonian (41.18) with E = 0.056; note the similarity to Figure 41.1. For E slightly greater than 0.16 it turns out that the 2-2 and 2-3 resonance zones are widely separated. (The 2-3 zone lies inside the 2-2 zone.) Figure 41.4, for E = 0.18, resembles a superposition of Figures 41.1 and 41.2, for the two resonances are, in effect, acting separately to perturb tori in different regions of phase space.

From arguments similar to those sketched above, Walker and Ford estimated that the 2-2 and 2-3 resonance zones should first overlap at E = 0.2095. Figure 41.5 shows the numerically generated (q_2,p_2) curves for this energy. Evidently a small zone of <u>unstable</u> motion has appeared in the region of resonance overlap. The motion in this region is not confined to tori. That is, the resonance overlap has destroyed the KAM tori.

The way in which tori are destroyed as the resonance zones approach overlap is quite intricate. Figure 41.6 shows numerically generated results for E = 0.20. [41.1] In addition to the 2-2 and 2-3 resonance zones one now observes a chain of five islands. And it turns out that there is also a detectable chain of seven islands near the chain of five. This suggests that in fact there is a whole "hierarchy of resonances" associated with a doubly resonant

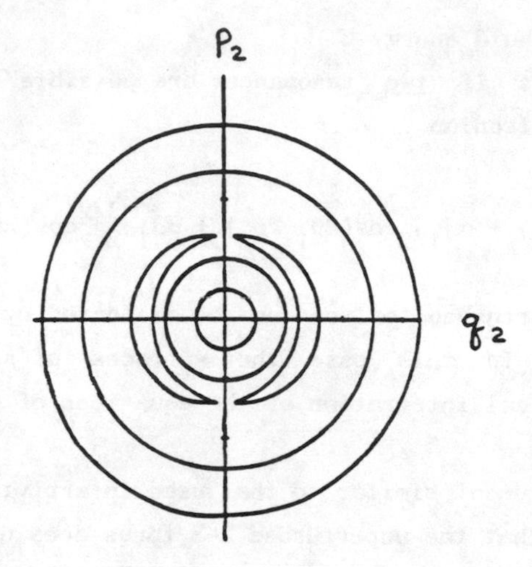

Figure 41.3 Surface of section for the Hamiltonian (41.18) with E = 0.056. From G.H. Walker and J. Ford, Reference [41.1], with permission.

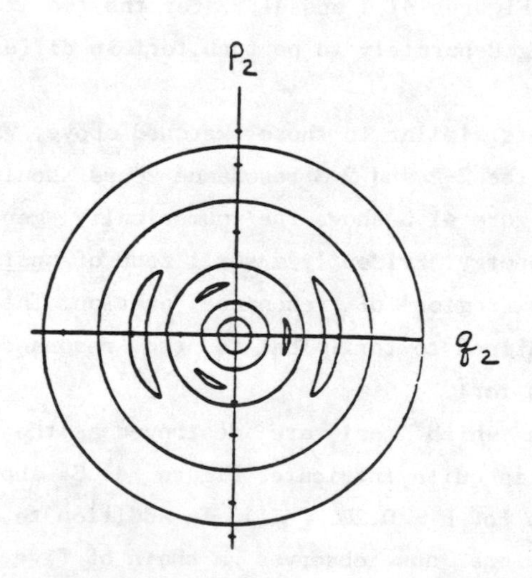

Figure 41.4 Surface of section for Hamiltonian (41.18) with E = 0.18. From G.H. Walker and J. Ford, Reference [41.1], with permission.

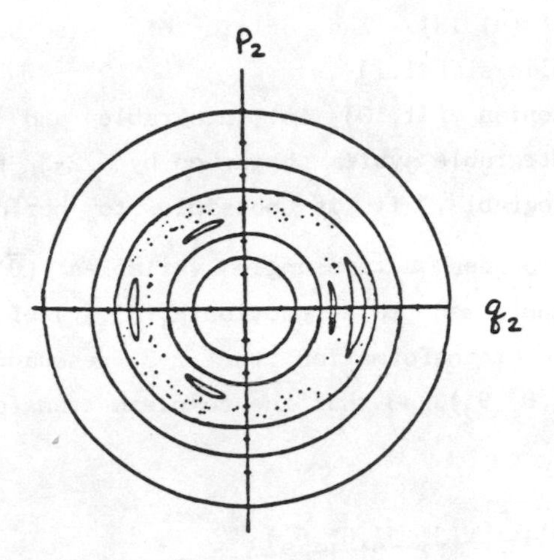

Figure 41.5 Numerically generated surface of section for Hamiltonian (41.18) with E = 0.2095, where the 2-2 and 2-3 resonance zones are predicted to begin overlapping. From G.H. Walker and J. Ford, Reference [41.1], with permission.

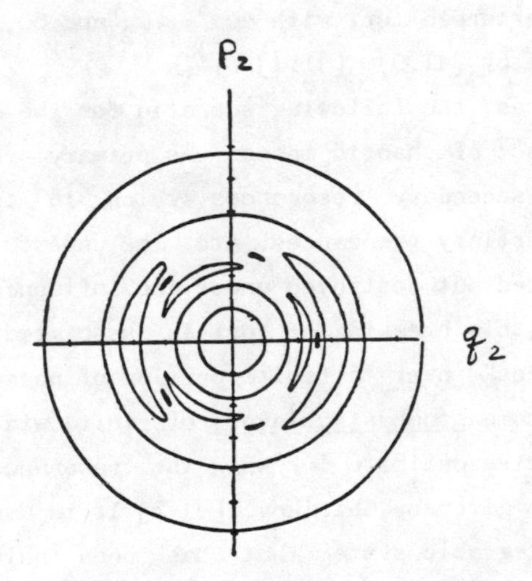

Figure 41.6 Numerically generated surface of section for Hamiltonian (41.18) with E = 0.20, showing a chain of five islands. From G.H. Walker and J. Ford, Reference [41.1], with permission.

Hamiltonian like (41.18). The origin of this hierarchy may be understood as follows. [41.1]

The Hamiltonian (41.10) is integrable and so we can regard (41.18) as an integrable system perturbed by a 2-3 resonance. Since (41.10) is integrable, it is possible to perform a canonical transformation to new action-angle variables $(\vec{\theta}', \vec{J}')$ such that (41.10) is transformed to a function $H_1(J_1', J_2')$ of action variables only. Under the transformation the 2-3 resonance becomes some function $V(J_1', J_2', \theta_1', \theta_2')$, so that the complete transformed Hamiltonian is

$$H = H_1(J_1', J_2') + V(J_1', J_2', \theta_1', \theta_2') \qquad (41.19)$$

The Fourier expansion of V will generally reveal "new" resonances that are not explicit in (41.18). These new resonances are called <u>secondary resonances</u> to distinguish them from the <u>primary resonances</u> (2-2 and 2-3) appearing explicitly in the Hamiltonian (41.18). In fact the five- and seven-island chains noted above occur in the vicinity of unperturbed tori with $4\omega_1 = 5\omega_2$ and $6\omega_1 = 7\omega_2$, where ω_1 and ω_2 are given by (41.9). [41.1]

One now has the following scenario for the destruction of KAM tori and the onset of chaotic motion. As primary resonances overlap they generate secondary resonances which in turn lead to the generation of tertiary resonances, etc. The unperturbed tori are then not only distorted but destroyed under the influences of all these resonances, and this breaking of tori is associated with the onset of "widespread chaos" over a finite portion of phase space. What were separatrices become <u>stochastic layers</u> of finite width.

A quantitative estimate for when the resonance overlap should appear has been given by Chirikov. [41.2] It is assumed that one has a nonlinear, integrable system that has been solved in terms of action-angle variables. In low-dimensional systems the Chirikov criterion has been found to predict to within a factor of about 2 the value of a perturbation parameter necessary for the onset of chaos.

We will later see examples of the use of the Chirikov criterion.

42. The Hénon-Heiles Model

The Hénon-Heiles system has become one of the paradigms for the study of chaos in Hamiltonian systems. It was considered by Hénon and Heiles as a test model for the motion of a star in the gravitational field of the other stars in its galaxy. [42.1] A cylindrically symmetric potential was assumed, so there are two constants of the motion, the energy and the z component of angular momentum. Since N = 3 the question is whether there is a third integral of the motion. The abstract of the Hénon-Heiles paper summarizes well the problem and the results: "The problem of the existence of a third isolating integral of motion in an axisymmetric potential is investigated by numerical experiments. It is found that the third integral of motion exists for only a limited range of initial conditions." (We note parenthetically that this "experimental" study was done at about the same time as that of Lorenz. [1.1] As anticipated early on by people like Fermi and Ulam, the computer had by this time already become a theorists' laboratory for testing ideas.)

Cylindrical symmetry reduces the problem effectively to motion in a plane. Hénon and Heiles chose the following specific Hamiltonian for numerical experimentation:

$$H = (1/2)(p_1^2 + p_2^2) + (1/2)(q_1^2 + q_2^2) + q_1^2 q_2 - (1/3)q_2^2 \qquad (42.1)$$

The particle is trapped in the well for energies $E < 1/6$. Trajectories of small energy were found to lie on invariant 2-tori in the four-dimensional phase space. The evidence for this was found in Poincare' maps whose points form smooth loops. As E was varied, however, Hénon and Heiles found trajectories that seemingly produce a random splattering of points in the Poincare' map, suggesting ergodic behavior (no third integral).

The Hamiltonian (42.1) defines a three-dimensional energy surface in the four-dimensional phase space (q_1, q_2, p_1, p_2). If another constant of the motion exists, then trajectories must lie on a two-dimensional surface. In this case the Poincare' map defined by $q_1 = 0$ and $\dot{q}_1 \geq 0$, for instance, will form a regular pattern, equivalent topologically to a plane cross section of a 2-torus. There are two possibilities for this regular motion:

(1) The frequencies ω_1 and ω_2 on the 2-torus are commensurate, i.e., the winding number ω_1/ω_2 (Section 15) is a rational number. In this case the trajectories on the 2-torus close on themselves and the Poincare' map consists of a discrete set of repeating points.

(2) The frequencies ω_1 and ω_2 on the 2-torus are incommensurate, i.e., the winding number ω_1/ω_2 is an irrational number. In this case trajectories on the 2-torus do not close on themselves, but cover the torus "ergodically." (Of course this is not ergodicity in the sense of covering the energy surface!) The Poincare' map then forms a smooth curve.

If the trajectories are not confined to 2-tori, however, the Poincare' map appears to be a random sprinkling of points over some area.

Figure 42.1 shows numerical results for the Hénon-Heiles model for E = 1/12. [41.1] These results indicate that all trajectories for this energy lie on invariant 2-tori. Figure 42.2, for E = 0.10629166, shows two eight-island chains. Figure 42.3, for the same energy, indicates the simultaneous appearance of an instability. Apparently a stochastic layer is forming where there was previously a separatrix. For E = 0.125 (Figure 42.4) there are still 2-tori, but the irregular motion makes up a significant portion of the available area; the random splatter of points, in fact, is produced by a <u>single trajectory</u>. At the disscociation energy E = 1/6 nearly all initial conditions produce irregular trajectories. (Figure 42.5) Again the splatter of points was generated by a single trajectory.

These results are nicely consistent with the resonance overlap

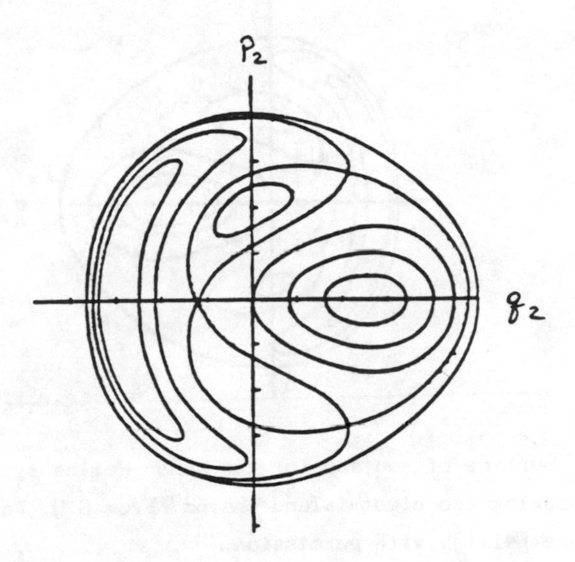

<u>Figure 42.1</u> Surface of section for the Hénon-Heiles system with E = 1/12. From G.H. Walker and J. Ford, Reference [41.1], with permission.

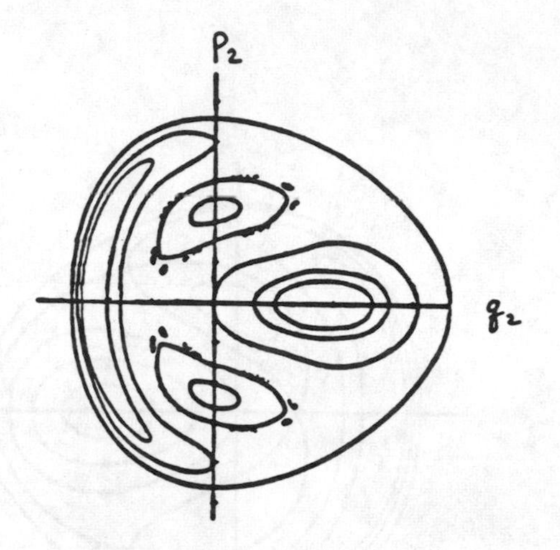

Figure 42.2 Surface of section for the Hénon-Heiles system with E = 0.10629166, showing two eight-island chains. From G.H. Walker and J. Ford, Reference [41.1], with permission.

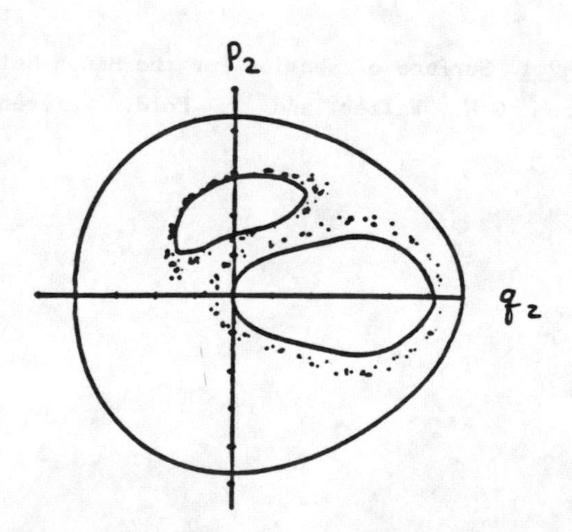

Figure 42.3 Surface of section for the Hénon-Heiles system for the same energy as in Figure 42.2. From G.H. Walker and J. Ford, Reference [41.1], with permission.

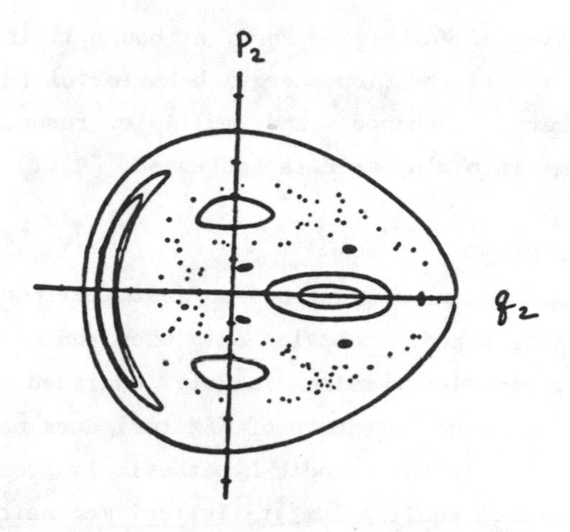

Figure 42.4 Surface of section for the Hénon-Heiles system with E = 0.125. From G.H. Walker and J. Ford, Reference [41.1], with permission.

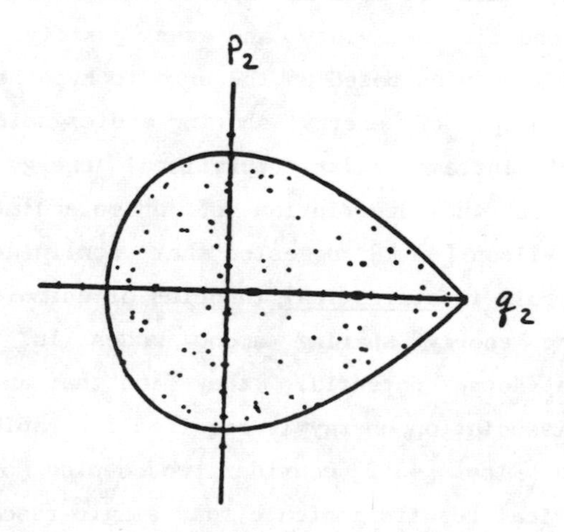

Figure 42.5 Surface of section for the Hénon-Heiles system with E = 1/6. From G.H. Walker and J. Ford, Reference [41.1], with permission.

interpretation of Walker and Ford. Although it is hardly obvious, it turns out that the low-energy behavior of (42.1) has in effect a single primary resonance, and multiple resonances and resonance overlap come into play as E is increased. [41.1]

43. Remarks

The preceding discussions suggest that nonlinear systems might generally show ergodic behavior only when the energy is large enough. Otherwise the motion remains largely confined to KAM tori. This suggests that the break-up of KAM tori does not "save" the ergodic hypothesis. And if the ergodic hypothesis is deemed necessary for the applicability of equilibrium statistical mechanics, we might expect statistical mechanics to fail at low energies, i.e., at low temperatures. But, no! Statistical mechanics appears to be quite adequate at all temperatures. This provides support for the view that the ergodic hypothesis is _not_ necessary for statistical mechanics.

In any case it does not appear that resonance overlap and the onset of chaotic behavior can ever justify the approach to statistical mechanics based on the ergodic hypothesis.

The concept of energy sharing and ergodicity is important in theories of intramolecular vibrational energy transfer, and in particular in the description of unimolecular processes. In 1961 Thiele and Wilson [43.1] suggested that nonlinearity must play an essential role in _statistical_ theories of unimolecular dissociation, where strong energy sharing among modes is taken for granted. Assuming a Morse potential, they find that an energy greater than half the dissociation energy is required for rapid energy exchange. Oxtoby and Rice [43.2] consider two coupled Morse oscillators, and their numerical results indicate that single resonance zones dominate the motion at low energies, and therefore that intramolecular vibrational transfer is slow at low energies. At higher energies there is overlap of resonance zones and _fast_ energy exchange. That is, statistical unimolecular dissociation theories could find justification in the overlap of resonance zones, if one accepts the validity of such classical models. We also refer the reader to the review by Brumer [43.3] of theories for the onset of statistical behavior in intramolecular energy transfer. Later we will consider very simple models of molecular vibrations and rotations in laser fields.

The question of just how relevant are these classical notions of ergodic behavior at the molecular level is still largely unanswered, even aside from the classical approximation. For one thing, <u>the chaotic trajectories that appear with the break-up of KAM tori are not generally ergodic</u>. And remember that the generic near-integrable systems we have been discussing are not C-systems or even K-systems. Rather, they exhibit "irregular" trajectories only in certain regions of phase space (i.e., for certain initial conditions), and regions of regular and irregular motion are generally interwoven in a very complicated way.

Walker and Ford [41.1] have noted that, " Since the unperturbed tori with commensurate frequencies which are destroyed by the perturbation are everywhere dense, it is remarkable indeed that KAM are able to show that the majority - in the sense of measure theory - of initial conditions lie on the preserved tori ... when [the perturbation] is sufficiently small." What about the "minority" of initial conditions for which the KAM theorem does not apply? It turns out that a set of initial conditions of measure zero can in effect wander quasi-ergodically over the energy surface. This brings us to consider briefly the phenomenon of <u>Arnold diffusion</u>.

By restricting our preceding discussions of resonance overlap to $N = 2$, we have stayed clear of Arnold diffusion, which occurs only for $N > 2$. The crucial difference between $N = 2$ and $N > 2$ is this: for $N = 2$ the surfaces of two-dimensional KAM tori divide the three-dimensional energy surface into distinct, closed surfaces, much like a (two-dimensional) plane is divided into distinct, closed areas by a set of (one-dimensional) lines. But for $N > 2$, N-dimensional KAM toroidal surfaces cannot divide up the $(2N-1)$-dimensional energy surface into separated regions, just as a set of lines in three-dimensional space cannot divide the space into separated volumes bounded by lines. Now for $N = 2$ the "irregular" trajectories not confined to tori are found in regions where the tori with commensurate frequencies have been destroyed. That is, they wander about between the preserved tori with incommensurate frequencies. This is indicated schematically in Figure 43.1, after Berry. [40.2] But for $N > 2$, because the tori do not divide the energy surface into disconnected regions, there are connected gaps between tori, as sketched in Figure 43.2. [40.2]

Thus the irregular trajectories can wander quasi-ergodically over a complex, interconnected <u>Arnold web</u>. This is Arnold diffusion. Unlike the case $N = 2$, the stochastic layers are not separated by regular trajectories, but are merged into a web that is everywhere dense on the energy surface (i.e., the Arnold web comes arbitrarily close to any point on the energy surface). Furthermore this permeation of the energy surface by the Arnold web persists even as the perturbation strength approaches zero, although for tiny perturbations the diffusion rate may be extremely small.

As in the onset of chaos for $N = 2$, chaos on the Arnold web is associated with overlapping resonances (or groups of overlapping resonances). Rather than proceeding further with general discussions about how chaos can arise in Hamiltonian systems, we next take up the question of how to identify and characterize this chaotic behavior. First, however, it is probably useful to summarize our discussion on the generic behavior of nonlinear Hamiltonian systems:

(1) For $N = 1$ (one degree of freedom) all Hamiltonian systems are integrable, and also ergodic.

(2) For $N = 2$ most systems will either be nonintegrable or will become nonintegrable if the energy is large enough. For near-integrable systems with sufficiently small perturbations, most trajectories remain confined to KAM tori which are slight distortions of the unperturbed tori. But the KAM tori can be destroyed when

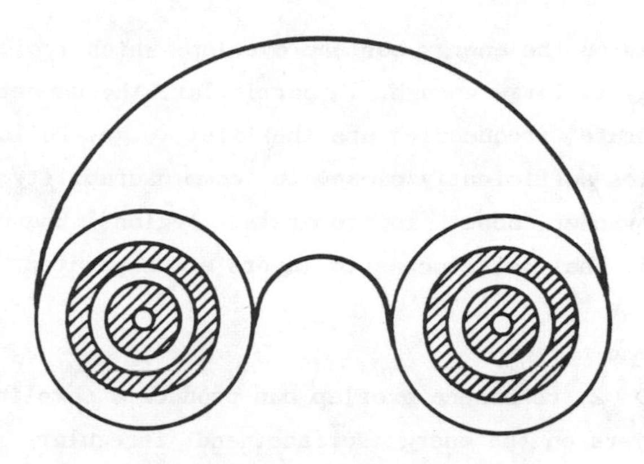

Figure 43.1 For N = 2 the two-dimensional KAM tori divide the three-dimensional energy surface into distinct, closed surfaces. The cross-hatched areas indicate irregular motion. After M.V. Berry, Reference [40.2].

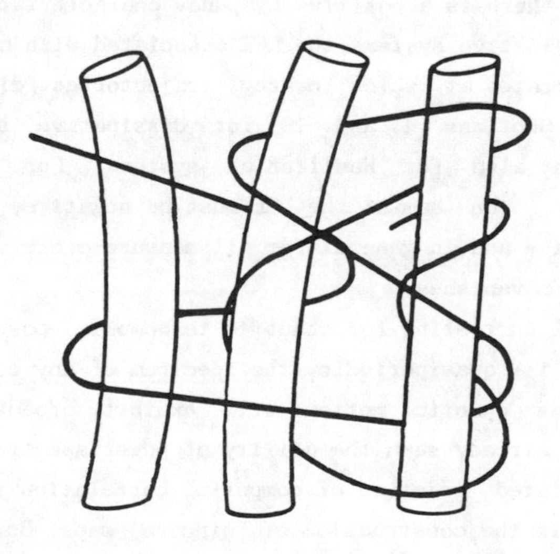

Figure 43.2 For N > 2 the KAM tori do not divide the energy surface into disconnected regions. After M.V. Berry, Reference [40.2].

resonance zones on the energy surface overlap, which typically occurs when the energy is large enough. In particular, the unperturbed tori with commensurate frequencies are the first to go, followed by tori with frequencies sufficiently close to commensurability. Irregular trajectories wander about in separated regions bounded by the preserved tori. That is, stochastic layers are separated by regular trajectories.

(3) For $N > 2$ resonance overlap can produce a complicated web of stochastic layers on the energy surface, and irregular trajectories can wander over essentially the whole energy surface.

44. Characterization of Chaotic Behavior

As noted earlier, chaos in Hamiltonian systems may be identified in the same way as in dissipative systems: chaos means exponential sensitivity to initial conditions and therefore occurs, by definition, if there is a positive Lyapunov characteristic exponent (LCE). As in dissipative systems the LCE associated with a trajectory give the average rates at which nearby trajectories diverge. The discussions in Sections 4 and 18 for dissipative systems are therefore relevant also for Hamiltonian systems. For dissipative systems, however, the sum of the LCE must be negative, whereas for Hamiltonian flows - and in general for all measure-preserving flows - the sum of the LCE vanishes.

Another tool in testing for chaos is to compute power spectra. If the motion is quasiperiodic the spectrum of any coordinate is discrete, whereas chaotic motion will exhibit broadband power spectra. We have already seen the utility of power spectra in Part I. The closely related practice of computing correlation functions is also useful, as is the construction of Poincare' maps. But perhaps it is worth emphasizing once more that <u>the sure way of identifying chaotic behavior is to compute the LCE, or at least the largest LCE</u>.

Classical Hamiltonian systems are deterministic in the sense that the state $(\vec{q}(t),\vec{p}(t))$ of a system is determined uniquely from some $(\vec{q}(0),\vec{p}(0))$. This determinism defines a dynamical system regardless of whether the flow is dissipative or conservative. We have discussed in Section 7 the fact that dynamical systems like the Bernoulli shift can nevertheless behave as "randomly" as coin tosses, and this notion may be carried over to chaotic Hamiltonian systems as well. It is therefore worthwhile now to discuss briefly what is meant by "randomness."

Suppose we have a sequence of N bits (i.e., a sequence of binary digits such as 0's and 1's - see Section 20). Can we give a prescription or rule for computing this sequence? Obviously we can program a computer to write out the sequence, and this program itself may be considered a rule for producing the sequence. The sequence may be called random if there is no rule significantly shorter than this trivial rule.

To be a bit more specific, let K_C be the smallest number of bits required for a program that generates our sequence of N bits on a machine C. K_C has two contributions - a machine-dependent part and a machine-independent part. The machine-independent part K_N is called the complexity of the sequence, and lies between 1 (any computer requires at least one command) and N. Then our sequence of bits may be said to be random if $K_N/N > 0$ as $N \to \infty$. In particular, there is a theorem to the effect that this condition for "positive complexity" is equivalent to the statement that our sequence of bits passes any computable test for randomness.

In this sense it may also be proven that a Bernoulli shift like $x_{n+1} = 2x_n \pmod 1$ generates a random sequence for almost all x_o. We noted in Sections 1 and 7 that the logistic map with $\lambda = 1$ is just such a Bernoulli shift, and so we can now assert that the (deterministic) logistic map can indeed produce a random sequence. Furthermore Hamiltonian systems can display similar behavior and so can generate randomness in spite of their deterministic character.

If a system has this quality of randomness we cannot, as a practical matter, predict its future exactly. If we have a set of differential equations with certain initial conditions and this system evolves chaotically, the best we can do, aside from approximations, is simply to "watch" the system evolve by numerical simulation. We cannot hope to invent a simpler way of describing the system, such as a closed-form analytical solution. In this sense it may be said that "The best description of a chaotic system is the system itself."

One striking manifestation of this randomness is seen in numerical computations. Suppose we start out from some point (initial condition) in phase space and integrate equations of motion out to some time, and then reverse the computation and integrate backwards out to the same time. If this time is large enough, roundoff error will catch up with us and we will not recover the initial conditions. This loss of "memory" occurs regardless of whether the evolution is regular or chaotic. But in chaotic evolution the initial conditions are lost much more quickly than in regular evolution. For a machine with N-bit precision, the time required for a chaotic system to lose memory of the initial conditions in this sense is $\approx O(N)$, whereas for regular systems it is $\approx O(2^N)$.

The recognition that such random behavior is possible in deterministic systems should help to remove certain "philosophical" prejudices about classical physics. In particular, we would like now to discuss briefly the following question, which historically has usually been answered in the affirmative.

45. Is Classical Physics Really Deterministic?

> And the first Morning of Creation wrote
> What the Last Dawn of Reckoning shall read
>
> Omar Khayyam

The notion of classical determinism, as opposed to quantum-mechanical indeterminacy, is often attributed to Laplace, who perhaps stated the idea most precisely: "An intelligence knowing, at a given instant of time, all forces acting in nature, as well as the momentary positions of all things of which the universe consists, would be able to comprehend the motions of the largest bodies of the world and those of the smallest atom in a single formula, provided it were sufficiently powerful to subject all data to analysis; to it, nothing would be uncertain, both future and past would be present before its eyes." [45.1] We might loosely say it this way: if we have unlimited computing power at our disposal, and know the correct equations of motion of any physical system, as well as the state of the system at some time $t = 0$, we can predict to any desired accuracy the state of the system at any time. But is it really true that the world-view of classical physics is deterministic? A strong case can be made that it is <u>not</u>.

We can take a more pragmatic point of view from that of Laplacian determinism and ask the following question: if the initial conditions of a system are known only to within some small error, can we nevertheless make reasonably accurate predictions about the system for any time $t > 0$? Maxwell (1873) recognized that in general we cannot: "There are certain classes of phenomena...in which a small error in the data only introduces a small error in the result...There are other classes of phenomena which are more complicated, and in which cases of instability may occur...influences whose physical magnitude is too small to be taken account of by a finite being, may produce results of the highest importance..." Maxwell also suggested that, if more attention were devoted to such things, "the promotion of natural knowledge may tend to remove that prejudice in favor of determinism..." [40.2] These were prescient remarks indeed!

Max Born also cautioned against a naive doctrine of classical determinism, and used an extremely simple example to illustrate his point. [45.2] Consider a ball moving between two walls at $x = 0$ and $x = L$, where it is elastically reflected. Given the initial values x_o and v_o of its position and velocity, we can easily determine $x(t)$ and $y(t)$ for any t, as indicated in Figure 45.1a. But if x_o and v_o are changed by Δx and Δv, then we obtain for any t an $x(t)$ differing form our original $x(t)$ by an amount $\Delta x(t)$. For $\Delta x = 0$ we have $|\Delta x(t)| = |\Delta v|t$ (Figure 45.1b), so that after a time $t = L/|\Delta v|$ the variation $\Delta x(t)$ is equal to the entire range (L) of possible values of x.

In particular, if $|\Delta x|$ and $|\Delta v|$ represent uncertainties in the initial conditions, our uncertainty $|\Delta x|$ in the position of the particle eventually becomes as large as the whole range over which x can vary. IF $|\Delta x|$ and/or $|\Delta v|$ are finite, however small, we cannot make (accurate) long-term predictions. It becomes operationally meaningless to say that the system is deterministic. It is no longer what Born called *determinable*. Born argued that we should therefore formulate classical physics in a *statistical* manner.

Classical indeterminability has an interesting implication *vis-à vis* quantum mechanics. We can argue that the "new" feature of quantum physics is not indeterminism *per se*. A classical chaotic system will, as a practical matter, be indeterminable in the long run because (a) we do not know initial conditions with infinite precision, or (b) we cannot handle an infinite string of digits in our computations. And aside from such practical considerations, such a system can be considered to have an *inherently* random character, as in the Bernoulli shift. (Systems have been investigated in which finite-bit arithmetic is not a limitation, where only integer "outputs" are possible. The outputs can nevertheless have a random character. [45.3])

Here again Born had an important message for us:

> It is misleading to compare quantum mechanics with deterministically formulated classical mechanics; instead, one should first reformulate the classical theory, even for a single particle, in an indeterministic, statistical manner. Then some of the distinctions between the two theories disappear, others emerge with great clarity. Amongst the first is the feature of quantum mechanics, that each measurement interrupts the automatic flow of events, and introduces new initial conditions (so-called "reduction of probability"); this is true just as

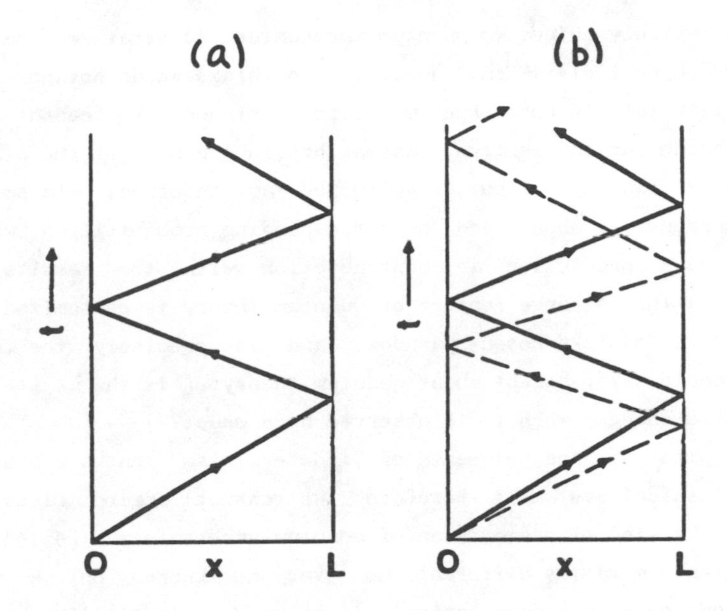

Figure 45.1 Max Born's simple illustration of indeterminability due to imprecise knowledge of initial conditions. (a) shows the trajectory of a ball bouncing between two walls. (b) shows the trajectories of two such balls with slightly different initial conditions. A slight uncertainty in the initial conditions for one ball can result in a large uncertainty in our predictions of the ball's location.

well for a statistically formulated classical theory...[45.2]

What is truly unique to quantum mechanics is that we cannot attribute objectively "real" qualities to things we do not observe. In a two-slit interference experiment, for instance, we cannot say that a photon (or an electron) passes through one slit or the other. (Of course we can say it, but it would be an incorrect statement. Were it correct, we should add the corresponding probabilities, which would give a prediction in contradiction with the results of experiment.) This strange feature of quantum theory is epitomized in the Einstein-Podolsky-Rosen paradox, and is precisely the thing Einstein could never accept about quantum theory. ("Is the state of the universe changed when it is observed by a mouse?")

In spite of the elements of "indeterminism" that are present even in classical systems, therefore, we cannot regard classical chaos as "a kind of premonition of quantum uncertainty." [45.4] The two theories are vastly different, as everyone knows. But a very interesting question does arise: if we have a classically chaotic system, and treat it quantum mechanically, does the classical chaos manifest itself in any way? Is there some "chaotic" feature of the quantum system that reflects the classical chaos, and would not be there if the classical system were not chaotic? Such questions of "quantum chaos" are broached later in these lectures.

46. The Kicked Pendulum and the Standard Mapping

Our remarks earlier about energy sharing among vibrational modes of a molecule, and the possible implications of nonlinearity and overlapping resonances for statistical approaches to unimolecular processes, certainly point to the importance of chaotic dynamics at the molecular level. Classical considerations also suggest that laser-molecule interactions are another area where notions of chaos are relevant: a molecule in the field of a laser is an example of a driven nonlinear oscillator, and chaos is commonplace in such systems. (Recall our discussion of the Duffing oscillator in Section

11.)

As discussed in Section 1, models with periodic kicking provide the most computationally convenient examples of driven nonlinear systems, because the integration of the equations of motion reduces to the iteration of a map. We showed in Section 3, for instance, how the logistic map could be obtained in this way. An important example of a periodically kicked system is the kicked pendulum (also known as the kicked rotator). [46.1] The Hamiltonian for this system is

$$H = p_\theta^2/2m\ell^2 - (m\ell^2\omega_o^2)\cos\theta \sum_{n=-\infty}^{\infty} \delta(t/T - n) \tag{46.1}$$

and the equations of motion are

$$\dot{p}_\theta = - (m\ell^2\omega_o^2)\sin\theta \sum_n \delta(t/T - n) \tag{46.2a}$$

$$\dot{\theta} = p_\theta/m\ell^2 \tag{46.2b}$$

m and ℓ are the pendulum mass and length, ω_o is the natural frequency for small displacements, and T is the period of the delta-function kicking. If $T \to 0$ the force term is on continuously. For $T \neq 0$ we can think of the gravitational force as being switched on and off to provide the kicks. Although the model is then very artificial, it turns out to play an important illustrative role not only classically but in "quantum chaos" as well.

From (46.2) we obtain by integration the discrete mapping

$$p_{n+1} = p_n - (m\ell^2\omega_o^2T)\sin\theta_n \tag{46.3a}$$

$$\theta_{n+1} = \theta_n + p_{n+1}T/m\ell^2 \tag{46.3b}$$

where p_n, θ_n are the values of p_θ and θ just before the n<u>th</u> delta-function kick. Writing $p_n = (m\ell^2/T)P_n$, and replacing θ_n by $\theta_n + \pi$, we have the <u>standard map</u> (or Chirikov map):

$$P_{n+1} = P_n + K\sin\theta_n \tag{46.4a}$$

$$\theta_{n+1} = \theta_n + P_{n+1} \tag{46.4b}$$

where $K \equiv (\omega_o T)^2$. For $K \to 0$ the gravitational potential is on continuously and of course (46.1) is integrable in this case. For small K we expect from the KAM theorem that most trajectories lie on invariant curves in the phase plane of the near-integrable system. Numerical experiments indicate that some of these KAM curves remain as $K \to 1$, but for $K >> 1$ they are all broken and most trajectories are chaotic.

The standard map, as its name implies, has been studied in considerable detail. [38.3] We will summarize a few pertinent points. First, the fixed points are easily found from (46.4):

$$P^* = 2\pi n, \ n = \text{integer} \tag{46.5a}$$

$$\theta^* = 0, \ \pi \tag{46.5b}$$

where it is understood that θ is given modulo 2π.

<u>Exercise</u>: Show that the fixed points with $\theta^* = 0$ are unstable, whereas those with $\theta^* = \pi$ are stable for $K < 4$.

The fixed points $(P^*, \theta^*) = (2\pi n, \pi)$ for $K < 4$, being stable fixed points of a Hamiltonian system, are centers. There are also n-cycles, $n \geq 2$. Some algebra reveals, for instance, that there are stable 2-cycles with $P = 2\pi(n + 1/2)$ and $\theta = 0, \pi$ for $K < 2$. Thus for small K we can expect the sort of phase curves sketched in Figure 46.1, which indicates the stable motion about the stable period-1 and period-2 fixed points, as well as the separatrices dividing these different regions of phase space.

Figure 46.2 shows numerical results for the standard map with $K = 0.5$ and 1.0. For $K = 0.5$ we see the sort of behavior sketched in Figure 46.1. At $K = 1.0$ it is clear that there are irregular trajectories, and an LCE computation confirms that these trajectories are indeed chaotic.

The onset of "widespread chaos" may be explained in terms of the overlap of resonance zones. For a rough approximation we consider the unperturbed Hamiltonian

$$H = p_\theta^2/2m\ell^2 - (m\ell^2\omega_o^2)\cos\theta = (m\ell^2/T^2)[P_\theta^2/2 - K\cos\theta] \tag{46.6}$$

where again $p_\theta = (m\ell^2/T)P_\theta$. Since the regions with $P_\theta \cong 2\pi n$ in Figure 46.1 are separated by $\Delta P_\theta = 2\pi$, and the maximum variation of P_θ about each region is $2\sqrt{K}$, we can expect a resonance overlap when $2(2\sqrt{K}) = 2\pi$, or

$$K = (\pi/2)^2 = 2.47 \tag{46.7}$$

(Note: Using the Poisson summation formula (47.12) below, it may be shown that in a rotating-wave approximation the Hamiltonian (46.1) consists of a sum of potentials of the form $\cos\theta_n$, to which resonance overlap ideas may be directly applied as in Section 41. See also Section 49.) Numerically, however, it is found that widespread chaos sets in for $K \cong 1$. Good agreement of the Chirikov overlap criterion with "experiment" may be obtained by including higher-order resonances. [38.3]

Of special interest to us is the energy gained by the pendulum as a result of the kicking. Figure 46.3 shows computed results for $\langle P_n^2 \rangle$ for different values of K, where the averaging in each case is

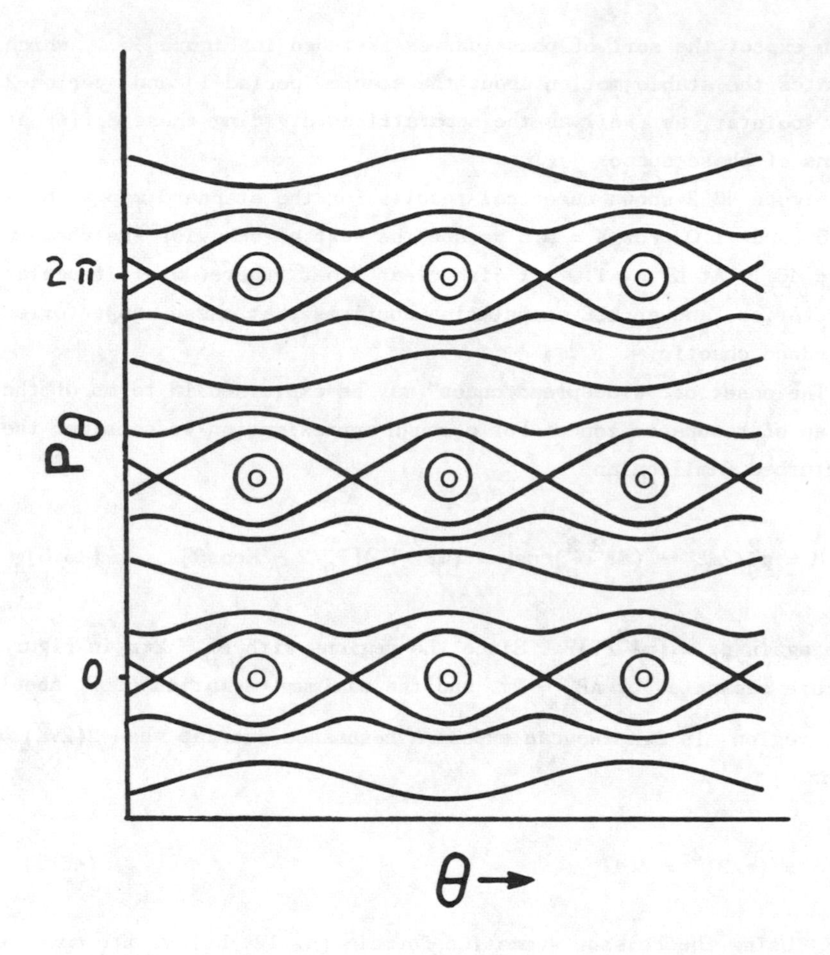

Figure 46.1 Phase curves for the kicked pendulum with small K.

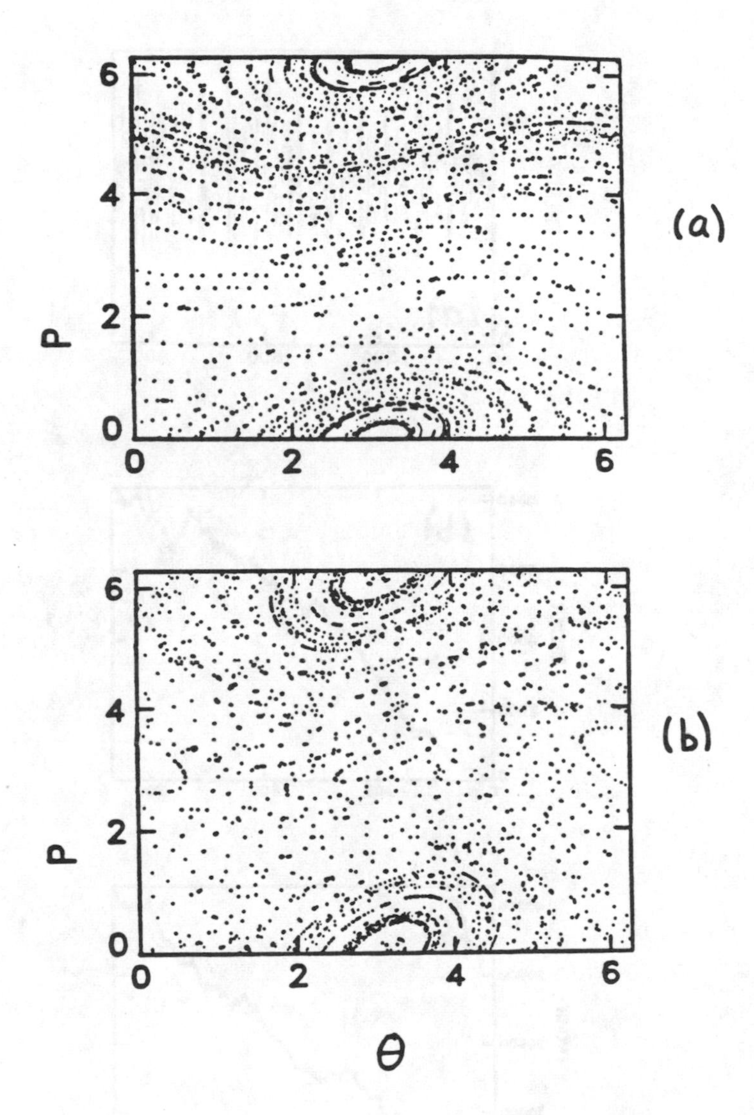

Figure 46.2 Computed phase curves for the kicked pendulum with (a) K = 0.5 and (b) K = 1.0.

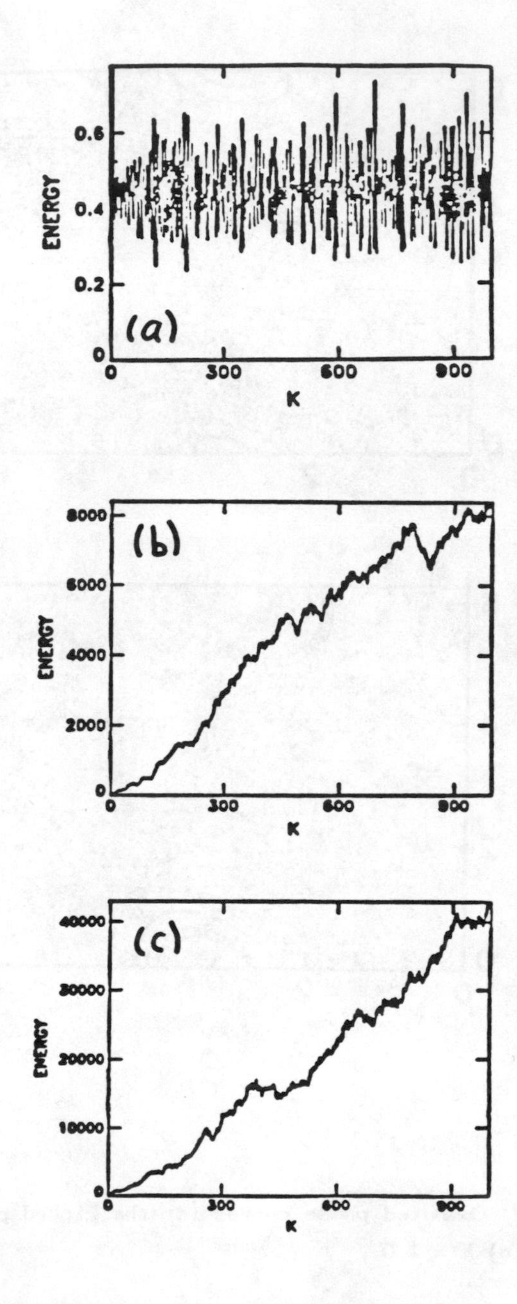

Figure 46.3 Energy in the kicked rotator for (a) $K = 0.5$, (b) $K = 4.0$, and (c) $K = 10.0$. In each case the plotted energy is an average over 40 uniformly distributed values of the initial angle θ_o.

taken over a set of 40 different values of θ_o. For large K - in the "chaotic regime" - the average of the square of the angular momentum (i.e., the average energy) grows in a random-walk, diffusive manner:

$$\langle P^2 \rangle \approx \tfrac{1}{2}K^2 t \tag{46.8}$$

Such results were first discussed by Casati, et al. [46.1] These authors gave the following intuitive explanation for the diffusive energy growth. First we note that from (46.4a) it follows that

$$P_n - P_o = K \sum_{j=0}^{n-1} \sin\theta_j \tag{46.9}$$

and therefore that

$$(P_n - P_o)^2 = K^2 \sum_{i=0}^{n-1} \sum_{j=0}^{n-1} \sin\theta_i \sin\theta_j \tag{46.10}$$

Now if the θ_i were uniformly distributed random variables, it would follow that the average of (46.10) over the θ_i is proportional to n^2. Thus we can see the role of chaos in giving rise to a linear dependence of average energy on time. And if from (46.9) we regard P_n as being a sum of random variables, it follows from the central limit theorem that the distribution function for P is Gaussian:

$$f(P) = (K\sqrt{\pi t})^{-1} e^{-P^2/K^2 t} \tag{46.11}$$

In fact this is the distribution found "experimentally." [46.1] An interpretation analogous to this appears in the following section.

47. Chaos in a Classical Model of Multiple-Photon Excitation of Molecular Vibrations

The multiple-photon excitation (MPE) and dissociation of polyatomic molecules with infrared lasers was studied extensively during the 1970s in connection with laser-controlled isotope separation. Early experiments of Isenor and Richardson [47.1] showed that molecules could be dissociated in laser fields of modest power. This was somewhat of a surprise, for one could have expected that the increasing mismatch between the laser frequency and the allowed transition frequencies as the excitation proceeds up the vibrational ladder would thwart the absorption of more than a few photons. Later experiments also revealed that over a range of conditions the MPE process depends mainly on the fluence of the laser field, i.e., the total energy in the pulse, but not on the detailed intensity variations within the pulse. [47.2] This trend towards fluence dependence has more recently been corroborated in the experiments of Simpson, et al., [47.3] who found increasingly strong fluence dependence as the number of atoms increased in the molecules they studied.

The currently accepted description of MPE in polyatomic molecules may be briefly summarized as follows. For the lower vibrational levels of a pumped vibrational mode, energy is absorbed by resonant, stepwise excitation, with the anharmonic energy defects being compensated for by the rotational energy levels. [47.4, 47.5] Due to the large number of vibrational-rotational levels, the density of states in a polyatomic molecule increases rapidly with increasing excitation, so that after sufficient excitation a "quasicontinuum" regime is reached. The large number of levels in this quasicontinuum regime allows resonant stepwise excitation to proceed without bottlenecking.

We have described a classical model for MPE of molecular vibrations by an infrared laser field. [47.6] The dynamics of this model were found to be chaotic, with the consequence that the MPE process was predicted to be fluence dependent. In particular, the energy absorbed from a constant laser field was found to grow, on average, linearly with time. We will first describe the model in quantum-mechanical terms, and then formulate it classically. The resulting classical model has been analyzed in some detail, [47.6] and here we follow rather closely that analysis.

We assume only one vibrational mode of a molecule interacts strongly with an applied field, and neglect interactions of the remaining, "background" modes with the field. The background modes, which can exchange energy with the laser-pumped mode by single-quantum vibrational energy transfer, are furthermore assumed for simplicity to be harmonic. In SF_6, for instance, there are two infrared-active normal modes (denoted ν_3 and ν_4) with a rather large separation ($\approx$ 300 cm^{-1}), so that a laser can interact selectively with one of the modes. In particular, the ν_3 mode lies near 10.6 μm and has been studied extensively with CO_2 laser radiation. [47.2] The nearest background mode to ν_3 has a small anharmonicity compared with that of ν_3, thus providing some justification for the treatment of the background modes as harmonic oscillators. Our idealized model for the MPE dynamics is then based on the following Hamiltonian:

$$\hat{H} = \Delta\hat{a}^\dagger\hat{a} - \chi(\hat{a}^\dagger\hat{a})^2 + \Omega(\hat{a} + \hat{a}^\dagger) + \sum_m(\Delta + \epsilon_m)\hat{b}_m^\dagger\hat{b}_m + \sum_m\beta_m(\hat{a}^\dagger\hat{b}_m + \hat{b}_m^\dagger\hat{a}) \tag{47.1}$$

in units in which $\hbar$ = 1. Here $\hat{a}$ and $\hat{b}_m$ are, respectively, the annihilation operators for the pumped mode and the m<u>th</u> background mode; the caret (^) is used to indicate a quantum-mechanical operator. Δ is the detuning of the laser from the pumped mode, and ϵ_m is the frequency of background mode m as measured from the pumped mode. χ is the anharmonicity of the pumped mode, corresponding to about 2 cm^{-1} in SF_6. Ω is the Rabi frequency associated with the applied field and the pumped mode; the field is treated as a prescribed, classical field, which is a superb approximation for the kinds of field strengths of interest here. The last term in the Hamiltonian (47.1) describes the vibrational exchanges between the pumped and background modes. For simplicity we ignore intramolecular vibrational transfer among the background modes.

Before proceeding further it may be useful to explain briefly the choice (47.1) for the Hamiltonian. We could quantize the field and begin with the Hamiltonian

$$\hat{H}' = \omega_o \hat{a}^\dagger \hat{a} - \chi(\hat{a}^\dagger \hat{a})^2 + D(\hat{c}^\dagger \hat{a} + \hat{a}^\dagger \hat{c}) + \sum_m (\omega_o + \epsilon_m) \hat{b}_m^\dagger \hat{b}_m + \sum_m \beta_m (\hat{a}^\dagger \hat{b}_m + \hat{b}_m^\dagger \hat{a}) + \omega \hat{c}^\dagger \hat{c} \qquad (47.2)$$

where $\hat{c}$ is the photon annihilation operator for a mode of the field of frequency ω, ω_o is the vibrational transition frequency of the pumped mode in the harmonic approximation, and D is the coupling constant between the field and the pumped mode. The first two terms of (47.2) give the usual sort of energy spectrum ($E_n = n\omega_o - \chi n^2$) for an anharmonic vibrator. It follows from the Hamiltonian (47.2) that the total excitation operator

$$\hat{C} \equiv \hat{a}^\dagger \hat{a} + \hat{c}^\dagger \hat{c} + \sum_m \hat{b}_m^\dagger \hat{b}_m \qquad (47.3)$$

is a constant of the motion. Subtracting ω times this constant of the motion from (47.2), we have

$$\hat{H}'' = \Delta \hat{a}^\dagger \hat{a} - \chi(\hat{a}^\dagger \hat{a})^2 + D(\hat{c}^\dagger \hat{a} + \hat{a}^\dagger \hat{c}) + \sum_m (\Delta + \epsilon_m) \hat{b}_m^\dagger \hat{b}_m + \sum_m \beta_m (\hat{a}^\dagger \hat{b}_m + \hat{b}_m^\dagger \hat{a}) \qquad (47.4)$$

Now when the field is treated classically, $\hat{c}$ and $\hat{c}^\dagger$ become ordinary (c-number) classical variables and $D(\hat{c}^\dagger\hat{a} + \hat{a}^\dagger\hat{c}) \rightarrow \Omega(\hat{a} + \hat{a}^\dagger)$. This replacement gives the Hamiltonian (47.1).

The Heisenberg equations of motion following from the Hamiltonian (47.1) are

$$\dot{\hat{a}} = -i(\Delta - \chi)\hat{a} + 2i\chi\hat{a}^\dagger\hat{a}\hat{a} - i\Omega - i\sum_m \beta_m \hat{b}_m \tag{47.5a}$$

$$\dot{\hat{b}}_m = -i(\Delta + \epsilon_m)\hat{b}_m - i\beta\hat{a} \tag{47.5b}$$

The rate at which the molecule absorbs energy from the field is determined by the rate of change of $\hat{a}^\dagger\hat{a}$ associated with the third term on the right-hand side of (47.5a):

$$\frac{d}{dt}(\hat{a}^\dagger\hat{a})_\Omega = i\Omega(\hat{a} - \hat{a}^\dagger) \tag{47.6}$$

Thus we define the expectation value

$$n(t) = -(2\Omega)\mathrm{Im}[\int_0^t dt'\langle\hat{a}(t')\rangle] \tag{47.7}$$

as the number of photons absorbed by the molecule. This number is determined explicitly only by the pumped-mode amplitude, because by assumption the background modes do not couple directly to the field.

Now we go over to a <u>classical</u> model by replacing all operators by their expectation values in (47.5):

$$\dot{a} = -i(\Delta - \chi)a + 2i\chi|a|^2a - i\Omega - i\sum_m \beta_m b_m \tag{47.8a}$$

$$\dot{b}_m = -i(\Delta + \epsilon_m)b_m - i\beta_m a \tag{47.8b}$$

where $a \equiv \langle \hat{a} \rangle$ and $b_m \equiv \langle \hat{b}_m \rangle$. In particular, we ignore the quantum-mechanical difference $\langle \hat{a}^{\dagger}\hat{a}\hat{a} \rangle - |a|^2 a$ in the anharmonic contribution to (47.8a). Thus our classical model is defined by (47.8) and the expression

$$n(t) = -(2\Omega)\mathrm{Im}\left[\int_0^t dt' a(t')\right] \tag{47.9}$$

for the number of absorbed photons.

There is a further approximation that is useful in both the classical and quantum-mechanical models, namely the so-called quasicontinuum approximation. In this approximation we assume there are an infinite number of background modes above and below the pumped mode in frequency, and furthermore we will assume that these background modes are evenly spaced by an amount ρ^{-1}, where ρ is the density of background modes. We also assume for simplicity that the background modes have equal couplings with the pumped mode. Thus we take

$$\epsilon_m = \Delta_o + m\rho^{-1}, \quad m = -\infty, \ldots, \infty \tag{47.10a}$$

$$\beta_m = \beta \quad \text{for all } m \tag{47.10b}$$

The approximation of letting m extend from minus to plus infinity may be justified following Bixon and Jortner, [47.7] who show that the exact eigenstates have a finite, Lorentzian overlap with the background levels. See also Reference [47.8].

The great utility of the quasicontinuum approximation is realized when we use the formal solution of (47.8b) in (47.8a):

$$\sum_m \beta_m b_m(t) = -i\beta^2 \sum_m \int_0^t dt' a(t') e^{i(\Delta+\Delta_o+m\rho^{-1})(t'-t)}$$

$$= -i\beta^2 \int_0^t dt' e^{i(\Delta+\Delta_o)(t'-t)} a(t') \sum_{m=-\infty}^{\infty} e^{im\rho^{-1}(t'-t)} \qquad (47.11)$$

According to the Poisson summation formula [47.9] we have

$$\sum_{m=-\infty}^{\infty} e^{im\rho^{-1}(t'-t)} = 2\pi\rho \sum_{m=-\infty}^{\infty} \delta(t'-t-2\pi m\rho) \qquad (47.12)$$

so that

$$\sum_{m=-\infty}^{\infty} \beta_m b_m(t) = -\, 2\pi i\beta^2\rho \int_0^t dt' e^{i(\Delta+\Delta_o)(t'-t)} a(t') \sum_{m=-\infty}^{\infty} \delta(t'-t+2\pi m\rho)$$

$$(47.13)$$

Defining

$$\gamma \equiv 2\pi\beta^2\rho \qquad (47.14a)$$

$$\tau_R \equiv 2\pi\rho \qquad (47.14b)$$

$$\phi \equiv (\Delta+\Delta_o)\tau_R \qquad (47.14c)$$

and using (47.13) in (47.8a), our classical model in the quasicontinuum approximation reduces to the equation

$$\dot{a}(t) = -i(\Delta-\chi)a(t) + 2i\chi|a(t)|^2 a(t) - i\Omega - \frac{\gamma}{2}a(t)$$

$$- \gamma \sum_{m=1}^{\infty} e^{-im\phi} a(t-m\tau_R)\Theta(t-m\tau_R) \qquad (47.15)$$

for the pumped-mode amplitude, together with equation (47.9) for the number of absorbed photons. In (47.15) Θ is the unit step function.

For numerical computations it is convenient to cast (47.15) in the form

$$\dot{a}(t) = -i(\Delta-\chi)a(t) + 2i\chi|a(t)|^2a(t) - i\Omega - \frac{\gamma}{2}a(t) - \gamma s(t) \tag{47.16a}$$

$$s(t) = e^{-i\phi}[s(t-\tau_R) + a(t-\tau_R)] \tag{47.16b}$$

In terms of the new independent variable $T = \gamma t$, and the scaled variables

$$A \equiv (\gamma/\Omega)a, \quad S \equiv (\gamma/\Omega)s, \quad \overline{\Delta} \equiv \Delta/\gamma, \quad \overline{\chi} \equiv \chi/\gamma$$

$$T_R \equiv \gamma\tau_R, \quad \alpha \equiv \chi\Omega^2/\gamma^3 \tag{47.17}$$

we have

$$\dot{A}(T) = -i(\overline{\Delta}-\overline{\chi})A(T) + 2i\alpha|A(T)|^2A(T) - i - \tfrac{1}{2}A(T) - S(T) \tag{47.18a}$$

$$S(T) = e^{-i\phi}[S(T-T_R) + A(T-T_R)] \tag{47.18b}$$

and (47.9) becomes

$$n(T) = -2(\Omega/\gamma)^2\mathrm{Im}[\int_0^t dt'A(T')] \tag{47.19}$$

Reasonable numerical values are $\Omega = 0.3\ \text{cm}^{-1}$, $\rho = 4$ cm, and $\alpha = 0.18$. [47.6] For these numbers the natural "recurrence" time τ_R is about 133 psec. It is interesting to see how the number of absorbed photons given by (47.19) depends on the parameter ϕ defined in (47.14). For $\phi \neq 0$ the number of absorbed photons appears to grow approximately linearly with time, with the slope depending on ϕ. [47.6] Thus the detailed predictions of the model depend on ϕ, which is determined by the detuning of the field and the nearest background mode from the pumped mode. Figure 47.1 shows the computed n(t) for $\phi = \pi/2$. We also show $|a(t)|^2$ and the power spectrum of a(t) determined numerically by a Fast Fourier Transform. The 500 intervals correspond to about 67 nsec.From the small values of $|a(t)|^2$ we can see that most of the absorbed energy goes into the background modes; the pumped mode acts only as an intermediary channel. The results shown in Figure 47.1 seem to be fairly typical.

Numerical experimentation indicates that the dynamical system (47.18) evolves chaotically for the sorts of parameters we have assumed; we believe these parameter values are realistic in terms of what one would expect for a real molecule. Figure 47.2, for instance, shows the results of a computation of the maximal Lyapunov exponent. The computation appears to be converging to a positive value indicative of chaos. The computation of power spectra and correlation functions supports this conclusion.

It is interesting to see how the chaos leads to an absorbed energy growing approximately linearly with time. Consider first the following expression for the total excitation in the background modes:

$$\sum_m |b_m|^2 = \gamma \int_0^t dt' \, |a(t')|^2 + 2\gamma \mathrm{Re} \sum_{m=1}^{\infty} \beta^m \Theta(t - m\tau_R) \int_{m\tau_R}^t dt' a^*(t' - m\tau_R) a(t') \tag{47.20}$$

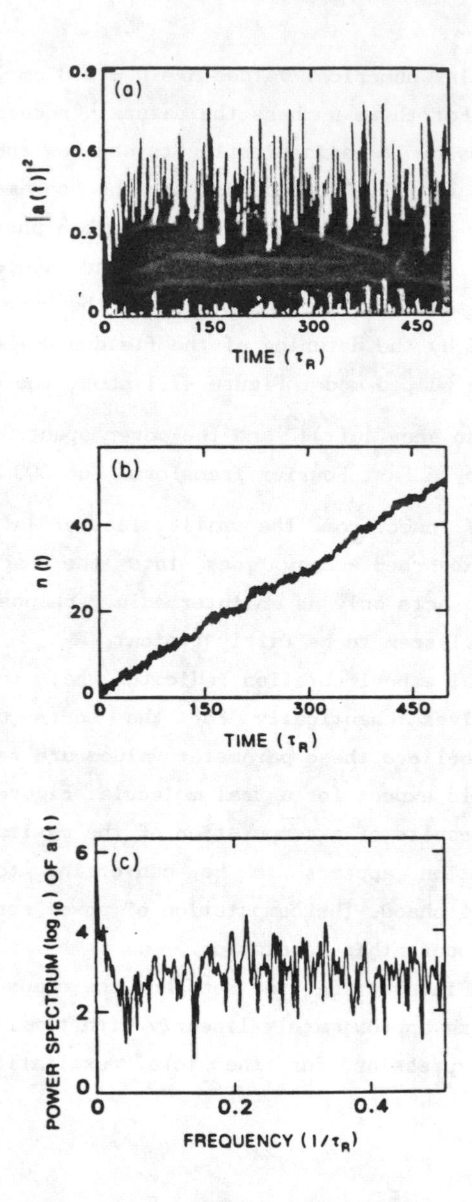

Figure 47.1 (a) Number of photons absorbed by the molecule for the case $\Omega = 0.3\ cm^{-1}$, $\beta = 0.2\ cm^{-1}$, $\rho = 4$ cm, $\Delta-\chi = \Delta_o = 0$, and the anharmonicity parameter $\alpha = 0.18$. The time is given in units of the recurrence time $\tau_R \approx 133$ psec. (b) Number of photons absorbed by the pumped mode. (c) Power spectrum of a(t).

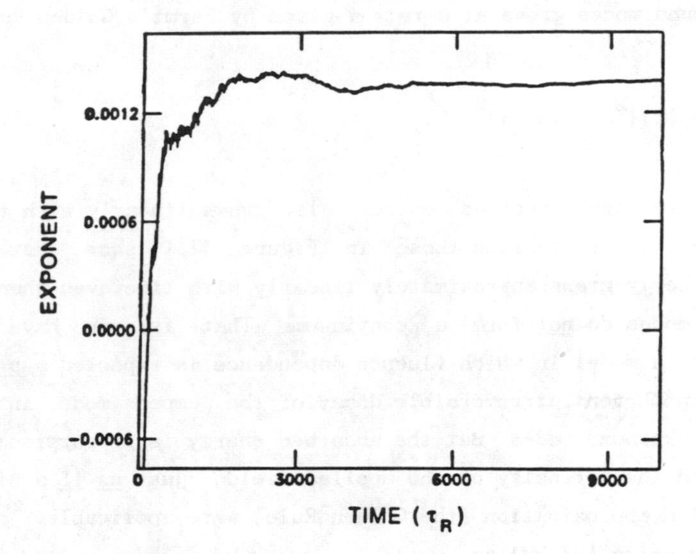

Figure 47.2 Computation of the largest Lyapunov exponent for the case of Figure 47.1.

In the limit $\tau_R \to \infty$ of a continuum of background modes, the second term would be absent. In this limit the number of photons absorbed by the background modes grows at a rate γ given by Fermi's Golden Rule:

$$\frac{d}{dt} \sum_m |b_m(t)|^2 = \gamma |a(t)|^2 \tag{47.21}$$

In this case the absorbed energy also grows linearly with time. [47.6] However, results like those in Figure 47.1 show that the absorbed energy grows approximately linearly with time even when the background modes do not form a continuum. That is, we have not constructed a model in which fluence dependence is expected a priori due to an incoherent, irreversible decay of the pumped mode into a bath of background modes. But the absorbed energy is proportional to the time and the intensity of the applied field - just as if a simple rate-equation approximation (the Golden Rule) were applicable.

We can write (47.20) as

$$\sum_m |b_m(t)|^2 = \gamma t \langle |a(t)|^2 \rangle_o + 2\gamma t \mathrm{Re} \sum_{m=1}^{\infty} \beta^m [X_m(t) \langle |a(t)|^2 \rangle_m + |\langle a(t) \rangle_m|^2] \tag{47.22}$$

where we introduce a "correlation function"

$$X_m(t) \equiv \frac{\langle a^*(t - m\tau_R) a(t) \rangle_m - |\langle a(t) \rangle_m|^2}{\langle |a(t)|^2 \rangle_m} \tag{47.23}$$

and define an "average" $\langle \ldots \rangle_m$ by

$$\langle f(t)\rangle_m \equiv (1/t)\Theta(t-m\tau_R) \int_{m\tau_R}^{t} dt'f(t') \qquad (47.24)$$

The approximately linear growth with time of the absorbed energy is a consequence of the fact that both terms multiplying t in equation (47.22) are approximately constant after many intervals. [47.6] In Figure 47.3 we show computed values of $X_1(t)$ and $X_{25}(t)$, and in Figure 47.4 the corresponding results for $\langle|a(t)|^2\rangle_m$ and $|\langle a(t)\rangle_m|^2$. The results for $X_m(t)$ are especially interesting, because they indicate the role played by chaos in establishing the linear growth with time, and therefore the fluence dependence, of the absorbed energy. As noted in Part I, quasiperiodic motion gives rise to oscillatory correlations, whereas it is typical of chaotic motion that correlations are decaying functions of the time difference. In particular, the fact that the $X_m(t)$ damp to constant values gives rise to the linear dependence of the absorbed energy with time, as is obvious from equation (47.22).

Note that, because the correlations decay to constant values, the linear dependence of the absorbed energy with time will persist for all times. That is, the energy growth will not saturate.

The quasicontinuum assumption of an evenly spaced, infinite ladder of background modes may seem highly artificial, but in fact it is not crucial for our results. We have performed computations with finite numbers of unevenly spaced background modes, and have obtained results in basically good agreement with those of the quasicontinuum model. [47.6]

To summarize, our classical model for MPE consists of an anharmonic pumped mode coupled to a background of harmonic, infrared active modes, and driven by a monochromatic applied field. For reasonable parameter values the model typically predicts approximate fluence-dependent behavior, more or less independently of the detailed assumptions made for the background modes. To the extent that the <u>classical</u> model is a reasonable one for a molecular vibrator

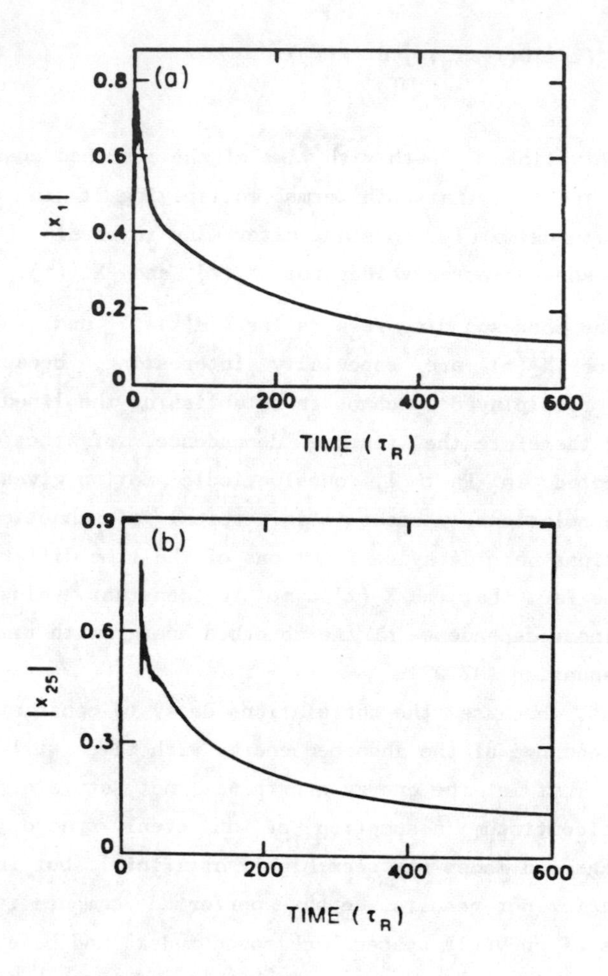

Figure 47.3 Correlation functions $X_m(t)$ for the case of Figure 47.1, for (a) m = 1 and (b) m = 25.

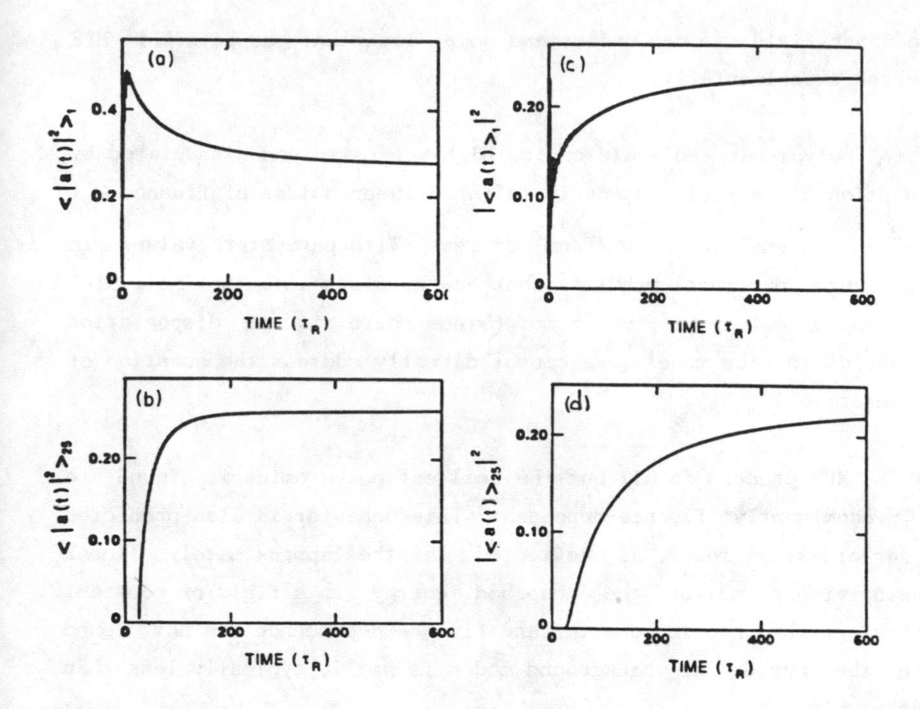

Figure 47.4 The average $\langle|a(t)|^2\rangle_m$ for the case of Figure 47.1 for (a) m = 1 and (b) m = 25, and $|\langle a(t)\rangle_m|^2$ for (c) m = 1 and (d) m = 25.

in a laser field, we can understand some important aspects of MPE observed experimentally:

(1) A polyatomic molecule can be highly excited and dissociated by absorption of tens of photons in infrared laser pulses of fluence ≈ 1 J/cm^2 and intensities ≈ 10 MW/cm^2 or less. With parameter values in this range, our model predicts that energy flows into the "molecule" at an approximately constant rate. (Since there is no dissociation threshold in the model , we cannot directly address the question of dissociation.)

(2) The MPE process in all but the smallest polyatomics is found to be predominantly fluence-dependent. This behavior is also predicted in our classical model, as reflected in the approximately linear growth with time of the absorbed energy in a field of constant intensity. The chaotic behavior and fluence dependence are not found when the number of background modes is small, typically less than five or six.

These features emerge without any a priori assumptions of statistical or irreversible behavior. In particular, we have explained the fluence dependence in terms of decaying correlations associated with the chaotic evolution of the pumped-mode amplitude.

Of course the model does have a serious, possibly fatal, shortcoming: it is completely classical. One might expect reasonably good agreement between the classical and quantum-mechanical predictions when the molecule is already highly pumped up the vibrational ladder, but our computations indicate that in fact the pumped mode does not get very highly excited. (We are unaware of experimental evidence for or against such low excitation of a driven mode in MPE, although it is frequently argued that the background modes act as a "sponge" for the absorbed photons, thereby casting doubt on the efficiency of isotopically selective excitation and dissociation. This conclusion depends, of course, on the specific value of the isotope shift.) We will later refer to results of fully quantum-mechanical computations for this model of MPE, after we have introduced some aspects of "quantum chaos." Before doing this, however, we turn our attention to a classical model of a <u>rotating</u> molecule in a laser field.

48. <u>Chaos in a Classical Model of a Rotating Molecule in a Laser Field</u>

We now consider the simplest model of a molecular rotator, namely a rigid-rotator harmonic oscillator. The Hamiltonian is

$$H = \Delta\vec{a}^{\dagger}\cdot\vec{a} + B_o\vec{J}^2 + \Omega\vec{\varepsilon}\cdot C\cdot(\vec{a} + \vec{a}^{\dagger}) \tag{48.1}$$

Here again $\vec{a}$ is the vector annihilation operator for the infrared-active normal mode of vibration of a molecule, with components referred to molecular body-fixed axes, $\vec{J}$ is the angular momentum operator, and C is the 3 × 3 orthogonal matrix (tensor) relating the laboratory and body frames. Δ, B_o, and Ω are respectively the laser detuning, the inverse of the molecular moment of inertia, and the Rabi frequency. $\vec{\varepsilon}$ is the polarization unit vector of the (linearly polarized) laser electric field, with components referred to the laboratory frame.

If $(\vec{\ell}_1,\vec{\ell}_2,\vec{\ell}_3)$ and $(\vec{f}_1,\vec{f}_2,\vec{f}_3)$ are respectively laboratory and body-frame coordinate unit vectors, then $C_{ij} = \vec{\ell}_i\cdot\vec{f}_j$,

$$J_i = i\sum_{s=1}^{3}\left(C_{sj}\frac{\partial}{\partial C_{sk}} - C_{sk}\frac{\partial}{\partial C_{sj}}\right) \tag{48.2a}$$

and

$$J_I = -i \sum_{s=1}^{3} (C_{Js} \frac{\partial}{\partial C_{Ks}} - C_{Ks} \frac{\partial}{\partial C_{Js}}) \qquad (48.2b)$$

are respectively the body-fixed and lab-fixed components of the molecular angular momentum, and i,j,k and I,J,K are cyclic in 1,2,3. We then have the following well-known commutation relations: [48.1]

$$[J_I, J_J] = i\epsilon_{IJK} J_K \qquad (48.3a)$$

$$[J_i, J_j] = -i\epsilon_{ijk} J_k \qquad (48.3b)$$

$$[J_i, J_I] = 0 \qquad (48.3c)$$

and

$$\vec{J}^2 = J_I J_I = J_i J_i \qquad (48.3d)$$

Here we use the Einstein summation convention for repeated indices, with ϵ_{ijk} the Levi-Civita symbol (= 1 if i,j,k is an even permutation of 1,2,3, - 1 if i,j,k is an odd permutation of 1,2,3, and 0 if any two of the i,j,k are equal). Furthermore it follows from (48.2a) and the definition of C that [48.1]

$$[J_i, C_{mj}] = -i\epsilon_{ijk} C_{mk} \qquad (48.4)$$

Then we easily obtain from these commutators and the Hamiltonian (48.1) the Heisenberg equations of motion

$$\dot{\vec{a}} = -i\Delta\vec{a} - i\Omega\vec{P} \qquad (48.5a)$$

$$\dot{\vec{P}} = -B_o(\vec{J} \times \vec{P} - \vec{P} \times \vec{J}) \qquad (48.5b)$$

$$\dot{\vec{J}} = \Omega\vec{P} \times (\vec{a} + \vec{a}^{\dagger}) \tag{48.5c}$$

where

$$\vec{P} \equiv \vec{\epsilon}\cdot C \tag{48.5d}$$

We will consider the classical approximation to (48.5), where we replace all the operators by c-numbers and solve the ODE system

$$\dot{\vec{a}} = -i\Delta\vec{a} - i\Omega\vec{P} \tag{48.6a}$$

$$\dot{\vec{P}} = -2B_o(\vec{J} \times \vec{P}) \tag{48.6b}$$

$$\dot{\vec{J}} = \Omega\vec{P} \times (\vec{a} + \vec{a}^{\dagger}) \tag{48.6c}$$

with the (arbitrary) initial conditions

$$\vec{a}(0) = (0,0,0) \tag{48.7a}$$

$$\vec{J}(0) = (0,0,J_o) \tag{48.7b}$$

$$\vec{P}(0) = (1/\sqrt{2})(1,0,1) \tag{48.7c}$$

Note that the vibrational and rotational motions are coupled through $\vec{P}$ in (48.6).

Numerical results obtained for the classical system (48.6) were described by Galbraith, Ackerhalt, and Milonni [48.2] and we follow their treatment here. First note that (48.6b) describes a precession of the polarization unit vector about $\vec{J}$ at the angular frequency

$2B_o|\vec{J}|$. If we take $\vec{J}$ to be constant (= $\vec{J}_o$) by ignoring (48.6c), we can easily solve (48.6a) and (48.6b) analytically. The second equation has the solution

$$\vec{P}(t) = (1/\sqrt{2})[\cos(2B_oJ_ot), \sin(2B_oJ_ot), 1] \qquad (48.8)$$

Using this solution in (48.6a), we find that $\vec{a}$ is driven at the three frequencies Δ and $\Delta \pm 2B_oJ_o$. (Reasonable values of B_o and Ω for molecules like SF_6 or CF_4 are 0.1 cm^{-1} and 1 cm^{-1}. In Reference [48.2] Δ was arbitrarily set to 0.5 cm^{-1}.) The explicit solution for $|\vec{a}(t)|^2$ is

$$|\vec{a}|^2 = \tfrac{1}{2}\Omega^2\{(\Delta+2B_oJ_o)^{-2}[1 - \cos(\Delta+2B_oJ_o)t] + (\Delta-2B_oJ_o)^{-2} \times$$

$$[1 - \cos(\Delta-2B_oJ_o)t] + \Delta^{-2}[1 - \cos\Delta t]\} \qquad (48.9)$$

The resonances at $\Delta = \pm 2B_oJ_o$ and 0 correspond respectively to the familiar P, R, and Q branches of rotational spectra.

It is interesting to solve the full system of equations (48.6) numerically, assuming (for instance) the initial conditions (48.7). The results shown in Figure 48.1 were obtained assuming $J_o = 5$. We show P_x vs. P_y and also the power spectrum of P_x. The spectrum suggests chaotic time evolution, and this is confirmed by a computation of the maximal Lyapunov exponent. [48.2]

Obviously these results show that $\vec{J}$ is far from constant. However, when J_o is large the motion is gyroscopically stabilized. [48.2] That is, there is still a precession of $\vec{P}$ about $\vec{J}$. Figure 48.2, for instance, shows how Figure 48.1 is modified when J_o is raised to 20. In this case the motion is not chaotic.

The conclusion to be drawn from this simple model is that the rotational nonlinearity arising from the laser-molecule coupling can

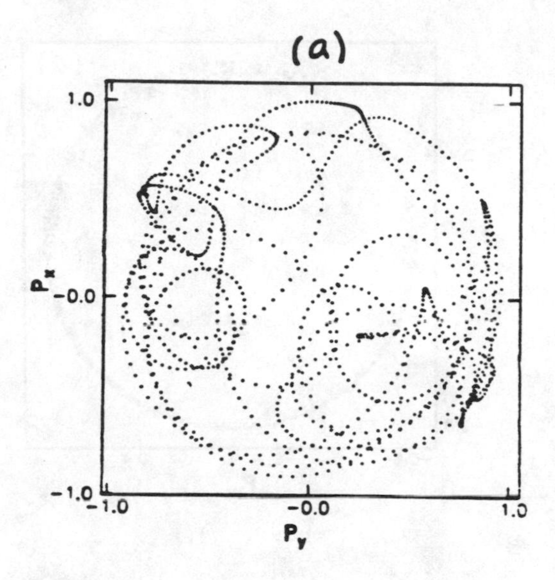

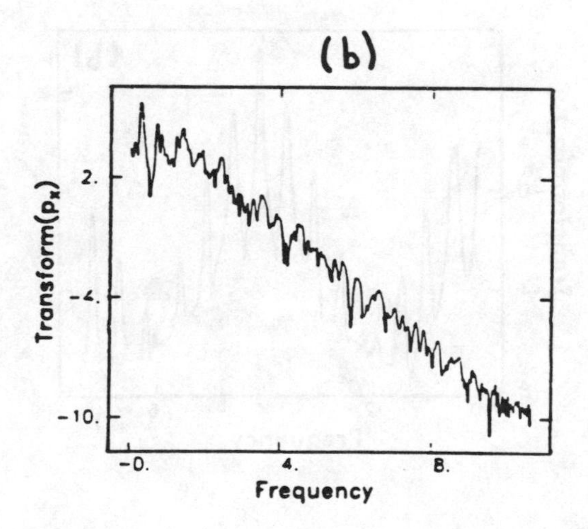

Figure 48.1 (a) P_x vs. P_y obtained by solving equations (48.6) with $J_o = 5$ and the initial conditions (48.7). (b) Power spectrum of P_x.

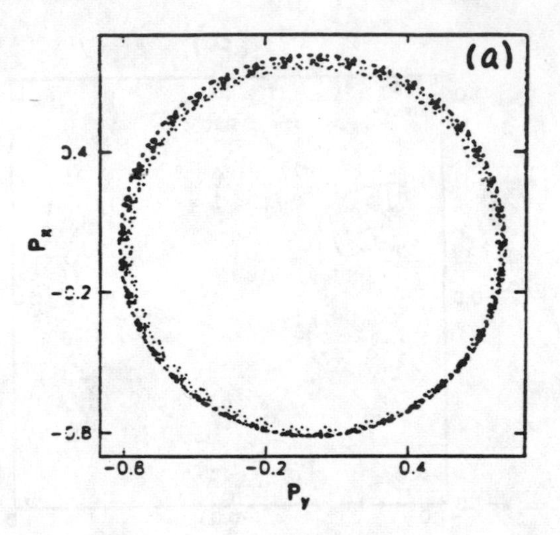

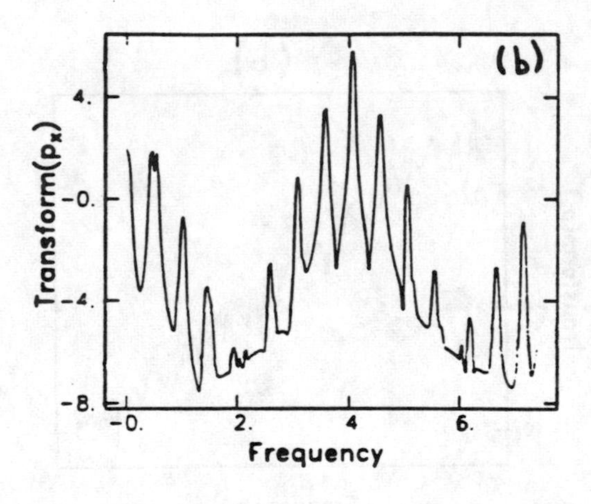

Figure 48.2 Results corresponding to Figure 48.1 when J_o is raised to 20.

give rise to chaotic dynamics. Chaos is dominant at the low angular momentum values, but the motion at large J_o is gyroscopically stabilized and regular (quasiperiodic). Furthermore these results indicate that we should not treat rotations cavalierly as a sort of inhomogeneous broadening of the vibrational motion. Once again we see that chaos arises even in very simple models of laser-molecule interactions. And once again the question arises as to the validity of the classical approximation. Before finally taking up some questions of "quantum chaos," we will briefly describe some other classical models of laser-molecule dynamics where chaotic behavior has been found.

49. Stochastic Excitation

We will focus our attention on studies concerned mainly with energy deposition in a molecule driven by a laser. In Section 47 we described a classical model of multiple-photon excitation (MPE) in which the experimentally observed "incoherence," or fluence dependence, was attributable to chaos and, in particular, a decay of correlations. However, that is not the only, nor even the first, model of MPE in which chaos plays a decisive role. We have more recently come across related classical models, and these will be briefly described in this section.

As noted in Section 47, it is interesting in the first place that a polyatomic molecule can absorb tens of (monochromatic) photons without bottlenecking. In the classical MPE models this is not difficult to understand: the chaotic behavior of the absorbing nonlinear system allows the phase space to be explored in an essentially diffusive manner, and so high levels of excitation (and dissociation) become possible. This is often referred to as stochastic excitation. (The kicked pendulum of Section 46 provides another example of stochastic excitation.) The loss of temporal coherence between the nonlinear oscillator (molecule) and the applied field also leads to the amount of absorbed energy depending on the field fluence (intensity times time) rather than intensity.

An early MPE model invoking stochastic excitation is due to Shuryak. [49.1] He begins with a one-dimensional nonlinear oscillator driven by a force varying sinusoidally in time. This system is approximated by the integrable Hamiltonian describing a pendulum. (See below) If a nonintegrable perturbation is added the system can become nonintegrable and chaotic as resonance overlap destroys the KAM tori. In Shuryak's model the nonintegrable perturbation is attributed to a nonlinear coupling between the pumped mode and background modes of the molecule. A critical intensity for stochastic excitation is estimated using Chirikov's resonance overlap criterion: it is assumed that stochastic excitation and dissociation can occur when the pumped-mode amplitude is large enough for the first resonance overlap. In this way Shuryak estimates a critical dissociation intensity ≈ 300 MW/cm^2 for SF_6. This is an order of magnitude larger than the threshold intensity observed experimentally by Ambartsumyan, et al. [47.4] Shuryak suggests the discrepancy may be attributed to quantum tunneling in the SF_6 molecule.

In the model described in Section 47 the coupling to background modes is linear, and furthermore there does not appear to be a threshold necessary for stochastic excitation. On the other hand the model describes only absorption, not dissociation, and experiments seem to indicate that there may be no threshold intensity necessary for MPE. [47.2]

In a model of Belobrov, et al. [49.2] a nonintegrable perturbation is supplied by the coupling of the molecules back to the field, in the spirit of laser theory and the Jaynes-Cummings model. In other words, the molecules interact with their own radiation fields. Belobrov, et al. suggest that "Allowance for the stochastic character of the behavior of the system may turn out to be essential in the mechanism whereby the molecules are excited into the region of high-lying states of the vibrational spectrum." To our knowledge this is the only MPE model, classical or quantum, that invokes the interaction of the molecules with their own radiated fields.

A classical MPE model with stochastic excitation has been very clearly described by Jones and Percival. [49.3] They consider a system with Hamiltonian

$$H = H_1 + H_{NL} - exE_o\cos(\Omega t+\phi_o) \qquad (49.1)$$

where H_1 is the Hamiltonian for a harmonic oscillator ($H_1 = p^2/2m + m\omega_o^2x^2/2$) and $H_{NL} = ax^3 + bx^4 + \ldots$ is the nonlinear part of the unperturbed (field-free) system. The transformation

$$p = -(2\omega_o mJ)^{1/2}\sin\theta \qquad (49.2a)$$

$$x = (2J/m\omega_o)^{1/2}\cos\theta \qquad (49.2b)$$

converts H_1 to action-angle form. In terms of the action-angle variables J,θ the last term in (49.1) becomes

$$H_{INT} = -(eE_o/2)(2J/m\omega_o)^{1/2}[\cos(\theta+\Omega t+\phi_o) + \cos(\theta-\Omega t-\phi_o)] \qquad (49.3)$$

It is assumed that the first term in brackets varies rapidly and may be averaged to zero (rotating-wave approximation). Then (49.1) is replaced by the approximate form

$$H(\theta,J,t) = H_o(J) - (eE_o/2)(2J/m\omega_o)^{1/2}\cos(\theta-\Omega t-\phi_o) \qquad (49.4)$$

where $H_o(J)$ is the Hamiltonian of the system without the applied field. Being one-dimensional, the field-free system is integrable.

Now define new variables

$$\psi = \theta - (\Omega t + \phi_o) \qquad (49.5a)$$

$$P = J - J_r \tag{49.5b}$$

where J_r is the value of J for which the nonlinear oscillator frequency $\omega = dH_o/dJ$ equals the driving frequency Ω. It is assumed that P is very small compared with J or J_r, and in the last term in (49.4) the approximation $J \cong J_r$ is used. Then if $H_o(J)$ is expanded up to second order in P we have in place of (49.4) the approximation

$$H_r(\psi,P) = P^2/2M - (eE_o/2)V(J_r)\cos\psi \tag{49.6}$$

with $M \equiv dJ/d\omega$ and $V(J_r) \equiv (2J_r/m\omega_o)^{1/2}$.

The Hamiltonian (49.6) describes a pendulum, and as such is integrable. To make the system nonintegrable, and so to allow the possibility of stochastic excitation, a nonintegrable perturbation must be added. Following Shuryak, Jones and Percival consider a nonlinear coupling between the pumped mode and background modes. In particular, in SF_6 the nearest mode to the pumped (ν_3) mode, denoted $\nu_2+\nu_6$, is about 14.7 cm^{-1} removed from the pumped mode. Jones and Percival suggest that a resonance overlap, and the onset of stochastic excitation, can occur when the resonance halfwidth

$$(\Delta\omega)_r = [2eE_o\omega'(J_r)V(J_r)]^{1/2} \approx 14.7 \text{ cm}^{-1} \tag{49.7}$$

where $\omega' \equiv d\omega/dJ$. The condition (49.7) is the Chirikov condition for resonance overlap. A derivation of (49.7) is not provided in [49.3], and so it may be worthwhile here to digress briefly and derive the expression for $(\Delta\omega)_r$.

Since $M \equiv dJ/d\omega$ we have

$$\Delta\omega = (d\omega/dJ)\Delta J = (1/M)\Delta J = (1/M)\Delta P \tag{49.8}$$

The maximum excursion of P allowed by (49.6) and energy conservation is (see also the discussion leading to (46.7))

$$\Delta P = 2[(MeE_o/2)V(J_r)]^{1/2} \tag{49.9}$$

and so

$$(\Delta\omega)_r = 2(1/M)[(MeE_o/2)V(J_r)]^{1/2} = [2eE_o\omega'(J_r)V(J_r)]^{1/2} \tag{49.10}$$

which is the expression used in (49.7).

To evaluate the field amplitude E_o needed to satisfy (49.7), it is necessary to have a specific model for $H_o(J)$. Jones and Percival use the Morse potential, for which the unperturbed Hamiltonian is

$$H_o = p^2/2m + D_e(1 - e^{-\alpha x})^2 \tag{49.11}$$

and

$$\omega_o = \alpha(2D_e/m)^{1/2} \tag{49.12a}$$

$$\omega(J) = dH_o/dJ = \omega_o - \omega_o^2 J/2D_e \tag{49.12b}$$

The last expression implies that $J_r = 2D_e\Delta/\omega_o^2$, where $\Delta = \omega_o - \Omega$ is the laser detuning. Equation (49.7) then becomes

$$2[(e\alpha E_o/m)(\Delta/2\omega_o)]^{1/2} = 14.7\ \text{cm}^{-1} \tag{49.13}$$

and, using parameter values appropriate to the ν_3 mode of SF_6, Jones and Percival obtain the expression

$$I = 9.5 \times 10^4\ [2(n + \tfrac{1}{2})\omega_o/\Delta]\ \text{W/cm}^2 \tag{49.14}$$

for the threshold intensity. Here the quantum number n arises via the expression $p^2/2m + m\omega_o^2x^2/2 = (n + \frac{1}{2})\hbar\omega_o$, which is used to estimate the maximum x and therefore the parameter α appearing in the Morse potential.

The expression (49.14) obviously depends strongly on the detuning Δ. The dependence of the threshold intensity for dissociation on detuning is not definitely established experimentally, although Ambartsumyan, et al. [47.4] apparently found no dependence in their experiments. Jones and Percival find that the average energy absorbed is independent of the laser intensity, and suggest that this unphysical result is due to the neglect of molecular rotations. In Shuryak's model the average absorbed energy turns out to be proportional to the square root of the fluence, in agreement with some early Russian experiments. The experiments of Judd, et al. [47.2] show a $(\text{fluence})^{2/3}$ dependence of the absorbed energy for a large number of molecules, whereas the model of Section 47 shows a linear dependence of the absorbed energy on the fluence.

Thus far we have described four models of molecular MPE. [47.1, 49.1-49.3] In each case the onset of chaotic behavior leads to stochastic, effectively diffusive growth of energy. These models can account for the high levels of excitation without bottlenecking, and the observed "incoherence" (fluence dependence) of the MPE process. But how realistic are these classical models?

The reason for classical treatments of the MPE problem, of course, is that realistic quantum calculations are too hard. In the quantum calculations one ordinarily has to contend with a very large number of basis states; in a molecule like SF_6 some 10^6 states/cm^{-1} are involved. A number of people, most notably Lamb, [49.4] have advocated classical models. In particular, Lamb notes that in many molecules quantum tunneling frequencies are slow on a time scale for dissociation. Jones and Percival [49.3] note the considerable success of classical theories in other areas of vibrational-rotational spectroscopy.

Of course when the density of states is large we can expect the classical approximation to work best, and the density of states <u>does</u> become large once a polyatomic molecule is pumped above the first few vibrational levels. Thus Bloembergen has suggested that the first few steps of the MPE process can be treated quantum mechanically, while a classical theory should suffice in the "quasicontinuum" regime. [49.5]

There do not appear to be a large number of numerical experiments investigating the classical-quantum correspondence for nonlinear systems. Early studies of Wilson, <u>et</u> <u>al</u>. [49.6-49.8] addressed the classical-quantum correspondence for coupled anharmonic oscillators and for a Morse oscillator with an initial Gaussian wave function. Comparing classical and quantum results for x(t), they find considerable discrepancies between the classical and quantum dynamics after a few cycles of oscillation. Walker and Preston [49.9] have studied a Morse oscillator driven by a sinusoidal force. They find that the classical and quantum predictions for the temporal behavior of average energy and displacement are in fairly good agreement, but that the classical description fails badly for multiphoton resonances. However, the good agreement in the predictions for average energy and displacement is hardly striking after a few tens of cycles.

One intuitively expects the best agreement between the classical and quantum predictions when the time elapsed is small. Based on the energy-time uncertainty relation, for instance, a short time interval implies a large ΔE; then our knowledge of the system, provided by the state vector, does not "resolve" the discrete, quantum nature of the energy spectrum, and the classical and quantum theories can be in excellent agreement. This is well borne out in the computations reported in References [49.6-49.9].

The computational simplicity of classical MPE theories is certainly very appealing, as is the general picture of resonance overlap and the onset of chaos and stochastic excitation. However, there are certain subtle features of the quantum theory of driven nonlinear systems that could not be anticipated in any of the classical models. In particular, it is possible for the stochastic excitation and diffusion predicted classically to be suppressed by a quantum localization effect analogous to Anderson localization. Furthermore it is not at all obvious how classical notions of chaos can carry over into the quantum domain. This brings us to the last, and perhaps most difficult, general topic of these lectures.

50. Quantum Chaos

Suppose we treat a system classically and find its dynamics to be chaotic. Does this classical chaos manifest itself in any way when the system is treated quantum mechanically? In other words, is the quantum behavior of the system fundamentally different from what it would be were the classical dynamics completely regular? Does the quantum dynamics change in a fundamental way when some parameter of the system exceeds a critical value necessary for classical chaos?

These questions form the basis of the subject that has come to be called "quantum chaos." In this section we will survey some approaches to the problem of quantum chaos. The subject is interesting, unsettled, and controversial, with some arguing that there is no such thing as quantum chaos, others defending their criteria for how best to characterize it.

For classical systems there is no controversy about how to define chaotic behavior: classical chaos means very sensitive dependence on initial conditions, and this property can be checked by computing the maximal Lyapunov exponent. In the case of quantum chaos there does not appear to be any such "hard number" that can be computed to unambiguously and incontrovertibly determine whether the system is "chaotic." Indeed the limits imposed by the uncertainty principle on simultaneous measurements of q's and p's make it impossible to talk about "initially close trajectories" in the same way as in classical dynamics. Thus a characterization of quantum chaos must rely on something other than exponentially separating trajectories. At the same time, one would hope that in the classical limit a system exhibiting "quantum chaos" would have the property of very sensitive dependence on initial conditions.

Consider first a system with discrete energy eigenvalues. One can prove a quantum recurrence theorem which Bocchieri and Loinger [50.1] state as follows: "Let us consider a system with discrete energy eigenvalues E_n; if $\psi(t_o)$ is its state vector at the time t_o and ϵ is any positive number, at least one T will exist such that the norm $\|\psi(T) - \psi(t_o)\|$ of the vector $\psi(T) - \psi(t_o)$ is smaller than ϵ."

The idea is very simple. [50.1] The state vector at time t has the form

$$\psi(t) = \sum_{n=0}^{\infty} |c_n| e^{-i(E_n t - \theta_n)} \phi_n \tag{50.1}$$

where the E_n are the (discrete) energy eigenvalues and the ϕ_n are the corresponding eigenvectors. Then

$$\|\psi(T) - \psi(t_o)\| = \sum_{n=0}^{\infty} |c_n|^2 (1 - \cos E_n \tau), \quad \tau \equiv T - t_o \tag{50.2}$$

and, since we can always find an N such that

$$\sum_{n=N}^{\infty} |c_n|^2 (1 - \cos E_n \tau) < \epsilon \tag{50.3}$$

the theorem holds if there is a τ such that

$$\sum_{n=0}^{N-1} |c_n|^2 (1 - \cos E_n \tau) < \epsilon \tag{50.4}$$

The last statement is in fact true, being a feature of the recurrence property of quasiperiodic functions mentioned in Section 5.

This recurrence theorem is the quantum-mechanical analogue of the classical Poincare' recurrence theorem, which states that any initial point $(\vec{q}, \vec{p})$ in the phase space of a system of finite volume will be revisited as closely (and as often) as desired if one waits long enough. But the quantum recurrence theorem is more far-reaching because, whereas nearby points in classical phase space may have quite different recurrence times, there can be many similar quantum states with similar recurrence times. [50.2]

Hogg and Huberman [50.3] have proven, using arguments similar to Bocchieri and Loinger, that "under any time-periodic Hamiltonian, a nonresonant, bounded quantum system will reassemble itself infinitely often in the course of time." According to them, "This in turn implies that no strict quantum stochasticity is possible, a result which disagrees with recent predictions." Their proof is mathematically correct, but in practical terms one can object that, as in the classical recurrence theorem, the recurrence times may be exceedingly large. Peres, for instance, has considered a simple example in which the recurrence time is greater than the age of the universe, and concludes that the recurrence argument against quantum chaos is of no practical concern unless there happens to be only a small number of incommensurate energy levels. [50.4] Thus it does not appear that recurrence arguments of this type can be used to generally rule out quantum chaos, any more than the Poincare' theorem can be invoked to argue against the possibility of classical chaos.

It may be, as first suggested in a seminal paper by Percival, [50.5] that the energy eigenvalues themselves become "irregular" when the corresponding classical system becomes chaotic. This is one of the more frequently used criteria for quantum chaos, and we therefore devote the following section to a discussion of eigenvalue distributions.

A large number of studies have focused on aspects of quantum <u>ergodicity</u>. (We should recognize, of course, that even in the classical limit the concept of ergodicity generally represents a weaker degree of "disorder" than chaos.) The early work of Nordholm and Rice [50.6] dealt with a "phase space" consisting of a zeroth-order basis of state vectors, and defined an ergodic state as one that uniformly overlaps the basis states. Thus an ergodic stationary state of the Hénon-Heiles system, according to them, is one consisting of a "nonlocalized" superposition of harmonic-oscillator stationary states. Noid, <u>et al</u>. [50.7] find in a study of a system of the Hénon-Heiles type (but with a 2:1 rather than 1:1

resonance) that wave functions seem to be localized when the classical motion is quasiperiodic, but more nonlocalized when the classical motion is chaotic.

Hutchinson and Wyatt [50.8] employed the Wigner distribution

$$\Gamma(\vec{q},\vec{p}) = (1/\pi\hbar)^N \int d\vec{z}\psi^*(\vec{q}+\vec{z})\psi(\vec{q}-\vec{z})e^{2i\vec{p}\cdot\vec{z}/\hbar} \tag{50.8}$$

to study quantum ergodic properties of the Hénon-Heiles system. (Recall that $\Gamma(\vec{q},\vec{p})$ is something like a joint probability distribution in $\vec{q},\vec{p}$, but of course it is impossible in quantum mechanics to have such a joint probability for conjugate variables like $\vec{q}$ and $\vec{p}$. Thus, although $\Gamma(\vec{q},\vec{p})$ is real, it is not positive-definite. Nevertheless it as close as one can come to a quantum-mechanical analogue of a classical phase-space distribution. For instance, one can obtain the probability distribution in $\vec{q}$ by integrating (50.8) over all $\vec{p}$.) Hutchinson and Wyatt consider the long-time average of the Wigner distribution for the Hénon-Heiles system, and characterize ergodicity in terms of the degree of uniformity of this long-time average phase space "density." They find a degree of ergodic behavior in this sense when the energy is increased, as is the case classically. However, the critical energy for this ergodic behavior is higher in the quantum case, and the degree of ergodicity is not as pronounced as classically. Hutchinson and Wyatt also argue that for such ergodic properties it may not suffice to consider only the configuration-space density $|\psi(\vec{q})|^2$: they find that the long-time average of $\Gamma(\vec{q},\vec{p})$ can be uniform while $|\psi(\vec{q})|^2$ is fairly localized.

In order to have some analogue of the classical separation of initially close trajectories, Hutchinson and Wyatt consider the separation measure

$$D(t) = 1 - |\langle\Psi(t)|\Psi'(t)\rangle|^2 \tag{50.9}$$

where $|\Psi(t)\rangle$, $|\Psi'(t)\rangle$ are the state vectors for two potentials differing by a small perturbation. They find that the long-term values of D(t) do appear to increase with energy. However, the separation increases only gradually, not with the abrupt transition to irregularity that occurs classically. Furthermore D(t) was found to be very sensitive to the strength of the perturbation, and did not depend sensitively on initial conditions. Also considered was the initial-state survival probability defined by

$$S(t) = |\langle\Psi(0)|\Psi(t)\rangle|^2 \tag{50.10}$$

Hutchinson and Wyatt found, in agreement with Brumer and Shapiro, [50.9] that with increasing energy the behavior of S(t) seemed to become more "irregular." However, they find this sort of behavior even in an integrable system for a few high-energy states, even though the long-time average of the Wigner distribution in such cases indicates nonergodic behavior. We refer the reader to the paper of Hutchinson and Wyatt for a very clear discussion of these points, and also to the discussion by Heller regarding the survival probability criterion. [50.10]

Feit and Fleck [50.11] have solved numerically the time-dependent Schrödinger equation for the Hénon-Heiles system, using Gaussian wave packets (coherent states) as initial wave functions. They characterized "chaotic" behavior of the quantum system in terms of (a) phase-space trajectories defined by expectation values, (b) the survival probability (50.10), and (c) the uncertainty product $\Delta x \Delta y \Delta p_x \Delta p_y$. Although these three tools gave consistent interpretations of the quantum dynamics, it was found that the corresponding classical dynamics "is not a reliable guide to regular or chaotic behavior in the quantum mechanical system." [50.11]

Heller has addressed questions of quantum chaos in the context of intramolecular dynamics and energy transfer. [50.10] He argues, among other things, that whether a molecule should be considered "stochastic" or "nonstochastic" depends on the specific experimental situation. He presents evidence to the effect that, if the overlap integrals $|\phi_n|^2$ of an initial (nonstationary) state with the eigenstates of the final potential surface vary smoothly with n, then the dynamics may be considered stochastic; if they fluctuate, the dynamics may be considered nonstochastic. Here "stochastic" means essentially "ergodic" : the coordinate-space probability is distributed smoothly over energetically allowed regions. In a molecule the $|\phi_n|^2$ are Franck-Condon factors, and so Heller's criterion can in principle be used to determine experimentally whether a system is stochastic or nonstochastic. As recognized by Heller, however, such criteria for quantum chaos will necessarily change with the choice of test states.

Quantum chaos in the context of intramolecular energy transfer has also been discussed by Reinhardt. [50.12] He argues that for many chaotic systems the dynamics are similar to quasiperiodic dynamics for times that are not too large, allowing the introduction of approximate constants of the motion. This leads to the notion of "vague tori" and semiclassical quantization on these vague tori. The vague tori are taken to define the modes of the molecule in the classically chaotic regime, and intermode energy transfer involves transitions between vague tori. Reinhardt criticizes the characterization of quantum chaos in terms of energy level distributions (see the following section) on the grounds that "any level spacing whatsoever may be obtained from a suitably chosen integrable Hamiltonian," a result which he attributes to C.R. Holt. This objection is discussed in the following section.

A different approach to quantum chaos, which makes no explicit reference to classical chaos, has been taken by Peres. [50.13] According to him, "the hallmark of quantum chaos is that simple dynamical variables (position, momentum, etc.) are represented by pseudorandom matrices when the Hamiltonian is diagonal" (i.e., in the energy representation). Since some discussion is required of the "reasonable" operators to which the definition applies, and of the notion of a "pseudorandom" matrix, we simply refer the reader to the clear discussion by Peres. Peres, et al. [50.13] illustrate these ideas by considering the examples of the Hénon-Heiles system and a pair of nonlinear coupled rotators. It is argued that the proposed definition is reasonable in that the quantum analogue of a classically chaotic system is "likely" to be chaotic in the sense of the definition.

Obviously we cannot do justice to the work of the authors above in such a brief overview. However, the papers cited in this section are very clearly written, and the reader who has made it this far into this volume should have little difficulty reading them.

The reader can perhaps sense even from this brief and very incomplete survey that the field at present is based largely on qualitative criteria fraught with caveats and exceptions to the rules. It appears, however, that a consensus is gradually emerging that quantum effects act to smooth out the degree of "disorder" possible in classical systems. This smoothing effect is perhaps best understood in terms of the summing over classical paths implicit in the quantum dynamics, i.e., in the path-integral formulation.

51. Regular and Irregular Spectra

In Section 36 we noted that the classical action is an adiabatic invariant and therefore can, according to the old quantum theory, be quantized. Let us briefly review some elementary aspects of the Bohr-Wilson-Sommerfeld quantization rule. In the limit $\hbar \to 0$ (see standard quantum mechanics texts for a more precise formulation of this classical limit) we can replace (36.17) by $\nabla S = \pm\sqrt{2m(E-V(x))} = \vec{p}$. If the wave function (36.15) is to be single-valued we must require that the difference ΔS between any two values of S be $n(2\pi\hbar) = nh$, where n is an integer. Thus we can write

$$\Delta S(\vec{x}) = \oint \nabla S\cdot d\vec{x} = \oint \vec{p}\cdot d\vec{x} = nh \qquad (51.1)$$

where the line integral is over any closed curve that begins and ends at $\vec{x}$.

More generally we can write

$$\psi(\vec{x}) = A(\vec{x})e^{iS(\vec{x})/\hbar} \qquad (51.2)$$

in place of (36.15), and the single-valuedness of ψ then demands that

$$\Delta S = [n + \frac{i\Delta \log A}{2\pi}]h = \oint \vec{p}\cdot d\vec{x} \qquad (51.3)$$

if we allow the possibility that both A and S might be multivalued. If $\Delta\log A = -i\pi$, for instance, then (51.3) implies that $\Delta S = (n + \frac{1}{2})h$ instead of nh, and this is the form of the semiclassical quantization rule that gives the correct energy levels for the harmonic oscillator. (Section 36) More generally still, we can replace (51.2) by [51.1]

$$\psi(\vec{x}) = \sum_k A_k(\vec{x})e^{iS_k(\vec{x})/\hbar} \qquad (51.4)$$

Here and above $\vec{x}$ can denote the full set of coordinates $q_1, q_2, \ldots, q_N$. The single-valuedness of ψ then leads to the generalization

$$\sum_j \oint p_j dq_j = [n + (i/2\pi)\oint \nabla\log A\cdot ds]h \qquad (51.5)$$

of (51.3), where ds is the differential element of length in the configuration space. When A is single-valued and the system is separable, so that $p_j = p_j(q_j)$, (51.5) reduces to the Bohr-Wilson-Sommerfeld quantization rule. Einstein's generalization (1917) removes the restriction that the system be separable, although integrability is required. (Einstein was attracted to this modification because of the fact that $\sum_j p_j dq_j$ is invariant under canonical transformations. [51.2] In Section 36 we noted that, by addressing the question of the (semiclassical) quantization of nonseparable systems, Einstein was in a sense inaugurating the subject of quantum chaos.) The modern version of the EBK (Einstein-Brillouin-Keller) quantization rule is

$$J_k = \oint p_k dq_k = (n_k + \alpha_k/4)h \tag{51.6}$$

where α_k is an integer called the "Maslov index." For a good review of the semiclassical theory we refer the reader to Percival. [51.2]

Percival [50.5] has invoked the correspondence principle to suggest that, in the semiclassical limit, the energy level spectrum of a quantum system consists of a regular part and an irregular part. Regular regions of classical phase space give rise to the regular spectrum, with the actions J_k given by EBK quantization conditions; the N J_k's correspond to classical motion on an invariant N-torus.

Irregular regions of phase space give rise to the irregular part of the spectrum. In these regions the semiclassical assignment of quantum numbers is complicated by the fact that invariant tori have been destroyed. Percival suggests that in this case the energy level distribution could appear to be random, and small changes in a nonlinear perturbation could produce large changes in the spectrum compared with the regular case.

Percival's conjecture was first tested with the Hénon-Heiles system by Pomphrey. [51.3] The eigenvalues of the Schrödinger equation were obtained for the potential $V(x,y) = \frac{1}{2}(x^2 + y^2) + \alpha(x^2 y - y^3/3)$, and the sensitivity of the spectrum to the parameter α was investigated by computing the second differences

$$\Delta_i = |[E_i(\alpha+\Delta\alpha) - E_i(\alpha)] - [E_i(\alpha) - E_i(\alpha-\Delta\alpha)]| \tag{51.7}$$

for each eigenvalue. The dissociation energy $D = 1/6$ of the Henon-Heiles model with $\alpha = 1$ becomes $D = 1/6\alpha^2$ for $\alpha \neq 1$; Hénon and Heiles found that invariant tori persist at energies up to about 0.68D. In the computations of Pomphrey it was found that for energies less than about 0.74D the second differences (51.7) are very small. But for larger energies, eigenvalues were found that were very sensitive to small changes $\Delta\alpha$ in the perturbation. Pomphrey's results, based on α values ranging from 0.086 to 0.090 in increments of 0.001, and the computation of 71 eigenvalues for each α, are shown in Figure 51.1a.

In the Hénon-Heiles model it is found that the relative area A_I of the surface of section covered by irregular, chaotic trajectories is given approximately by

$$A_I(E) = 0 \qquad E < 0.68D$$

$$= 3.125(E/D) - 2.125 \qquad E > 0.68D \tag{51.8}$$

Pomphrey compared the total relative area covered by irregular trajectories up to energy E, namely

$$I(E) = \int_0^E A_I(E)dE \tag{51.9}$$

to the quantity

$$S(E) = (1/D)\sum_i^E n_I(E_i)\langle\Delta E_i\rangle \tag{51.10}$$

determined by the energy spectrum computed from the Schrödinger equation. Here $n_I(E_i) \equiv 1$ if E_i is very sensitive to small changes in the perturbation, $n_I(E_i) \equiv 0$ otherwise, and $\langle\Delta E_i\rangle$ is defined as the

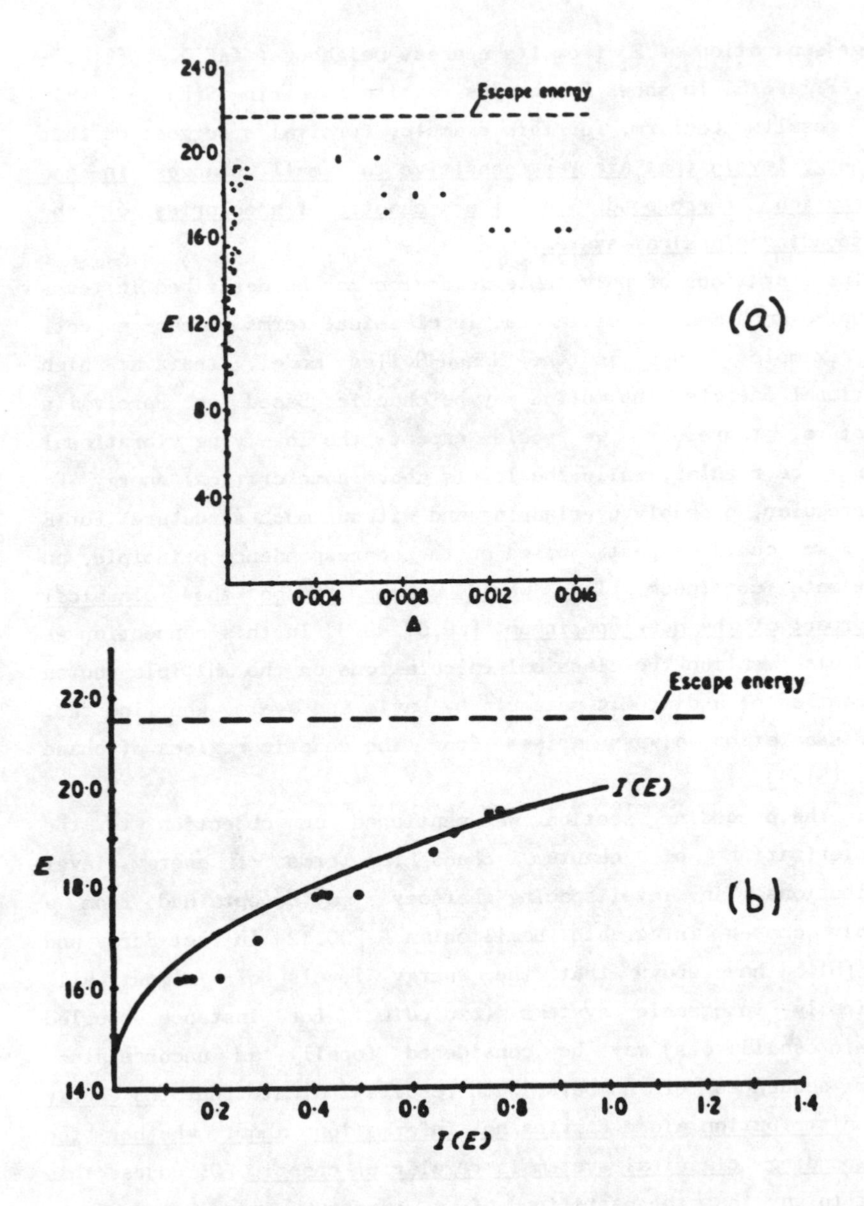

Figure 51.1 (a) The second differences (51.7) as a function of E, and (b) comparison of I(E) (solid curve) with S(E) (dots) computed by Pomphrey. From N.Pomphrey, Reference [51.3], with permission.

average separation of E_i from its nearest neighbors: $\langle \Delta E_i \rangle = \frac{1}{2}(E_{i+1} - E_{i-1})$. Figure 51.1b shows Pomphrey's results comparing S(E) to I(E). These results confirm, for this example, Percival's suggestion that the energy levels that are very sensitive to small changes in the perturbation correspond to the chaotic trajectories of the corresponding classical system.

The vibrations of polyatomic molecules may be described in terms of coupled anharmonic oscillators. In classical terms we can expect, from examples such as the Hénon-Heiles model, that at high vibrational energies the motion may be chaotic. Based on Percival's conjecture, therefore, we would expect the low-lying vibrational levels to be regular, while the levels above some critical energy E_c are irregular, probably overlapping and without much structure. For $E \gg E_c$ we could expect, based on the correspondence principle, an approximate continuum. This would appear to be the classical counterpart of the quasicontinuum. [50.5, 49.3] In this connection we should also mention the classical calculations on the multiple-photon dissociation of a diatomic molecule by Davis and Wyatt, who find that the dissociation always arises from the chaotic regions of phase space. [51.4]

In the preceding section we mentioned an objection to the characterization of quantum chaos in terms of energy level distributions: "any level spacing whatsoever may be obtained from a suitably chosen integrable Hamiltonian." [50.12] In fact Berry and Tabor [51.5] have shown that the energy levels of "almost all" classically integrable systems (excluding, for instance, coupled harmonic oscillators) may be considered locally as uncorrelated, random sequences of numbers. Such results indicate that the energy level distribution alone carries no information about whether the corresponding classical system is regular or chaotic. Of course this is not to say that the partition of a spectrum into regular and irregular portions is invalid, nor that the onset of classically chaotic behavior does not produce in the quantum system a transition from a regular to an irregular spectrum. It simply means that one cannot deduce from a table of energy levels alone whether the quantum dynamics is "chaotic." However, a strong sensitivity of these levels to the value of some anharmonicity parameter does appear to provide evidence for a different sort of quantum behavior when the classical motion is chaotic, as in the numerical experiments of Pomphrey. [51.3]

It must also be mentioned that energy levels and spacings may themselves be more orderly when the classical motion is chaotic than when it is regular, [51.6 - 51.8] in the sense that adjacent levels "repel" each other and the spectrum has a "rigid" character. It has been predicted that in the chaotic regime the distribution of successsive levels is peaked at a nonzero value, whereas in integrable cases there is a clustering of levels and a maximum of the distribution at zero separation. [51.5, 51.9 - 51.11]

Confirmation of this prediction has been reported by McDonald and Kaufman. [51.12] They consider a "stadium" problem in which a particle is confined to move within an area consisting of two semicircles connected by straight-line segments. If the straight-line segments have zero length, so that the stadium reduces to a circle, then the classical motion is integrable, whereas any deviation from the perfect circle makes the classical motion chaotic. McDonald and Kaufman solved the time-independent Schrödinger equation (in this case the Helmholtz equation with the boundary condition $\psi = 0$ at the walls) for the two cases. Figure 51.2 shows the distribution of energy level spacings for the circle, compared with the distribution found for the case in which the straight-line segments have length equal to the diameter of the semicircles. In the stadium case small spacings are seen to be improbable, and there is evidently a level "repulsion" as predicted. The distribution of spacings is also well approximated by the Wigner distribution ($X\exp[-\pi X^2/4]$) predicted theoretically. It might be noted parenthetically that some of the more recent ideas concerning energy level statistics have their origins in nuclear theory.

It seems fair to say, based on such results, that there is such

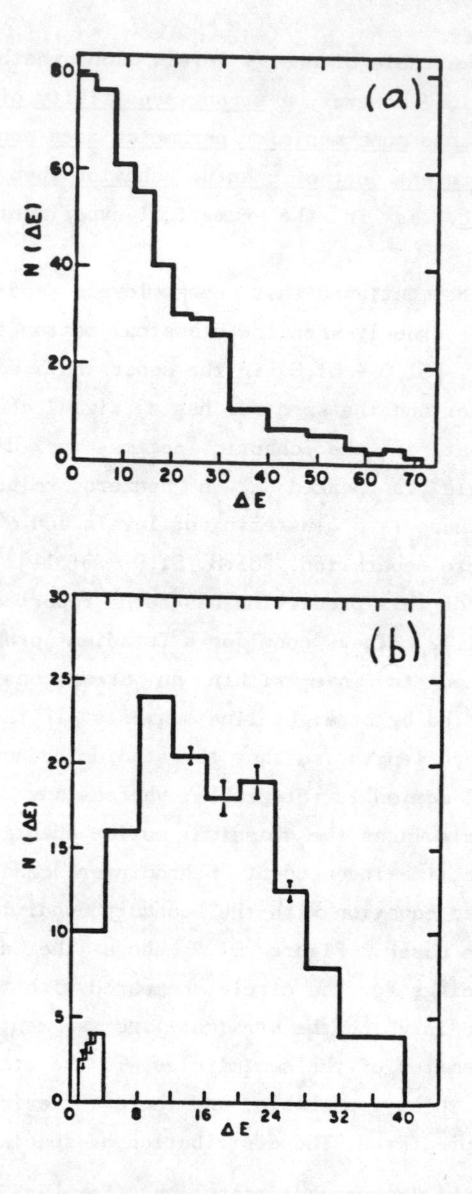

Figure 51.2 Distribution of eigenvalue spacings ΔE computed by McDonald and Kaufman for (a) the circular stadium and (b) a noncircular stadium. In each case the eigenvalues were found for wavefunctions of a particular ("odd-odd") parity. From S.W. McDonald and A.N. Kaufman, Reference [51.12], with permission.

a thing as "quantum chaos," in the sense that wave functions, [51.10, 51.12] matrix elements, [51.7] and the sensitivity of energy eigenvalues to changes in a perturbation [50.5, 51.3] can all reflect the chaotic character of the corresponding classical motion. The reader wishing to probe more deeply into these matters is referred especially to the papers of Berry on the properties of semiclassical wave functions.

52. The Kicked Two-State System

Systems "kicked" periodically by delta-function impulses are easy to compute with because the integration of equations of motion reduces to the iteration of a discrete mapping. In Section 46 we discussed the classical kicked pendulum, and later we will treat the kicked pendulum quantum mechanically. In this section we will consider the very simple example of a kicked two-state system. [52.1] In spite of its simplicity, this model sheds some light on what kinds of behavior are possible in <u>driven</u> quantum systems.

Consider first the general problem of a quantum system described by the Hamiltonian

$$H = H_o + A(x)F(t) \sum_{-\infty}^{\infty} \delta(t/T-n) \tag{52.1}$$

Let $|\Psi(k)\rangle$ be the state vector just prior to the $k\underline{th}$ δ-function impulse. Just after the $k\underline{th}$ kick the state vector is $\exp[-iA(x)F(kT)T/\hbar]|\Psi(k)\rangle$, and between kicks the free evolution of the state vector is governed by the time evolution operator $\exp[-iH_o t/\hbar]$. Thus

$$|\Psi(k+1)\rangle = e^{-iH_o T/\hbar} e^{-iA(x)F(kT)T/\hbar} |\Psi(k)\rangle \tag{52.2}$$

Writing

$$|\Psi(k)\rangle = \sum_m a_m(k)|\phi_m\rangle \tag{52.3}$$

where $H_o|\phi_m\rangle = E_m|\phi_m\rangle$, we obtain from (52.2) the quantum map

$$a_n(k+1) = \sum_m V_{nm}(k)a_m(k) \tag{52.4a}$$

with

$$V_{nm}(k) \equiv \langle\phi_n|e^{-iA(x)F(kT)T/\hbar}|\phi_m\rangle e^{-iE_nT/\hbar} \tag{52.4b}$$

In the case $F(t)$ = constant, in which we have a periodically kicked quantum system, V_{nm} is independent of k.

The case of a two-state system is particularly simple. In this case $H_o = (\hbar\omega_o/2)\sigma_z$, where σ_z is the Pauli spin-1/2 matrix in the standard representation and ω_o is the transition frequency. For the perturbation we take $A(x) = -dE\sigma_x = \hbar\Omega\sigma_x$. For a two-state "atom" in an electric field, d is the transition dipole moment, E is the amplitude of the electric field, and Ω is the Rabi frequency. For such a perturbation we have

$$e^{-iA(x)F(kT)T/\hbar} = e^{i\Omega F(kT)T\sigma_x} = \cos[\Omega(k)T] + i\sigma_x\sin[\Omega(k)T] \tag{52.5}$$

with $\Omega(k) \equiv \Omega F(kT)$. Defining $c_n(k) = a_n(k)e^{ikE_nT/\hbar}$, n = 1, 2, we may write (52.4) in the form

$$c_1(k+1) = \cos[\Omega(k)T]c_1(k) + i\sin[\Omega(k)T]e^{-ik\omega_oT}c_2(k) \tag{52.6a}$$

$$c_2(k+1) = i\sin[\Omega(k)T]e^{ik\omega_oT}c_1(k) + \cos[\Omega(k)T]c_2(k) \tag{52.6b}$$

Consider first the case of purely periodic kicking, i.e., $\Omega(k) \to \Omega$ independent of k. In Figure 52.1 we plot the upper-state probability $|c_2(k)|^2$ versus the kick number k for $\omega_o T = 2\pi$ and $\Omega T = 0.60$. (In each case we start with the initial condition $c_1(0) = 1$ and therefore $c_2(0) = 0$.) It is interesting to compare such results with the solution of the Schrödinger equation for a two-state atom in a monochromatic field $E = E_o \cos\omega t$. In Bloch form this Schrödinger equation takes the form (see equations (22.15))

$$\dot{x} = -\omega_o y \tag{52.7a}$$

$$\dot{y} = \omega_o x + \bar{\Omega} z \cos\omega t \tag{52.7b}$$

$$\dot{z} = -\bar{\Omega} y \cos\omega t \tag{52.7c}$$

where $\bar{\Omega} \equiv 2dE_o/\hbar$. The atom is most strongly driven when the resonance condition $\omega \cong \omega_o$ is met; driving frequencies far removed from resonance do not have much of an effect by comparison. Therefore we can replace $\cos\omega t$ in (52.7) by

$$\tfrac{1}{2}\sum_{-\infty}^{\infty} e^{in\omega t} = \tfrac{1}{2}\sum_{-\infty}^{\infty} e^{2\pi int/T} = \tfrac{1}{2}\sum_{-\infty}^{\infty} \delta(t/T - n) \tag{52.8}$$

where $T = 2\pi/\omega$. We expect the mapping (52.6) to approximate well the solution of the Bloch equations whenever $\omega \cong \omega_o$ and $\bar{\Omega} = 2\Omega$. In Figure 52.1 we also show the solution of the Bloch equations (52.7) for the upper-state probability $\frac{1}{2}(1+z)$, assuming $\omega = \omega_o$ (i.e., $\omega_o T = 2\pi$) and $\bar{\Omega}T = 1.2$. The solution is quite complicated (but not chaotic!), and we plot only the values at times $t = nT$ in order to compare directly with the solution of the discrete mapping. It is seen, as expected, that the mapping provides an excellent approximation to the solution

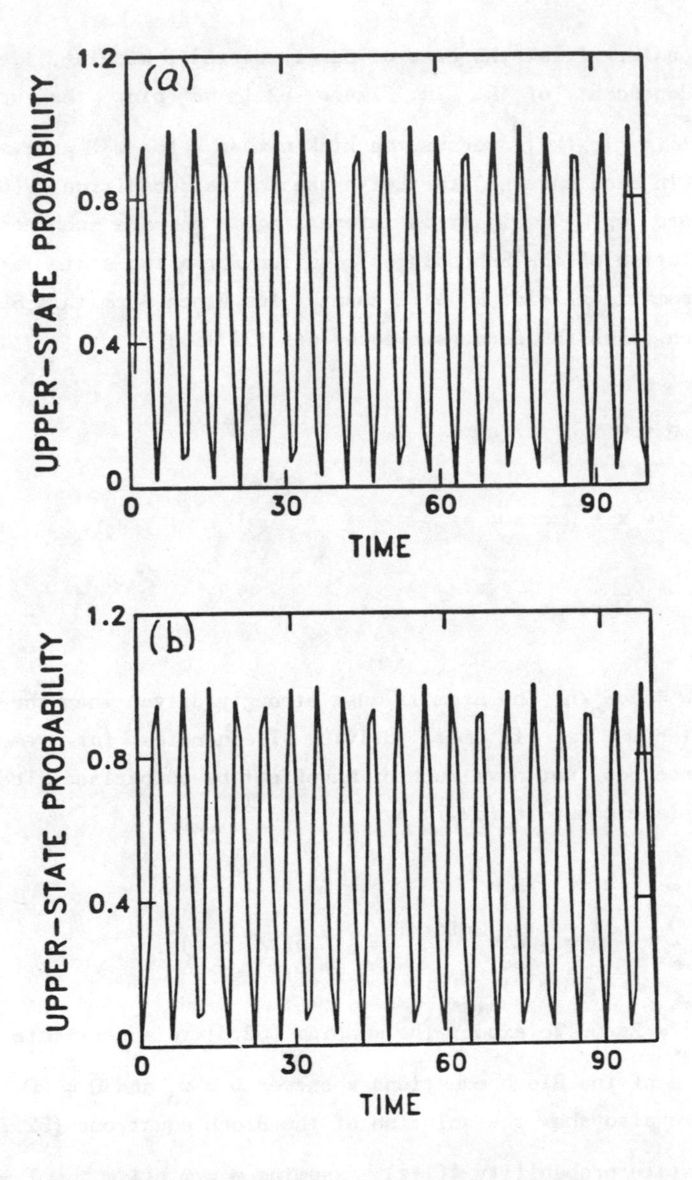

Figure 52.1 (a) Solution of (52.6) for the upper-state probability $|c_2(k)|^2$ for the case $\omega_o T = 2\pi$, $\Omega T = 0.60$, and $c_1(0) = 1$.
(b) Solution of the Bloch equations (52.7) for the upper-state probability $\frac{1}{2}(1+z)$ for the case $\omega = \omega_o$, $\bar{\Omega}T = 1.2$, and $z(0) = -1$.

of the Bloch differential equations (52.7). The agreement is best for small values of ΩT, where the effects of all the overtone frequencies implicit in the discrete mapping are small. For small ΩT the mapping is also accurate considerably far from resonance.

Consider the autocorrelation function of the state vector, which we define by

$$C(\tau) = \lim_{T\to\infty}(1/T) \int_0^T dt\langle\Psi(t)|\Psi(t+\tau)\rangle \tag{52.9}$$

For the kicked two-state system we consider

$$|C(k)| = \lim_{N\to\infty}(1/N)\left|\sum_{n=0}^{N} [c_1^*(n)c_2(n+k) + c_2^*(n)c_1(n+k)e^{-ik\omega_o T}]\right| \tag{52.10}$$

We recall from Sections 6 and 47 that one of the consequences of chaos is a decay of correlations. Numerical estimates indicate that for a periodically kicked two-state system the correlations are quasiperiodic rather than decaying. And of course, since the mapping (52.6) is linear, its iterates cannot evolve chaotically in the sense of a positive Lyapunov exponent.

However, we find that $|C(\tau)|$ can decay rapidly in the case of a <u>quasi</u>periodically kicked two-state system. Consider, for instance, the kicking with $F(t) = \cos\omega' t$, in which case

$$\Omega(k) = \Omega\cos(\omega' kT) = \Omega\cos(2\pi k\omega'/\omega) = \Omega\cos(2\pi kx) \tag{52.11}$$

where $x \equiv \omega'/\omega$ is the ratio of the two driving frequencies. A rational value of x means the two frequencies are commensurate. In such cases we obtain nondecaying, quasiperiodic autocorrelations. When x is irrational, however, the autocorrelations are found to decay rapidly when ΩT is large. The correlations do not go exactly to zero; there are small but finite correlations even for very large values of τ, as well as occasional peaks as high as around 0.4 in $|C(\tau)|$. But the behavior of the autocorrelation function is dramatically different from the case of rational x. In Figure 52.2 we plot $|c_2(k)|^2$ and $|C(k)|$ for $\Omega T = 500$ and an irrational value of x. (For machine computations we can take x to be the ratio of two large primes. Following Pomeau, et al. [52.2] we present here results for x = 4637/13313 as our "irrational" frequency ratio.) The decay of $|C(k)|$ suggests "chaotic" time evolution of the state vector, and moreover the power spectrum of the time series of Figure 52.2a is broadband, as shown in Figure 52.3. (This is not surprising, since the power spectrum and autocorrelation function are related by Fourier transformation.) These indications of quantum chaos are consistent with the conclusions of Pomeau, et al., [52.2] who integrated the Bloch equations for a bichromatic driving field and found decaying correlations and broadband power spectra.

In terms of the three real Bloch variables

$$x(k) = a_1(k)a_2^*(k) + a_1^*(k)a_2(k) \tag{52.12a}$$

$$y(k) = i[a_1^*(k)a_2(k) - a_1(k)a_2^*(k)] \tag{52.12b}$$

$$z(k) = |a_2(k)|^2 - |a_1(k)|^2 \tag{52.12c}$$

the mapping (52.6) takes the form

$$x(k+1) = x(k)\cos(\omega_o T) - y(k)\sin(\omega_o T)\cos[2\Omega(k)T]$$

$$- z(k)\sin(\omega_o T)\sin[2\Omega(k)T] \tag{52.13a}$$

$$y(k+1) = x(k)\sin(\omega_o T) + y(k)\cos(\omega_o T)\cos[2\Omega(k)T]$$

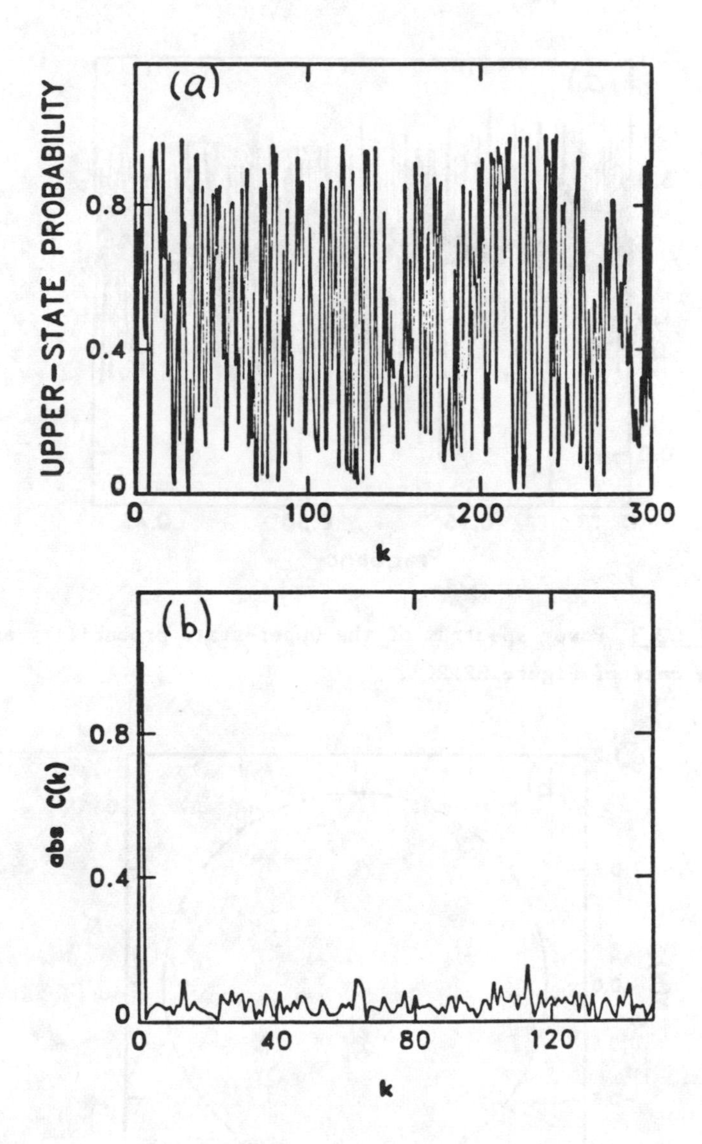

Figure 52.2 (a) Upper-state probability and (b) absolute value of the autocorrelation function $C(\tau)$ for $\Omega T = 500$, $\omega_o T = 3.0$, x irrational.

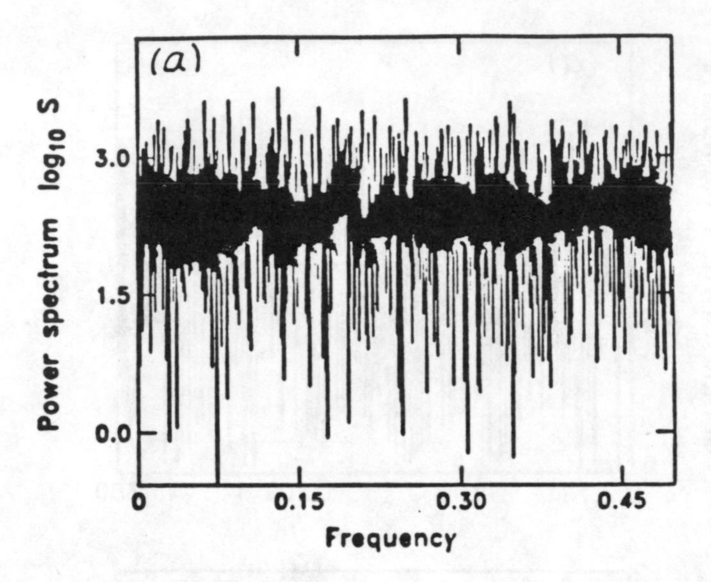

Figure 52.3 Power spectrum of the upper-state probability amplitude for the case of Figure 52.2.

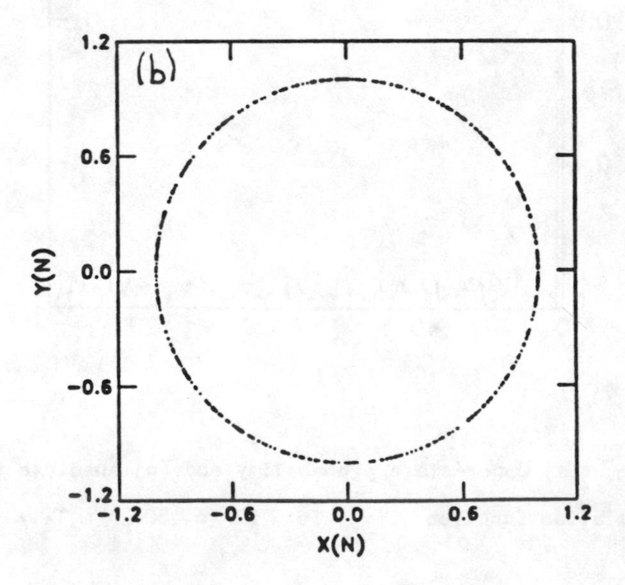

Figure 52.4 Surface of section in the xy plane for the case of Figure 52.2.

$$+ z(k)\cos(\omega_o T)\sin[2\Omega(k)T] \qquad (52.13b)$$

$$z(k+1) = z(k)\cos[2\Omega(k)T] - y(k)\sin[2\Omega(k)T] \qquad (52.13c)$$

In the limit $T \to 0$ we recover the Bloch equations (52.7). Since the motion is confined by conservation of probability to the Bloch sphere $x^2(k) + y^2(k) + z^2(k) = 1$, (52.13) is basically a mapping relating the azimuthal and polar angles on the Bloch sphere from kick to kick. The form (52.13) is convenient for computing the surface of section (Poincare' map) of points x(k), y(k) such that $z(k+1) = 0$ and $z(k) > 0$ (for instance). This provides a mapping on a circle, i.e., a mapping on the equator of the Bloch sphere.

In Figure 52.4 we show the results of this Poincare' map for the case of Figure 52.3. The points appear to the eye to fill in the circle in an erratic sequence; a similar result is obtained for the surface of section in the xz plane. The surface of section in the yz plane fills only the semicircle $y^2 + z^2 = 1$, $y > 0$. This is a simple consequence of the form of the Bloch equations: $\dot{x} < 0$ implies $y > 0$. In other words, the projection in the xy plane of trajectories on the Bloch sphere must always spiral in a counterclockwise sense.

For an intuitive appreciation of the origin of this "chaotic" behavior, consider the factors $\cos[\Omega(k)T] = \cos[\Omega T\cos(2\pi kx)]$ and $\sin[\Omega T\cos(2\pi kx)]$ appearing in the Schrödinger equation (52.6). For large values of ΩT, and irrational values of x, these functions vary erratically with k. We plot $\cos[\Omega T\cos(2\pi kx)]$ in Figure 52.5, together with its autocorrelation function, for the case of Figure 52.3. Note the rapid decay of the autocorrelation function, which does not occur at smaller values of ΩT, nor at rational values of x. For large ΩT and irrational x, therefore, the probability amplitudes are being driven "chaotically" and therefore evolve "chaotically." It is also worth noting that the angles $\theta_k = 2\pi kx$ satisfy the "circle map" $\theta_{k+1} = \theta_k + x$, and it may be proven that for irrational x the circle is filled in densely (ergodically).

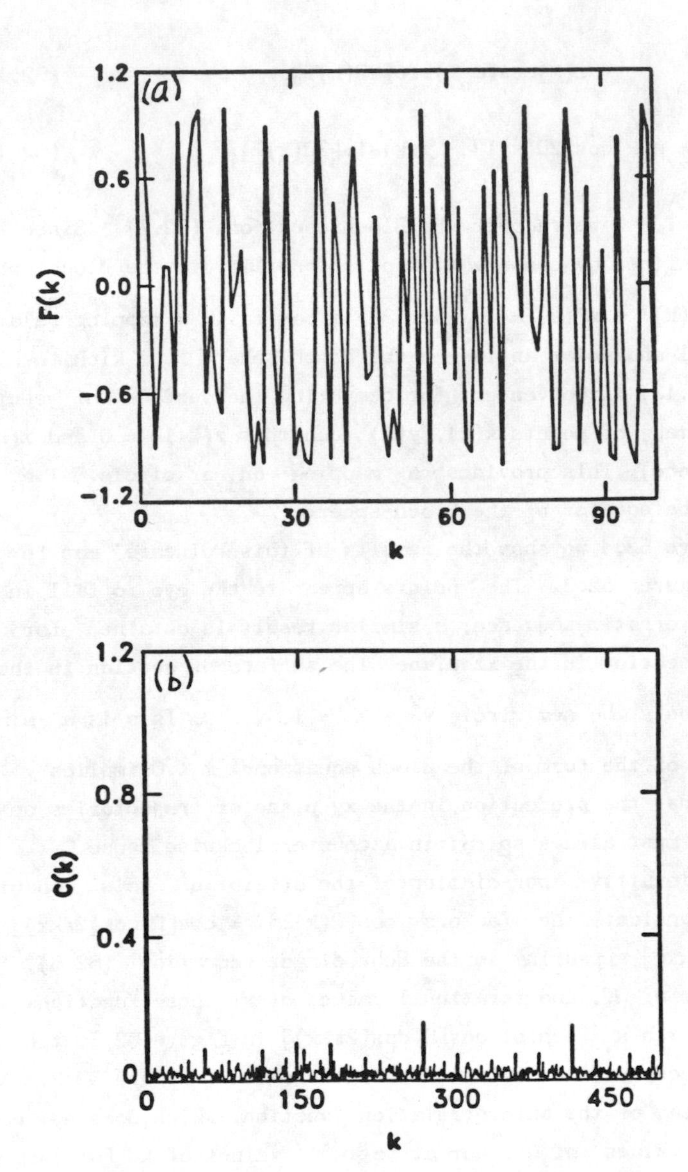

Figure 52.5 (a) The function $\cos[\Omega\cos(2\pi kx)]$ and (b) its autocorrelation function for the case of Figure 52.2.

We can summarize our results for the quasiperiodically kicked two-state system as follows. For commensurate driving frequencies the time evolution is regular, in the sense that the state vector remains strongly correlated in time and the motion on the Bloch sphere is nonergodic. For incommensurate driving frequencies and large Rabi frequencies, however, the time evolution is "chaotic" in the sense that (a) the autocorrelation function of the state vector decays, (b) the power spectrum of the state vector is broadband, and (c) the motion of the state vector on the Bloch sphere is ergodic. The properties (a) - (c) appear to be reasonable criteria for "quantum chaos" of externally driven systems, in that they also characterize classically chaotic behavior. In the classical case, of course, there is an unambiguous criterion for chaos - a positive Lyapunov exponent, implying very sensitive dependence on initial conditions. <u>This hallmark of classical chaos is not shared by bounded quantum systems</u>: a perturbation of the state vector cannot grow exponentially because the state amplitudes satisfy continuous, linear equations.

In Section 44 we noted that in truly chaotic systems we lose "memory" of initial conditions in the sense that a backward integration of the equations of motion will not generally reproduce the initial conditions. We have confirmed numerically that the backward mapping for the kicked two-state system does reproduce the initial state, thus providing another example of the greater degree of stability enjoyed by quantum systems.

The kicked two-state system is a very simple example, but it illustrates several important points:

(1) Ergodicity and chaos are not the same. In this example the dynamics can be ergodic on the Bloch sphere but not chaotic in the sense of a positive Lyapunov exponent.

(2) Quantum systems can show features like broadband power spectra and decaying correlations, which are consequences of chaos in classical systems, without being chaotic in the classical sense.

These features may be among the strongest possible manifestations of any sort of "quantum chaos".

(3) Quasiperiodically driven quantum systems can display a qualitatively different type of behavior than periodically driven systems. This will be exploited in our discussion of the kicked pendulum in Section 54.

53. Chaos in the Jaynes-Cummings Model

The Bloch equations (52.7) describe a two-state atom in an applied monochromatic field $E_0 \cos\omega t$. These equations may be derived by taking expectation values of fully quantum-mechanical Heisenberg equations of motion, provided the field is assumed to be in a Glauber coherent state. [53.1] The discrete mapping of the preceding section for a kicked two-state system may likewise be obtained if the field modes at the different frequencies $n\omega$ are assumed to be described by coherent states. Thus, under the assumption of a coherent driving field, these equations are fully quantum-mechanical, with no semiclassical approximation of treating the applied field classically.

Consider now a two-state atom contained in a resonant cavity allowing only a single field mode. In this case we describe the atom and the radiation field self-consistently. That is, the field determines the two-state atom's dynamics through the Bloch equations, while the atom acts as a (dipole) source for the field in the Maxwell wave equation. This atom-field system, called the Jaynes-Cummings model, is one of the fundamental theoretical paradigms of quantum optics. [53.2] In the Jaynes-Cummings model the field dynamics is determined by the source atoms (and vice versa), and so in this case a classical treatment of the field is only an approximation, though it may be quite a good one.

One of the virtues of the Jaynes-Cummings model is that, within the rotating-wave approximation (RWA), it is exactly solvable, regardless of whether the field is treated classically or quantum mechanically. A generalization of this model to include N two-state atoms interacting with a single mode of the field was studied by Tavis and Cummings. [53.3] Again, within the RWA this model is exactly solvable.

The RWA is used almost universally in quantum optics, and its validity is seldom questioned. An exact treatment, if possible, typically leads in resonance problems to very small quantitative differences from the RWA. An example is the Bloch-Siegert shift. [22.1] Tavis and Cummings note that the breakdown in the RWA occurs only for very intense fields. In fact it is just in this regime where the atomic system can generate strong fields that chaos is possible. [53.4, 53.5] In this section we consider this chaotic behavior following first the approach of Reference [53.5].

If the density of atoms is denoted N, the single-mode field of frequency ω may be assumed to satisfy the Maxwell equation

$$\ddot{E} + \omega^2 E = -4\pi N d\dot{x} \tag{53.1}$$

As in References [53.4] and [53.5] we will treat the field as a classical variable, although we continue to treat the atoms quantum mechanically. In other words, we will treat the atom-field system semiclassically; this point is discussed later. In the semiclassical approximation here the field is assumed to have some small initial value, representing spontaneous emission, that triggers the atom-field interaction. The field satisfying (53.1) couples to the two-state atoms via the Bloch equations

$$\dot{x} = -\omega_o y \tag{53.2a}$$

$$\dot{y} = \omega_o x + (2d/\hbar)Ez \tag{53.2b}$$

$$\dot{z} = -(2d/\hbar)Ey \tag{53.2c}$$

Let $\tau = \omega_o t$ and $\bar{E} = (2d/\hbar\omega_o)E$. Then with τ as the independent variable we may write the system (53.1) plus (53.2) in the form

$$\dot{x} = - y \tag{53.3a}$$

$$\dot{y} = x + \bar{E}z \tag{53.3b}$$

$$\dot{z} = - \bar{E}y \tag{53.3c}$$

$$\ddot{\bar{E}} + \mu^2\bar{E} = \beta\dot{y} \tag{53.3d}$$

where we have introduced the dimensionless parameters

$$\mu = \omega/\omega_o \tag{53.4a}$$

$$\beta = 8\pi N d^2/\hbar\omega_o \tag{53.4b}$$

Taking $d = 1$ Debye and $\omega_o = 10^{15}$ Hz, we have $\beta = 2.4 \times 10^{-23}N$, where N is the atomic number density in units of cm^{-3}.

Numerical computations for the system (53.3) have revealed chaotic behavior. We first focus our attention on the results presented in Reference [53.5]. We consider the case of exact resonance ($\mu = 1$) between the atom and the field, and the initial condition $x(0) = y(0) = 0$, $z(0) = 1$. That is, we assume the atoms are all initially in the excited state.

Figure 53.1 shows $z(\tau)$, together with its power spectrum, for $\beta = 0.01$ and $\beta = 1.0$. As β increases the spectrum becomes increasingly broadband. A computation of the maximal Lyapunov exponent reveals that in fact <u>the system is chaotic for $\beta \approx 1$ or larger</u>.

The dynamical system (53.3) evolves in a regular, nonchaotic fashion for $\beta = 0$. In this case the atoms are uncoupled from the field, which evolves freely as a harmonic oscillator. Figure 53.2

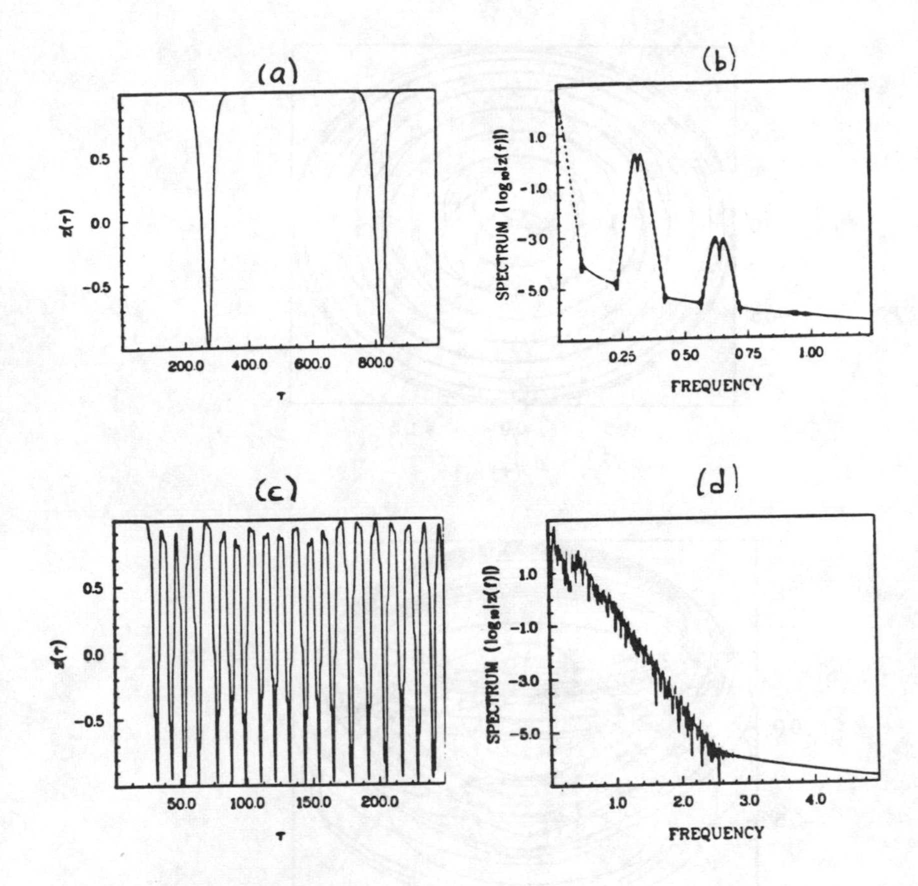

Figure 53.1 (a) Population inversion $z(\tau)$ for $\beta = 0.01$ and (b) its power spectrum. The corresponding results for $\beta = 1.0$ are shown in (c) and (d).

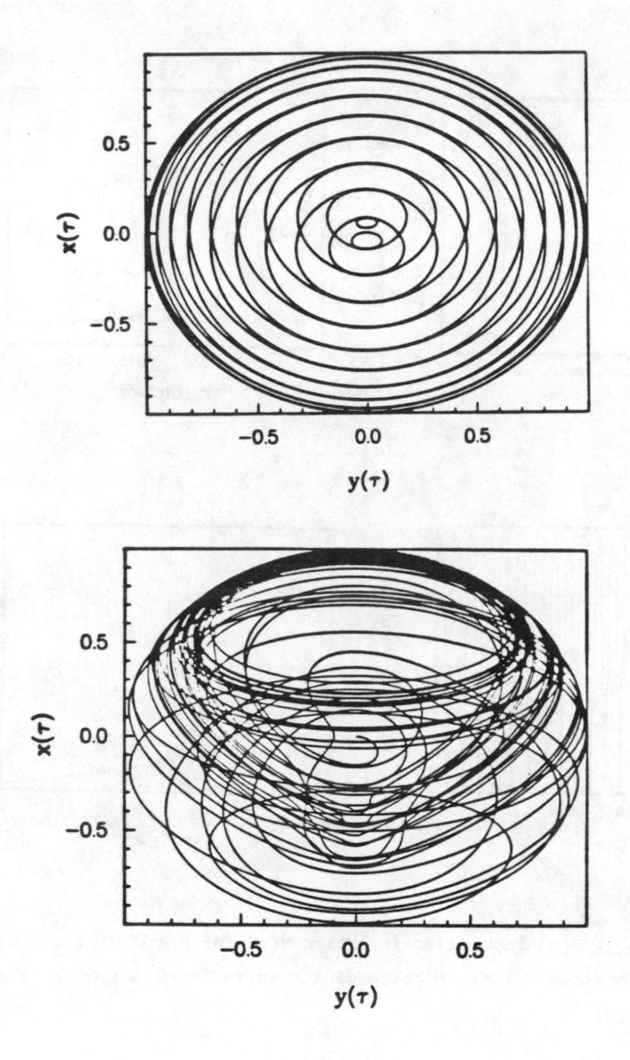

Figure 53.2 $x(\tau)$ vs. $y(\tau)$ for (a) $\beta = 0$ and (b) $\beta = 1.0$.

shows a plot of $x(\tau)$ vs. $y(\tau)$ for this case, compared with the corresponding result for $\beta = 1.0$. Obviously the orderly pattern for $\beta = 0$ is destroyed when β is raised to 1.0.

In the RWA we write

$$x(t) = u(t)\cos\omega t - v(t)\sin\omega t \tag{53.5a}$$

$$y(t) = u(t)\sin\omega t + v(t)\cos\omega t \tag{53.5b}$$

$$z(t) = w(t) \tag{53.5c}$$

$$E(t) = \varepsilon(t)\cos\omega t \tag{53.5d}$$

and assume that u, v, w, and ϵ are slowly varying compared with $\cos\omega t$. Then the RWA version of (53.1) and (53.2) is found to be (Section 22)

$$\dot{u} = -\Delta v \tag{53.6a}$$

$$\dot{v} = \Delta u + (d/\hbar)\epsilon w \tag{53.6b}$$

$$\dot{w} = -(d/\hbar)\epsilon v \tag{53.6c}$$

$$\dot{\epsilon} = (2\pi N d\omega)v \tag{53.6d}$$

This system has two integrals: $u^2 + v^2 + w^2 = 1$ and $N\hbar\omega w + \epsilon^2/4\pi =$ constant, representing conservation of probability and energy, respectively, and these integrals reduce the dimensionality of the autonomous system (53.6) to two. Thus the RWA can give only regular time dynamics, and of course analytic solutions of these equations have been known for a long time. The conclusion that there is no chaos in the RWA holds also if we include a slowly varying field phase, as in Section 22.

Note, however, that these integrals break down when damping terms are included. In fact the system (53.6) with damping terms added is equivalent to the Lorenz model, as we saw in Section 23.

The fact that the RWA precludes chaos in this example suggests that caution should be exercised in other quantum-optical problems where the possibility of chaos arises. As noted by Belobrov, et al., [53.4] "The neglect of the [non-RWA] terms is so weakly based because of the formal difficulties which occur, that the generally accepted use of the [RWA] has changed into a sort of symbol of faith." They conclude that the emergence of chaos in the non-RWA system "shows that the analysis of the interaction of radiation with matter under conditions where there is a strong coupling must as a matter of principle take into account the unremovable statistical nature of the motion which in quantum optics so far has not been considered."

The chaotic behavior of the (non-RWA) Jaynes-Tavis-Cummings model was first discovered by Belobrov, et al. [53.4] and later and independently by Milonni, et al. [53.5] Belobrov, et al. note that the loss of coherence due to chaos for $\beta \approx 1$ can occur more quickly than spontaneous emission, which is neglected above. It is worth noting that large atomic densities are necessary to realize the chaotic regime: the values $\beta = 0.01$ and 1.0 correspond respectively to $N = 4.17$ and 417×10^{20} cm^{-3} for $d = 1$ D and $\omega_o = 10^{15}$ sec^{-1}. [53.5] Belobrov, et al. give the estimate $N \approx 4 \times 10^{21}$ cm^{-3} for $d \approx 3$ D and ω_o corresponding to a 250 μm rotational transition in HF. Because of the onset of incoherence associated with chaotic evolution, such number densities may set an upper limit on the density at which superradiance may be observed.

The equations used by Belobrov, et al. and Milonni, et al. are actually slightly different. The difference stems from the fact that the latter authors use the $\vec{d}\cdot\vec{E}$ form of the interaction Hamiltonian, whereas the former use the $\vec{A}\cdot\vec{p}$ form. Since this point is not directly relevant to matters of chaos, we will not discuss it here. A discussion of precisely this point may be found in Reference [53.6].

Both Belobrov, et al. and Milonni, et al. treated the field classically in the Jaynes-Tavis-Cummings model. If the field as well as the atoms are treated quantum mechanically, we obviously have an interesting model for quantum chaos. In the past few years various authors have studied this model of quantum chaos, and in the remainder of this section we summarize some results of this work.

First we note that the question of quantum chaos in the Jaynes-Cummings model can be addressed either for a single atom or a collection of atoms (all confined to a region small compared with the wavelength of the field). For a single atom, chaos can be found if the atom-field coupling constant is made artificially large, just as chaos arises in the multi-atom model above if there is a sufficiently large density of atoms. Graham and Höhnerbach [53.7] begin with the single-atom Hamiltonian

$$\hat{H} = \hbar\omega_o\hat{\sigma}_z + \hbar\omega\hat{a}^\dagger\hat{a} + \hbar g\hat{\sigma}_x(\hat{a} + \hat{a}^\dagger) \tag{53.7}$$

in the present notation, where $\hat{a}$, $\hat{a}^\dagger$ are the photon annihilation and creation operators, respectively, for the single field mode of frequency ω, and g is the atom-field coupling constant. In the RWA (53.7) reduces to the usual Jaynes-Cummings Hamiltonian

$$\hat{H}_{RWA} = \tfrac{1}{2}\hbar\omega_o\hat{\sigma}_z + \hbar\omega\hat{a}^\dagger\hat{a} + \hbar g(\hat{\sigma}^\dagger\hat{a} + \hat{a}^\dagger\hat{\sigma}) \tag{53.8}$$

where $\hat{\sigma}$, $\hat{\sigma}^\dagger$ are atomic lowering and raising operators, $\hat{\sigma} = \tfrac{1}{2}(\hat{\sigma}_x - i\hat{\sigma}_y)$. [53.1]

An obvious basis set consists of the states $|m,n\rangle$, with

$$\hat{\sigma}_z|m,n\rangle = m|m,n\rangle, \quad m = \pm 1 \tag{53.9a}$$

$$\hat{a}^\dagger\hat{a}|m,n\rangle = n|m,n\rangle, \quad n = 0,1,2,\ldots \tag{53.9b}$$

Thus the state $|m,n\rangle$ is one in which the atom is in the upper state ($m = +1$) or lower state ($m = -1$), with n photons in the field. (Of course these states are not eigenstates of the coupled atom-field system.) Since the "parity" operator

$$\hat{P} \equiv \exp[i\pi(\hat{a}^\dagger\hat{a} + \tfrac{1}{2}\hat{\sigma}_z + \tfrac{1}{2})] \tag{53.10}$$

is a constant of the motion, the states $|m,n\rangle$ of parity $p = \exp[i\pi(n + m + \frac{1}{2})] = +1$ are uncoupled from the states of parity $p = -1$. Writing

$$|\Psi\rangle = \sum_{n=0}^{\infty} \sum_{m\pm1} C(m,n)|m,n\rangle \tag{53.11}$$

the Schrödinger equation for the positive-parity states takes the form

$$[E_+ - (2\ell-1)\omega - \tfrac{1}{2}\omega_o]C(1,2\ell-1) = g[\sqrt{2\ell}\, C(-1,2\ell) + \sqrt{2\ell-1}\, C(-1,2\ell-2)] \tag{53.12a}$$

$$[E_+ - 2\ell\omega + \tfrac{1}{2}\omega_o]C(-1,2\ell) = g[\sqrt{2\ell+1}\, C(1,2\ell+1) + \sqrt{2\ell}\, C(1,2\ell-1)] \tag{53.12b}$$

where $\ell = 0, 1, 2, \ldots$ and we have set $\hbar = 1$.

Graham and Höhnerbach determine the eigenvalues and eigenstates of the coupled (non-RWA) atom-field system by numerical diagonalization of (53.12). Figure 53.3 shows their results for the first few (positive-parity) energy levels as a function of ω/ω_o, both for $g/\omega_o = 0.01$ and $g/\omega_o = 1.0$. In the former case the semiclassical dynamics is regular, whereas in the latter it is chaotic. Narrow and

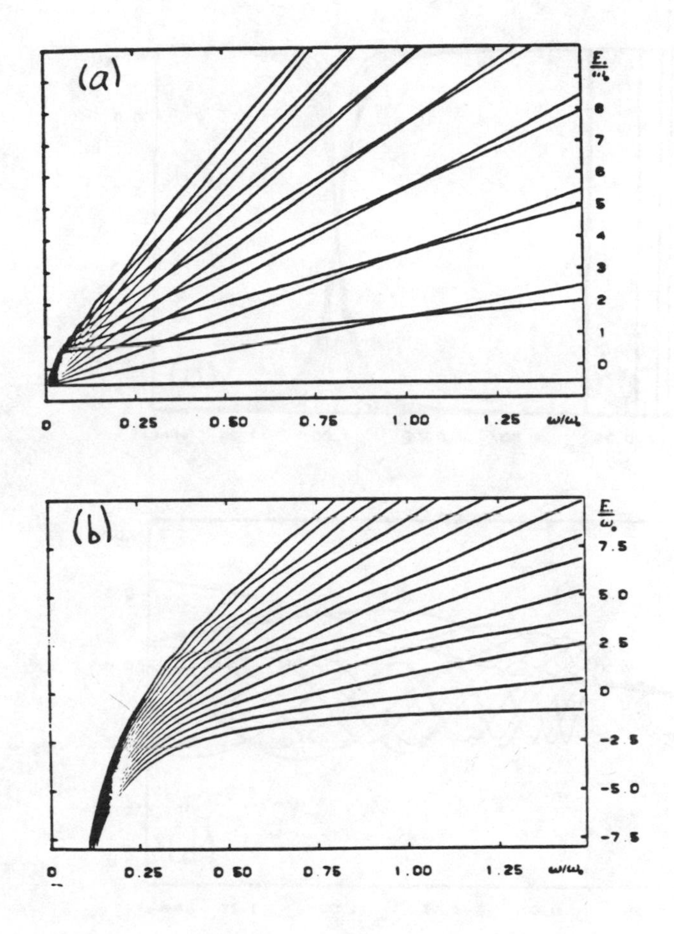

Figure 53.3 Lowest (positive-parity) energy levels as a function of ω/ω_0 for (a) $g/\omega_0 = 0.01$ and (b) $g/\omega_0 = 1.0$. From R. Graham and M. Höhnerbach, Reference [53.7], with permission.

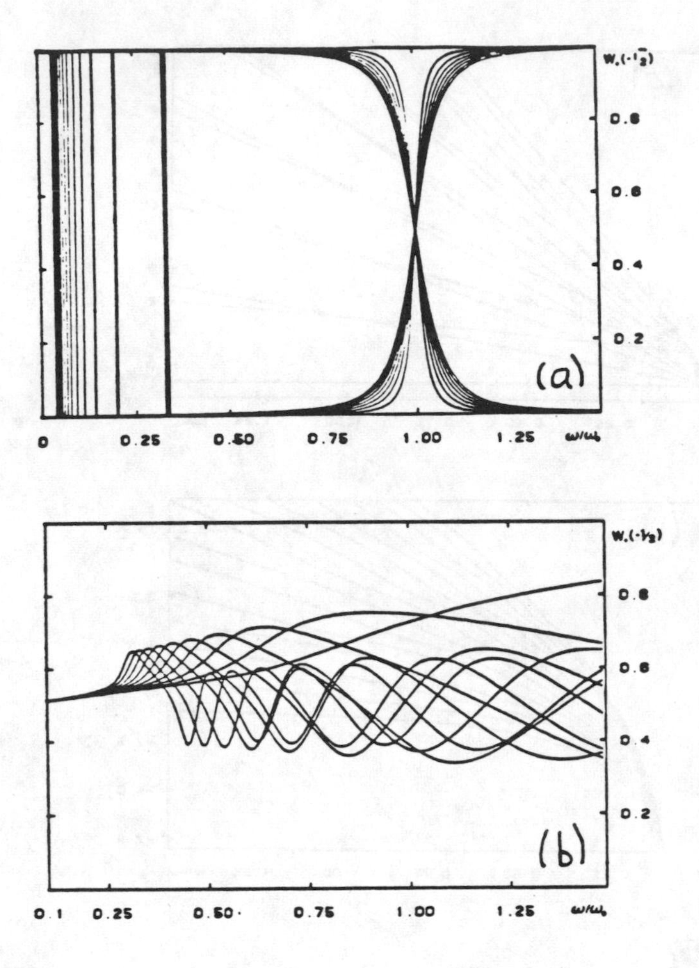

Figure 53.4 Lower-state probability for the lowest (positive-parity) energy levels as a function of ω/ω_0 for (a) $g/\omega_0 = 0.01$ and (b) $g/\omega_0 = 1.0$. From R. Graham and M. Höhnerbach, Reference [53.7], with permission.

well separated avoided crossings in the case of weak coupling (Figure 53.3a) may be associated with the multiphoton resonances $\omega = \omega_o/(2k+1)$, $k = 0, 1, 2, \ldots$; the restriction to odd multiphoton resonances is due to the electric-dipole approximation in the atom-field coupling. In the strong coupling case (Figure 53.3b), however, the avoided crossings are broad and overlapping, and can evidently no longer be associated with well-defined multiphoton resonances. [53.7] Thus the energy levels appear to have different characteristics depending on whether the corresponding semiclassical dynamics is regular or chaotic.

This difference is further reflected in the plots shown in Figure 53.4 for the occupation probability

$$W = \sum_{n=0}^{\infty} |C(-1,n)|^2 \tag{53.13}$$

of the lower atomic state. In the case of weak coupling (Figure 53.4a) the sharp transitions between the upper and lower states occur at the multiphoton resonances $\omega = \omega_o/(2k+1)$. In the case of strong coupling (Figure 53.4b), however, the transitions are much less sharply defined and cannot be associated with well-defined resonances. [53.7]

Kus [53.8] has investigated the statistical properties of the energy eigenvalues of the non-RWA Jaynes-Cummings model. In particular, he finds that second differences of the type considered by Pomphrey (Section 51) do not seem to be very sensitive to changes in parameters in either the semiclassically regular or chaotic regimes.

It should also be noted that Fox and Eidson [53.9] have discussed the origin of chaos in the semiclassical Jaynes-Cummings model, exploiting the well-known pendulum analogy to explain certain aspects of the observed power spectra. Their analysis also serves to clarify the approach of Belobrov, _et al_. [53.4] They emphasize that broadband spectra and decaying correlations do not necessarily imply chaos in the sense of a positive Lyapunov exponent. (Recall the discussion in Section 52.) Eidson and Fox have also studied in detail the continuous spectrum found by Pomeau, et al. [52.2] for the quasiperiodically driven two-state system, and have confirmed that the spectrum can indeed become truly continuous when the driving frequencies are incommensurate.

In the past few years there has been a resurgence of interest in the fully quantized, RWA Jaynes-Cummings model in connection with the "collapse" and "revival" properties discussed by Eberly, et al. [53.10] These properties, which are characteristic of regular rather than chaotic dynamics, have recently been observed experimentally. [53.11]

54. Quantum Theory of the Kicked Pendulum

For the remainder of our discussion of "quantum chaos" we will be concerned with systems driven by an externally applied force. The principal physical system we have in mind in this context, of course, is an atom or molecule in the field of a laser. In Sections 47 and 49 we discussed classical models of a vibrating molecule driven by a laser, and found that there could be a "stochastic excitation" in which the chaotic dynamics of the driven molecule leads to a diffusive growth of the absorbed energy. In particular, the average absorbed energy can be proportional to the time. (Section 47) An important aspect of "quantum chaos" is whether such predictions survive a fully quantum-mechanical treatment. Can we have stochastic excitation in quantum theory?

In this section we return to the kicked-pendulum paradigm, this time treating the problem fully quantum mechanically. Recall that in Section 46 we discussed the classical theory of the kicked pendulum, leading to the standard mapping, and we saw that for large values of K there was a diffusive growth of the pendulum energy, with the average energy growing approximately linearly with time. The quantum theory of the kicked pendulum is straightforward, as we will see. This allows us to test the validity of the classical prediction of stochastic excitation of the kicked pendulum.

For the periodically kicked pendulum we may write the Hamiltonian (46.1) in the form (52.1) with

$$H_o = p_\theta^2/2m\ell^2 \tag{54.1a}$$

$$A(\theta) = -(m\ell^2\omega_o^2)\cos\theta \tag{54.1b}$$

$$F(t) = 1 \tag{54.1c}$$

The eigenstates of the unperturbed Hamiltonian H_o are simply

$$\psi_n(\theta) = (2\pi)^{-1}e^{in\theta} \tag{54.2a}$$

and the corresponding energy eigenvalues are

$$E_n = n^2\hbar^2/2m\ell^2 \tag{54.2b}$$

with n = 0, ±1, ±2, ... Thus the quantum map (52.4) takes the form

$$a_n(k+1) = \sum_m V_{nm}a_m(k) \tag{54.3}$$

with

$$V_{nm} = \langle\psi_n|e^{im\ell^2\omega_o^2T\cos\theta/\hbar}|\psi_m\rangle e^{-in^2\hbar T/2m\ell^2}$$

$$= e^{-in^2\tau/2}(1/2\pi)\int_0^{2\pi} d\theta e^{i(m-n)\theta}e^{i(K/\tau)\cos\theta} \tag{54.4}$$

Here

$$\tau \equiv \hbar T/m\ell^2 \tag{54.5a}$$

and

$$K \equiv (\omega_o T)^2 \tag{54.5b}$$

Note that K is identical to the parameter appearing in the classical standard map (46.4). In the quantum map, however, an additional parameter (τ) appears, which vanishes in the classical limit $\hbar \to 0$.

We can write (54.4) as

$$V_{nm} = (1/2\pi)e^{-in^2\tau/2}\int_0^{2\pi} d\theta e^{i(m-n)\theta}\sum_{s=-\infty}^{\infty} b_s(K/\tau)e^{is\theta} \tag{54.6}$$

with

$$b_s(y) \equiv i^s J_s(y) \tag{54.7}$$

where J_s is the Bessel function of the first kind of order s. Thus

$$V_{nm} = e^{-in^2\tau/2}\sum_{-\infty}^{\infty} b_s(K/\tau)\ (\tfrac{1}{2\pi})\int_0^{2\pi} d\theta e^{i(m-n+s)\theta}$$

$$= e^{-in^2\tau/2}\sum_{-\infty}^{\infty} b_{n-m}(K/\tau) \tag{54.8}$$

and the quantum map (54.3) becomes

$$c_n(k+1) = \sum_m b_{n-m}(K/\tau)e^{-in^2\tau/2}c_m(k) \tag{54.9}$$

where

$$c_n(k) \equiv a_n(k)e^{in^2\tau/2} \tag{54.10}$$

The expectation value of the pendulum energy is obtained by iterating the map (54.9) and using the expression

$$\langle E(k)\rangle = (\tau^2/K)\sum_{-N}^{N} n^2|c_n(k)|^2 \tag{54.11}$$

in units of $m\omega_o^2\ell^2/2$. Here N is an integer, typically ≈ 400, chosen large enough that the total probability is conserved at each iteration. Results for ⟨E(k)⟩ were first reported by Casati, Chirikov, Izrailev, and Ford, [46.1] and here we report similar results.

In Figure 54.1 we show ⟨E(k)⟩ for $\tau = 1$ and $K = 10$. Comparing with the classical results shown in Figure 46.3, we see that there is a substantial quantum suppression of the classical diffusive energy growth: after some time the energy expectation value does not increase approximately linearly with time. Instead there is a "saturation" of the pendulum energy, or at least a much slower growth than at short times (where the best agreement with the classical theory would be expected, as noted in Section 49). This quantum suppression effect, which was first found in the numerical experiments of Casati, *et al*., does not depend on our particular choice for τ and K.

The quantum suppression of the classically predicted stochastic excitation has been nicely interpreted by analogy with Anderson localization. In the following section we discuss this localization, and also point out a possible way to beat it. We conclude this section by mentioning two other aspects of the kicked pendulum.

Generically, the classical kicked pendulum exhibits mainly chaotic behavior for large K and regular behavior for small K.

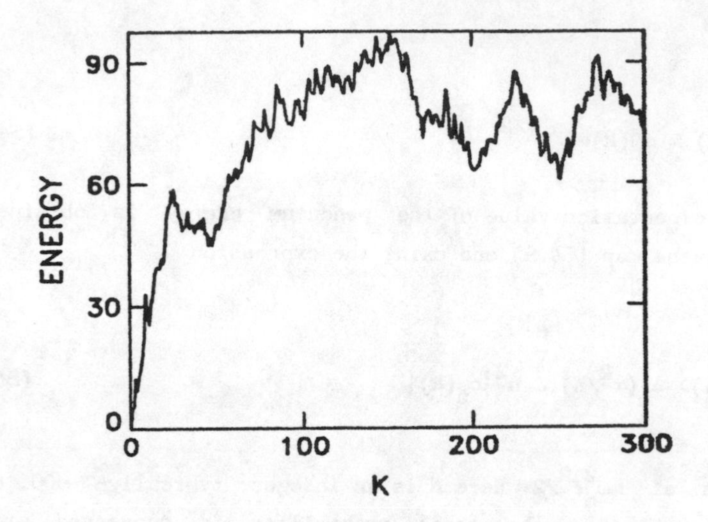

Figure 54.1 $\langle E(k)\rangle$ for K = 10 and τ = 1, illustrating the quantum suppression of the classical diffusive energy growth.

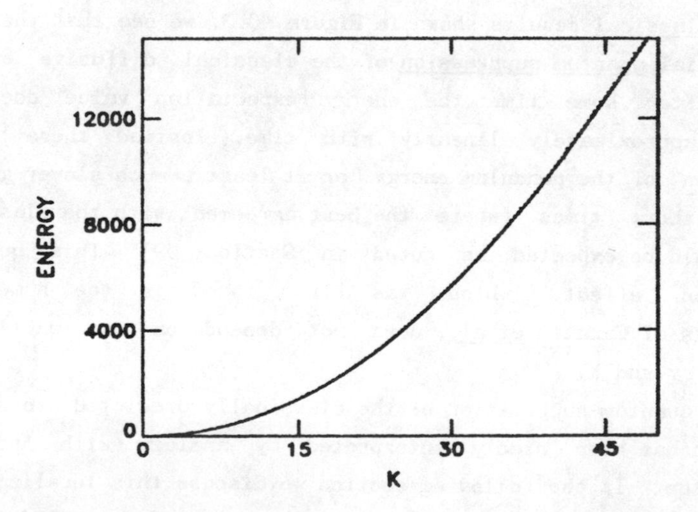

Figure 54.2 $\langle E(k)\rangle$ for K = 1 and the "quantum resonance" value $\tau = 4\pi$, showing energy growth $\sim t^2$.

However, for $K \cong 2\pi n$ there are certain initial conditions for which trajectories are continually accelerated, with energy increasing as t^2. For $(\theta_o, P_o) = (\pi/2, 0)$ and $K = 2\pi$, for example, we obtain from the standard mapping (46.4) the trajectory $(\theta_n, P_n) = (\pi/2, -2\pi n) = (\pi/2, -Kn)$, and so $P_n^2 = K^2 n^2$. Such an initial point in phase space is called an accelerating point, and the region near such a point is called an <u>accelerator mode</u>. [54.1] Accelerator modes give rise to energy growth much faster than that predicted from a diffusion coefficient

$$D \equiv \lim_{n\to\infty}[\langle(P_n - P_o)^2\rangle]/2n \tag{54.12}$$

In fact if any part of the ensemble in (54.12) belongs to an accelerator mode the limit does not exist. Thus, if a small fraction of an initial ensemble of trajectories lies within an accelerator mode, the average energy may be dominated at long times by the accelerator modes and grow as t^2 on average. For shorter times the average energy growth may only be proportional to t because most of the initial ensemble belongs to non-accelerating regions.

In the quantum kicked pendulum the very small accelerator regions of phase space may not give rise to such behavior if they are small on a scale of $\hbar$, or if there is substantial quantum tunneling out of these regions. Hanson, <u>et al</u>. [54.2] have studied the effects of tunneling in mitigating the effect of the accelerator modes.

Energy growth proportional to t^2 can also occur for special values of τ in the quantum kicked pendulum. This occurs when the <u>quantum resonance</u> condition $\tau = 4\pi r/s$ is satisfied, where r and s are relatively prime integers. [46.1, 54.3] Figure 54.2, for instance, shows $\langle E(k)\rangle$ for the case $\tau = 4\pi$. (Note that the factor $\exp(-in^2\tau/2)$ is unity when $\tau = 4\pi r$.). From (54.5a) it is easily checked that such values of τ imply the resonance condition $r\omega = s\Delta E$, where $\omega \equiv 2\pi/T$ and $\Delta E \equiv \hbar^2/2m\ell^2$, $E_n = n^2\Delta E$.

55. Localization

Random walks in which the hopping probabilities are themselves random variables can exhibit dramatic localization compared with the familiar diffusion occurring when the probabilities are not randomly varying. [55.1] In quantum dynamics there is the Anderson localization of a particle moving in a one-dimensional random potential. Consider the "tight-binding model" described by the Schrödinger equation

$$i\hbar\dot{a}_n = E_n a_n + \sum_{m\neq n} V_{nm} a_m \tag{55.1}$$

where a_n is the probability amplitude, and E_n the energy, for an electron at the nth lattice site. Stationary states of energy E are described by the corresponding time-independent Schrödinger equation

$$(E-E_n)a_n = \sum_{m\neq n} V_{nm} a_m \tag{55.2}$$

In the tight-binding model of localization the site energies E_n are taken to be independent random variables with a finite set of possible values, each described by a probability P_n. This is not the only model for localization. In the "Lloyd model," for instance, the random variables E_n are distributed according to a continuous, Lorentzian distribution function (the Cauchy probability distribution). In such models it is found that the eigenstates are exponentially localized in space, i.e., there is no quantum diffusion. (The title of Anderson's original paper in 1958 was "Absence of Diffusion in Certain Random Lattices." [55.2])

Anderson localization is generally regarded as a purely quantum effect associated with destructive interference of probability amplitudes. However, one may construct classical analogues. Equation (55.1), for instance, can describe a set of coupled harmonic oscillators in a slowly-varying amplitude approximation. In any case we will simply accept the existence of Anderson localization. The interested reader is referred especially to the reviews by Ishii [55.3] and Thouless. [55.4]

Grempel, Prange, and Fishman [55.5] have related the quantum suppression of diffusion in the kicked pendulum to Anderson localization. They obtain from the Schrödinger equation for the kicked pendulum an equation of the same form as (55.2). The random diagonal terms E_n of the tight-binding model correspond in the kicked pendulum equation to a "pseudorandom" sequence, and the lattice sites of the tight-binding model correspond to the integer values of quantized angular momentum in the kicked pendulum. We begin by deriving this correspondence. [55.5]

The Hamiltonian (52.1) with $F(t) = 1$ is periodic with period T, and so Floquet's theorem is applicable. Thus we may write the state vector as

$$|\Psi(t)\rangle = e^{-i\omega t}|\psi(t)\rangle \tag{55.3}$$

with $|\psi(t)\rangle$ periodic: $|\psi(t+T)\rangle = |\psi(t)\rangle$. The ω's, which are analogous to the k-vectors in the energy band theory of solids, define the <u>quasienergy spectrum</u>. In (55.3) ω may be regarded as a new quantum number, but for simplicity we will not bother to indicate explicitly the ω-dependence of $|\Psi\rangle$ and $|\psi\rangle$. States associated with different ω's may be shown to be orthogonal. [55.6]

Let $|\Psi(k)\rangle_-$ and $|\Psi(k)\rangle_+$ denote respectively the state vectors just before and just after the k<u>th</u> kick. In the notation of Section 52 we have

$$|\Psi(k)\rangle_+ = e^{-iA(x)T/\hbar}|\Psi(k)\rangle_- \tag{55.4a}$$

$$|\Psi(k)\rangle_- = e^{-iH_oT/\hbar}|\Psi(k-1)\rangle_+ \tag{55.4b}$$

or, using (55.3),

$$|\psi\rangle_+ = e^{-iA(x)T/\hbar}|\psi\rangle_- \tag{55.5a}$$

$$|\psi\rangle_- = e^{i\omega T}e^{-iH_oT/\hbar}|\psi\rangle_+ \tag{55.5b}$$

Note that, since $|\psi(t+T)\rangle = |\psi(t)\rangle$, we have $|\psi(k+1)\rangle_\pm = |\psi(k)\rangle_\pm \equiv |\psi\rangle_\pm$. From (55.5a),

$$|\psi\rangle \equiv \tfrac{1}{2}(|\psi\rangle_+ + |\psi\rangle_-) = \tfrac{1}{2}(1 + e^{-iA(x)T/\hbar})|\psi\rangle_-$$

$$= e^{-iA(x)T/2\hbar}\cos[A(x)T/2\hbar]|\psi\rangle_-$$

$$= [1-iW(x)]^{-1}|\psi\rangle_- \tag{55.6}$$

where $W(x)$ is the Hermitian operator defined by

$$W(x) = -\tan[A(x)T/2\hbar] \tag{55.7a}$$

or

$$e^{-iA(x)T/\hbar} = \frac{1+iW(x)}{1-iW(x)} \tag{55.7b}$$

Similarly

$$|\psi\rangle = [1+iW(x)]^{-1}|\psi\rangle_+ \tag{55.8}$$

where $|\psi\rangle$ is the value of $|\psi(t)\rangle$ at any of the times $t = kT$. From (55.5b) we then have

$$[1+iW(x)]|\psi\rangle = e^{-i(\omega-H_o/\hbar)T}[1-iW(x)]|\psi\rangle \tag{55.9}$$

Now writing

$$|\psi\rangle = \sum_n c_n|\phi_n\rangle \tag{55.10}$$

where $H_o|\phi_n\rangle = E_n|\phi_n\rangle$, we have from (55.9),

$$c_n + i\sum_m W_{nm}c_m = e^{-i(\omega-E_n/\hbar)T}[c_n - i\sum_m W_{nm}c_m] \tag{55.11}$$

This result may be rewritten in the form

$$(W_{nn}+T_n)c_n = -\sum_{m\neq n} W_{nm}c_m \tag{55.12}$$

with

$$T_n \equiv \tan[\tfrac{1}{2}(\omega - E_n/\hbar)T] \tag{55.13}$$

For the kicked pendulum

$$W_{nm} = (1/2\pi)\int_0^{2\pi} d\theta e^{-i(n-m)\theta}W(\theta) \equiv W_{n-m} \tag{55.14}$$

and (55.12) takes the form derived by Grempel, et al. [55.5]:

$$(E-T_n)c_n = \sum_{r\neq 0} W_r c_{n+r} \tag{55.15}$$

with $E \equiv - W_0$.

Equation (55.15) is of the form (55.2) with E_n in the latter equation replaced by (recall equations (54.2b) and (54.5a))

$$T_n = \tan[\tfrac{1}{2}(\omega-E_n/\hbar)T] = \tan[\tfrac{1}{2}(\omega T-n^2\tau)] \qquad (55.16)$$

The case $\tau = 4\pi m$, m an integer, corresponds to the "quantum resonance" mentioned at the end of the preceding section. If τ is an irrational multiple of 4π, however, $\{T_n\}$ is effectively a random sequence, having decaying correlations and a broadband power spectrum, much like the sequence $\{\sin^2\pi\theta_n\}$, $\theta_{n+1} = 2\theta_n$, considered in connection with the logistic map. (Section 1) This observation is the basis of the Anderson localization analogy established by Grempel, Prange, and Fishman for the kicked pendulum.

The formal mapping relating the kicked pendulum to the tight-binding model implies there will be no quantum diffusion over the space of unperturbed eigenstates of the kicked pendulum. <u>This explains the quantum suppression of diffusive energy growth discussed in the preceding section</u>.

Some obvious questions arise in connection with the stochastic excitation process found in <u>classical</u> models of multiple-photon excitation. (Sections 47-49) Some computations by Wyatt and Brunet [55.7] suggest that there is a quantum supression of the classically predicted diffusive energy growth in the model considered in Section 47. There is also evidence of such quantum suppression in the theory of atomic hydrogen in a microwave field; this is discussed in Section 56.

Using (55.5), the quasienergy spectrum may be defined by the equation

$$e^{i\omega T} = S \equiv e^{i[H_o-A(x)]T/\hbar} \qquad (55.17)$$

More generally, the quasienergy spectrum is defined as the spectrum of the (Hermitian) operator Ω such that the Floquet operator $S = \exp(i\Omega)$, with S defined by $|\Psi(t+T)\rangle = S|\Psi(t)\rangle$. This definition is applicable to any periodically driven quantum system, regardless of whether it is kicked or driven continuously. It may be shown that a discrete ("pure point") quasienergy spectrum implies recurrent behavior of the state vector, whereas a continuous quasienergy spectrum implies a spreading of the state vector over the set of unperturbed eigenstates. The latter case is associated, for instance, with a quantum resonance when $\tau/4\pi$ is rational.

Casati and Guarneri [55.8] have proven that there is a dense set of irrational (but "close" to rational) values of $\tau/4\pi$ producing a continuous quasienergy spectrum, and they conclude that "quantum mechanics seems to place no intrinsic limitation on the display of stochasticity." However, based on numerical experiments, it appears that nonlocalized behavior for irrational $\tau/4\pi$ is certainly nongeneric.

Schuster [55.9] has shown that a discontinuity in the kicking potential can result in diffusive energy growth, corresponding in the tight-binding model to hopping matrix elements W_n that are long-ranged. He suggests that this may provide a mechanism for quantum chaos.

The simplest generalization of purely periodic driving is, of course, quasiperiodic driving. Based on the results in Section 52 for a quasiperiodically kicked two-state system, we have performed numerical experiments for a quasiperiodically kicked pendulum, [52.1] an example that was considered earlier in a paper by Shepelyansky. [55.10] In the case $F(t) = \cos\omega' t$ considered in Section 52, the parameter K appearing in (54.9) for the kicked pendulum is replaced by $K(k) = K\cos(2\pi kx)$, where $x = \omega'/\omega$ is the ratio of the two driving frequencies, as in Section 52. For rational values of x we find the localization behavior characteristic of purely periodic kicking. For irrational values of x, however, there is evidently a diffusive energy growth as in the classical kicked pendulum. Figure 55.1, for instance, shows the energy expectation value $\langle E(k)\rangle$ for $K = 10$, $\tau = 1$, and x irrational. There is no evidence of any quantum suppression (localization) of the classically predicted diffusion, even after a large number of kicks. Shepelyansky [55.10] has also found that quasiperiodic kicking can greatly extend the time scale over which

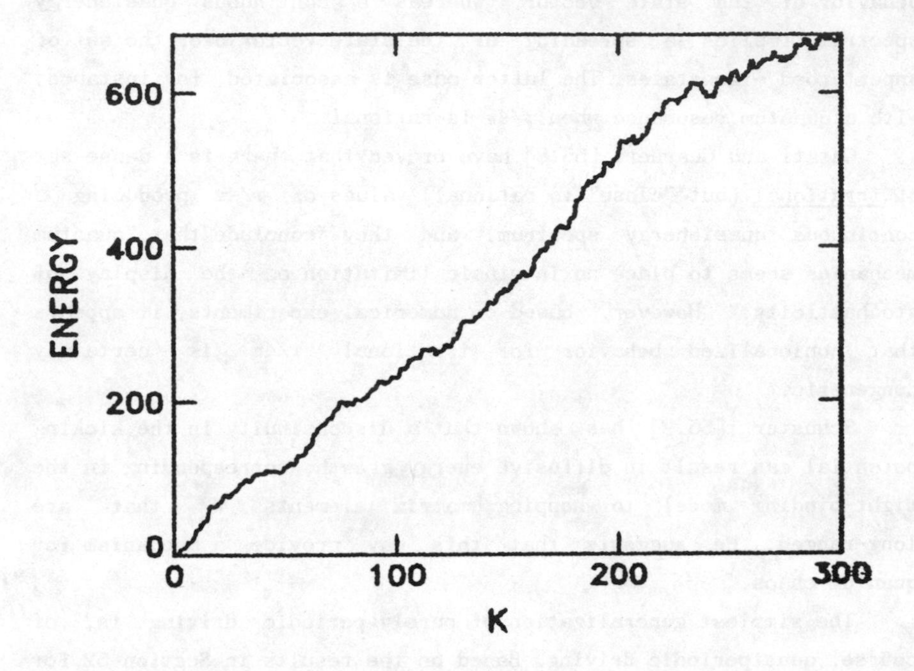

Figure 55.1 $\langle E(k) \rangle$ for $K = 10$, $\tau = 1$, and x irrational, illustrating average energy growth ~ t in the case of quasiperiodic kicking with incommensurate frequencies.

there is diffusion, and has suggested that the diffusive time scale in this case increases exponentially with K.

Based on these results, it does not appear that diffusive energy growth is generally ruled out in driven quantum systems. Let us emphasize again, however, that the quasiperiodically kicked quantum systems considered here and in Section 52 are not chaotic in the classical sense of a positive Lyapunov exponent. Nevertheless they can have properties, like broadband power spectra, decaying correlations, and diffusive energy growth, that are typically _consequences_ of chaos in classical driven systems.

56. Classical and Quantum Calculations for a Hydrogen Atom in a Microwave Field

In 1974 Bayfield and Koch [56.1] reported experimental results on the ionization of highly excited hydrogen atoms by a microwave field. Hydrogen atoms with principal quantum numbers $63 \leq n \leq 69$ were produced by charge transfer from xenon using an 11 keV proton beam, and then passed through a microwave cavity. Using a Faraday cup and amplification electronics, Bayfield and Koch measured the ionization probability as a function of the "adiabatic tunneling parameter" $\gamma = \omega/nF$, where ω and F are the microwave field (angular) frequency and amplitude, respectively. (See below) For the microwave frequency of 9.9 GHz they found substantial ionization above a critical field strength $\approx$ 20 V/cm. Note that the transition $n = 66 \rightarrow 67$ has a resonance frequency

$$\nu_o = (13.6 \text{ eV/h})[1/66^2 - 1/67^2] = 22 \text{ GHz} \qquad (56.1)$$

and so the applied field frequency is only about 40% of the resonance frequency. The ionization is found to depend strongly on the microwave power but not on the frequency. Furthermore the ionization energy ($13.6 \text{ eV}/66^2$) is about 76 times the field photon energy $\hbar\omega$.

Obviously the ionization process involves a large number of photons. This has made standard quantum-mechanical analyses very difficult, since we are faced with high-order perturbation theory and complicated intermediate resonances. Indeed the most successful theoretical analyses to date, beginning with the Monte Carlo approach of Leopold and Percival, [56.2] have been based on classical dynamics. These classical studies indicate that the microwave ionization of hydrogen is a stochastic excitation process associated with the destruction of KAM tori. (Section 49)

Consider the classical equations of motion

$$\dot{\vec{r}} = \vec{p}/m \tag{56.2a}$$

$$\dot{\vec{p}} = -(e^2/r^2)\hat{r} + F_{max}\hat{z}\cos\omega t \tag{56.2b}$$

for an electron in the field of the nucleus plus an externally applied sinusoidal field. Define the energy, force, and frequency parameters

$$E_{at} = e^2/a_{at} \tag{56.3a}$$

$$F_{at} = E_{at}/a_{at} = m\omega_{at}^2 a_{at} \tag{56.3b}$$

$$\omega_{at} = v_{at}/a_{at} \tag{56.3c}$$

where a_{at} and v_{at} are respectively the semimajor axis and root-mean-square velocity of an orbiting electron. Using these definitions in (56.2), we may write the following equation for $\vec{R} \equiv \vec{r}/a_{at}$:

$$\ddot{\vec{R}} = -R^{-2}\hat{R} + (F_{max}/F_{at})\hat{z}\cos(\omega\tau/\omega_{at}) \tag{56.4}$$

where now the independent variable is $\tau \equiv \omega_{at}t$. Thus the classical dynamics of the hydrogen atom in a monochromatic field depend only on the dimensionless ratios ω/ω_{at} and F_{max}/F_{at}. [56.2] The parameter γ mentioned above in connection with the Bayfield-Koch experiment is defined as

$$\gamma = \frac{\omega/\omega_{at}}{F_{max}/F_{at}} = \frac{P_{at}}{F_{max}/\omega} \tag{56.5}$$

where $P_{at} = F_{at}/\omega_{at}$ is the linear momentum of the electron in its unperturbed orbit and F_{max}/ω is the maximum momentum imparted by the field in the absence of the nucleus. [56.2] γ may also be written in the form

$$\gamma = \hbar\omega/(F_{max} n a_o) \tag{56.6}$$

for an electron with principal quantum number n, where a_o is the Bohr radius. This is the form used by Bayfield and Koch. [56.1] (They use units in which m = e = ħ, and so they write $\gamma = \omega/nF$.)

For an electron in a Bohr orbit with n = 66 the electric field due to the nucleus is ≈ 275 V/cm, which is about an order of magnitude larger than the amplitude of the microwave field. From (56.6) we estimate $\gamma = 5.9$ for a 9.9 GHz field of peak amplitude 20 V/cm. Leopold and Percival report calculations with $\gamma = 6$ and 7, corresponding to the ratios $F_{max}/F_{at} = .072$ and .061, respectively, of applied field to atomic field. Initial conditions for the integration of the classical equations of motion were chosen by a Monte Carlo method from a (classical) microcanonical distribution corresponding quantum mechanically to an equal distribution of population over degenerate (ℓ,m) states for n = 66. Leopold and Percival computed the "compensated energy"

$$E_c = \tfrac{1}{2}(p_x^2 + p_y^2) + \tfrac{1}{2}[p_z - (F_{max}/\omega)\sin\omega t]^2 - 1/r \tag{56.7}$$

in natural units. E_c is useful because it removes the rapid oscillations of the energy due to the oscillating applied field; in the absence of the Coulomb field it is constant. A compensated energy $E_c > 0$ indicates that ionization has occurred, whereas the energy E can be positive without ionization.

Leopold and Percival identified four types of trajectories: (1) trajectories confined to invariant tori; (2) trajectories that rapidly ionize; (3) trajectories excited to very high levels with subsequent ionization; and (4) trajectories excited to very high levels without subsequent ionization. The ionization probability as a function of time was fitted well by the formula

$$P_{ion}(t) = (1 - Q_T)[1 - e^{-\beta(t)t}] \tag{56.8}$$

where Q_T is the estimated probability that the motion is confined to an invariant torus, and β is nearly constant until ionization is nearly complete. For $\gamma = 7$ and 6 Leopold and Percival computed ionization probabilities of 40-50% and 62-80%, respectively, compared to the values 50% and 62% measured by Bayfield and Koch. This impressive agreement of classical calculations with experiment has been strongly upheld by more recent analyses, as we will see shortly.

In the experiments the microwave field is effectively turned on and off gradually as the atomic beam enters and exits the microwave cavity. Leopold and Percival [56.3] model this effect by taking the applied field to be of the form $A(t)F_{max}\cos\omega t$, where (Figure 56.1)

$$\begin{aligned} A(t) &= e^{\lambda(t-t_i)} && 0 \leq t \leq t_i \\ &= 1 && t_i \leq t \leq t_f \\ &= e^{-\lambda(t-t_f)} && t_f \leq t \leq T \end{aligned} \tag{56.9}$$

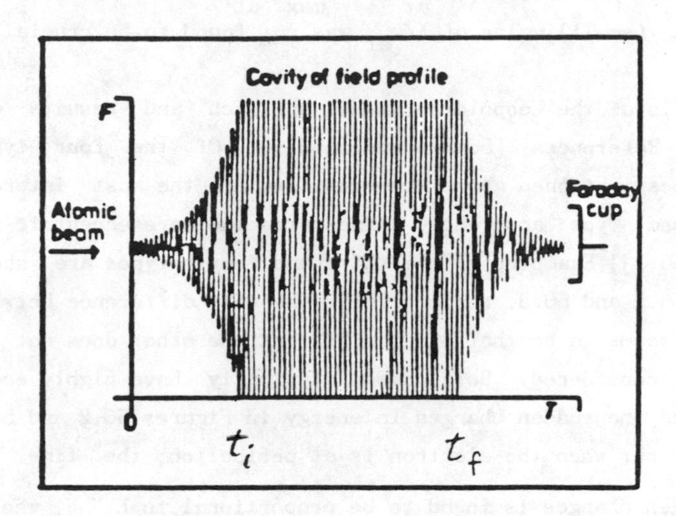

Figure 56.1 **Electric field profile assumed in numerical experiments of Leopold and Percival. From J.G. Leopold and I.C. Percival, Reference [56.3], with permission.**

This introduces the parameter λ/ω_{at} in addition to the other two dimensionless parameters (ω/ω_{at} and F_{max}/F_{at}) of the classical model. The exact (small) value of λ/ω_{at} was not found to be crucial to the results.

Details of the Leopold-Percival approach and results may be found in References [56.3] and [56.4]. Of the four types of trajectories mentioned above, the last two are the most interesting. (The second type apparently arises from the more eccentric initial orbits. [56.3]) Examples of the third and fourth types are shown in Figures 56.2 and 56.3, respectively. The only difference between the two types seems to be that one ionizes but the other does not during the time considered. Both cases typically have highly eccentric orbits, and the sudden changes in energy in Figures 56.2 and 56.3 are found to occur when the electron is at perihelion; the time between these sudden changes is found to be proportional to $E^{-3/2}$, where E is the compensated ionization energy. Thus the atom is <u>most stable</u> for the <u>weakest</u> binding. Of course such highly excited atoms would be very sensitive to collisions or stray fields. [56.3]

Remarkable agreement between experiment and classical calculations has been reported by van Leeuwen, <u>et al</u>. [56.5] In their experiments a static electric field of 0, 2, 5, or 8 V/cm was superimposed on the 9.9 GHz microwave field and the atoms were prepared in levels n = 32, 40, and 51 - 74 before entering the cavity. Ionization curves as functions of the microwave field strength were measured, and classical Monte Carlo calculations (which included the static field) of the field strength $F_o(10\%)$ at which 10% of the atoms were ionized were compared with measured values. Both one- and two-dimensional classical models showed excellent agreement with experiment, with the two-dimensional treatment showing generally better agreement. Figure 56.4 shows the classically scaled values of $F_o(10\%)$ $(F_o/(e/r^2) \sim n^4F_o)$ versus the classically scaled microwave frequency $(\omega/\omega_{at} \sim n^3\omega)$ determined from both theory and experiment. The ability of the classical theory to reproduce even fine details of

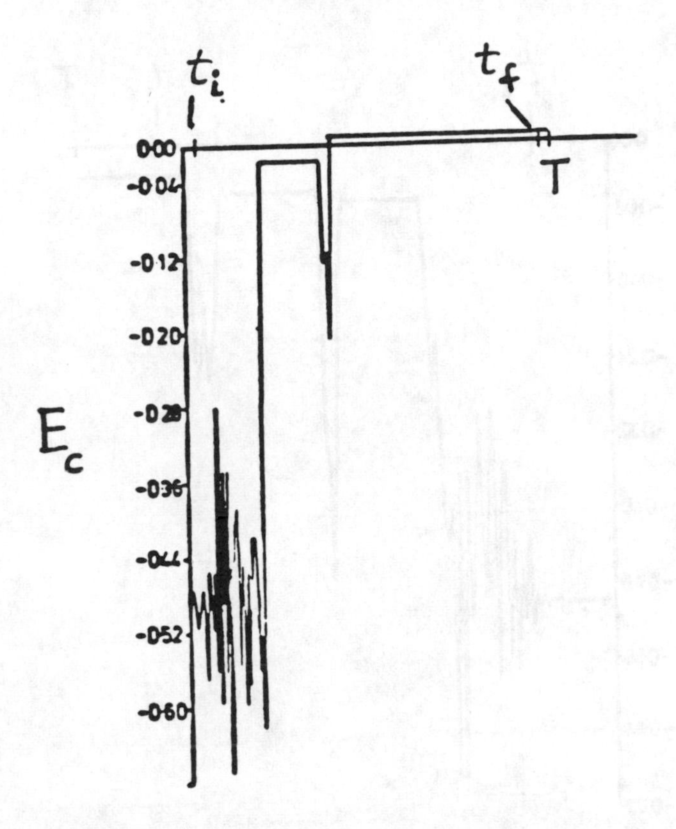

Figure 56.2 Compensated energy E_c vs. time for a trajectory that reaches a high level of energy with subsequent ionization. From J.G. Leopold and I.C. Percival, Reference [56.3], with permission.

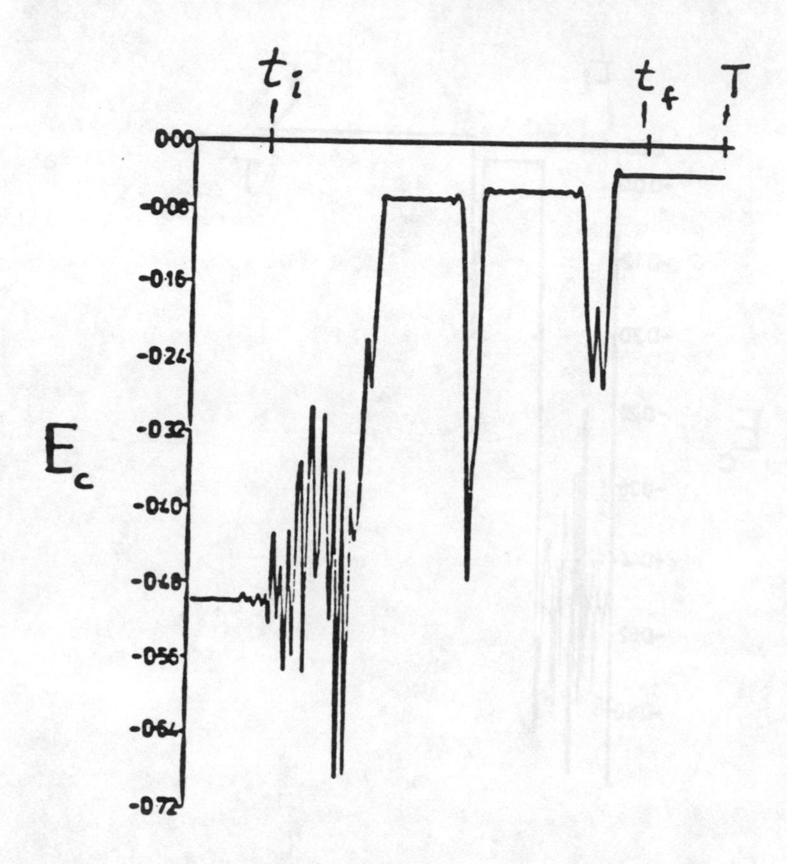

Figure 56.3 Compensated energy E_c vs. time for a trajectory that reaches a high level of energy without subsequent ionization. From J.G. Leopold and I.C. Percival, Reference [56.3], with permission.

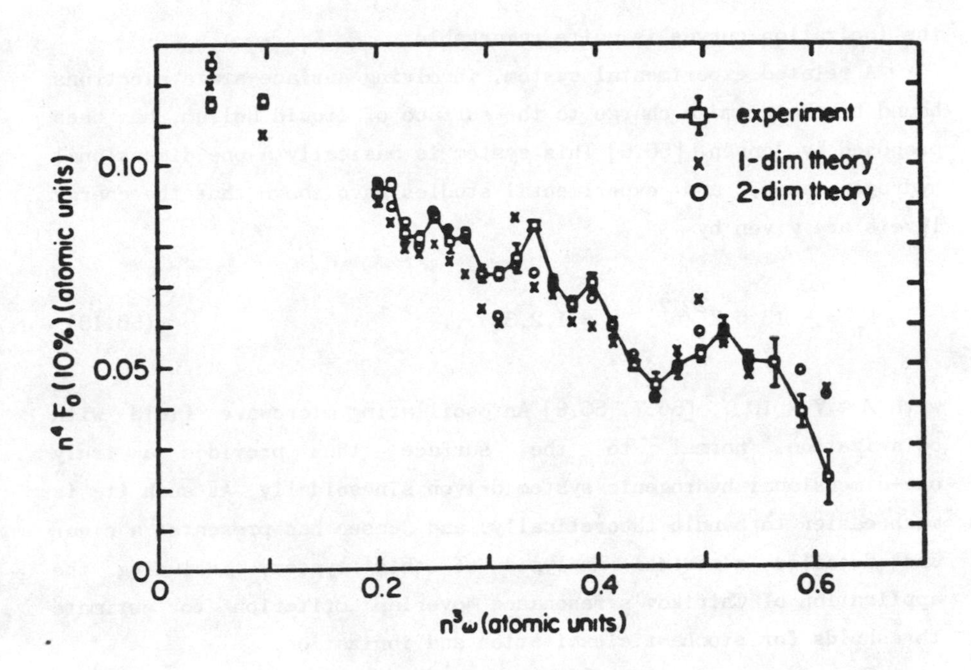

Figure 56.4 **Classically scaled F_o(10%) vs. classically scaled microwave frequency, showing classical (one- and two-dimensional) theoretical predictions vs. experimental results on the microwave ionization of Rydberg hydrogen atoms. From K.A.H. van Leeuwen, et al., Reference [56.5], with permission.**

the ionization curves is quite remarkable.

A related experimental system, involving surface-state electrons bound by their image charge to the surface of liquid helium, has been proposed by Jensen. [56.6] This system is basically a one-dimensional hydrogen atom, and experimental studies have shown that the energy levels are given by

$$E_n = - 13.6\ Z^2/n^2, \quad n = 1,2,3, \ldots \tag{56.10}$$

with $Z \approx 7 \times 10^{-3}$. [56.7, 56.8] An oscillating microwave field with polarization normal to the surface thus provides a truly one-dimensional hydrogenic system driven sinusoidally. As such it is much easier to handle theoretically, and Jensen has presented a clear and detailed classical analysis of this system, including the application of Chirikov's resonance overlap criterion to estimate thresholds for stochastic excitation and ionization.

These sinusoidally driven hydrogenic systems with $\omega \approx \omega_{at}$ are quite analogous to the models for multiple-photon excitation of molecular vibrations considered in Sections 47 and 49. In each case the absorption of many photons in the atom or molecule is associated with the onset of chaos and stochastic excitation: chaotic trajectories eventually migrate to high energy levels and possibly into the continuum. Obviously the hydrogenic systems are cleaner to analyze theoretically and to make unambiguous comparisons with experiments. The striking agreement between the classical theories and the experiments is perhaps surprising in light of the quantum localization effect that might have been anticipated from the discussion in the preceding section.

Quantum-mechanical calculations for a one-dimensional hydrogenic system in a monochromatic field have been carried out by Casati, Chirikov, and Shepelyansky. [56.9] In these computations the initially excited level had n = 45, 56, or 66, and a total of typically 200 levels was included in the range $20 \leq n \leq 200$. Comparisons were made with classical trajectory results corresponding to the same initial n. To gauge the degree of diffusion over the unperturbed eigenstates, the second moment of the distribution function $f_n(t)$ was computed:

$$M_2 \equiv \langle (n - \langle n \rangle)^2 \rangle / n_o^2 = \langle \Delta n^2 \rangle / n_o^2 \tag{56.11}$$

where n_o is the value of n chosen for the initial state.

Casati, <u>et al</u>. [56.9] reported that the most interesting result of their computations was a quantum suppression of the classically predicted diffusion, confirming the expectations based on simple models like the kicked pendulum. For short times M_2 was close to the classically predicted value, but then oscillated about a constant value. Classically, however, M_2 continued to grow with time in the chaotic regime. The difference between the classical and quantum results is illustrated in Figure 56.5 for $n_o = 66$, $n_o^3\omega = 1.2$, and $n_o^4 F_{max} = 0.03$ and 0.04. Note that one of the two quantum-mechanical cases shown in Figure 56.5 does show continued growth of M_2 with time (curve 3), but at a much slower rate than in the corresponding classical case (curve 4).

Figure 56.6 shows an averaged distribution function $\bar{f}_n$ computed by Casati, <u>et al</u>. for the case of curves 1 and 2 of Figure 56.5. The average is over 40 values of $\tau = \omega t/2\pi$ in the interval $80 < \tau \leq 120$. The classical distribution (dashed line) is fairly broad, whereas the quantum distribution (solid line) is peaked at the initial value $n_o = 66$. The peaks in the "plateau" in such quantum distributions were associated with multiphoton resonances.

These results obviously suggest something very different from those of Reference [56.5] One thing that should be noted, however, is that these quantum-mechanical computations assume a completely <u>discrete</u> set of unperturbed eigenfunctions. Furthermore the physical system of interest here is not a bounded quantum system with a purely discrete spectrum, and recurrence and/or localization arguments against quantum chaos are not applicable if the continuum plays an

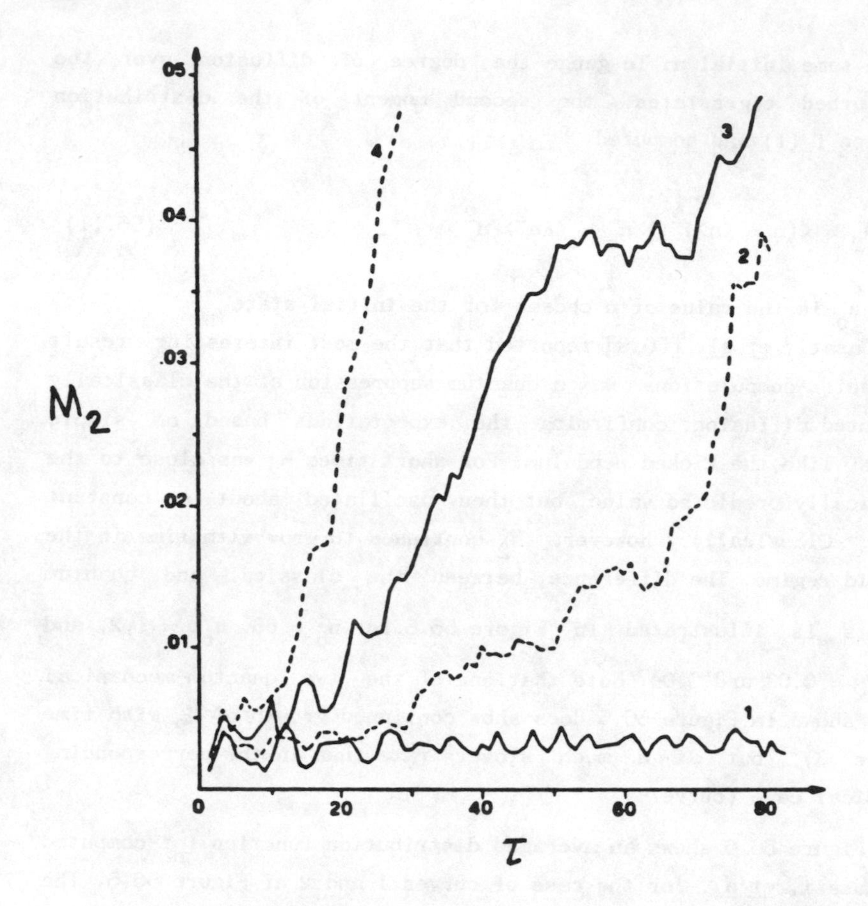

Figure 56.5 Dependence of M_2 on time $\tau = \omega t/2\pi$. Curves 1 and 3 are obtained with quantum calculations for $n_o = 66$ and $n_o^4F_{max} = 0.03$ and 0.04, respectively. Curves 2 and 4 are obtained by corresponding classical calculations. From G. Casati, et al., Reference [56.9], with permission.

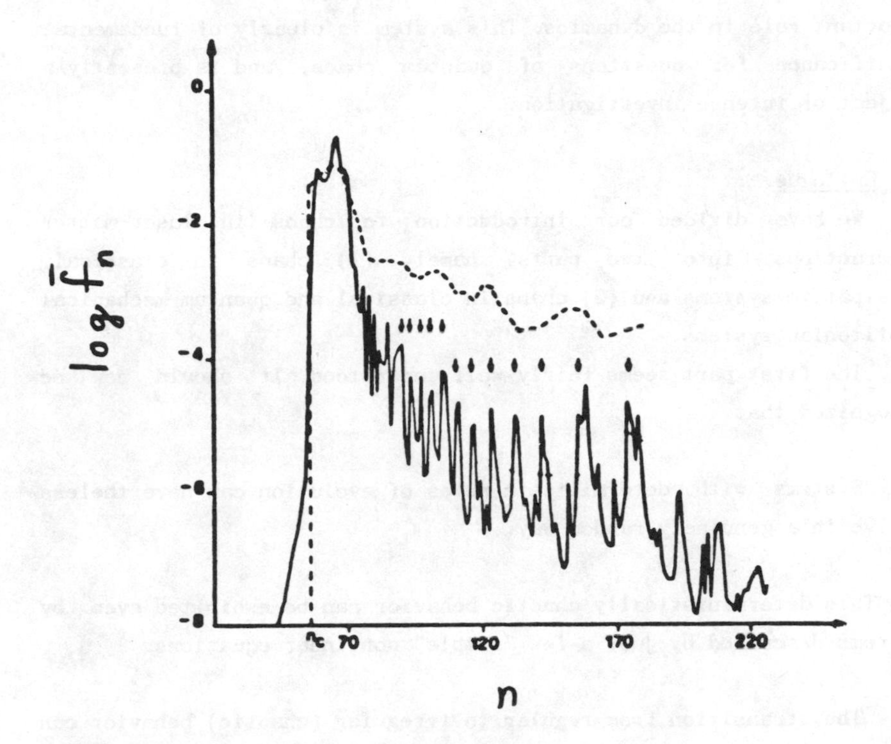

Figure 56.6 The distribution function $\bar{f}_n$, obtained by averaging $f_n(t)$ over 40 values of $\tau = \omega t/2\pi$ for the case of curves 1 and 2 of Figure 56.5. The quantum and classical results are given by the solid and dashed lines, respectively. n_c gives the transition point for classical chaos, and the arrows are drawn with equal (on the energy scale) spacing $\Delta E = \omega$. From G. Casati, et al., Reference [56.9], with permission.

important role in the dynamics. This system is clearly of fundamental significance for questions of quantum chaos, and is presently a subject of intense investigation.

57. Epilogue

We have divided our introduction to chaos in laser-matter interactions into two parts, namely (1) chaos in classical, dissipative systems and (2) chaos in classical and quantum-mechanical Hamiltonian systems.

The first part seems fairly well understood. It should now be recognized that

(1) Systems with deterministic rules of evolution can nevertheless evolve in a genuinely random way.

(2) This deterministically chaotic behavior can be exhibited even by systems described by just a few "simple" nonlinear equations.

(3) The transition from regular to irregular (chaotic) behavior can proceed in well-characterized, universal ways (e.g., by period doubling).

With regard to laser-matter interactions, chaotic behavior has been found in systems as "simple" as a single-mode gas laser. The three most ubiquitous routes to chaos have been observed. Chaos can arise in many other active and passive optical devices, and many observations have been made of chaotic behavior in lasers and/or optically bistable devices. It is not clear whether any applications will come out of all this, but certainly these studies have sharpened our understanding of optical instabilities.

On the Hamiltonian front the situation is different. From the pioneering work of Ford, Chirikov, and others we have some qualitative pictures of how chaos can develop in these systems. But it is not so easy to think of laboratory experiments that can be performed to study things like chaos and ergodicity in nondissipative, Hamiltonian systems. At the microscopic level of atoms and molecules there are clear indications, from classical studies, that chaos may play a fundamental role. Although there appear to be elements of truth in these classical models, quantum effects may smooth out, and in some cases completely suppress, any manifestations of classically predicted chaos. Many numerical studies are being made but the problem, to paraphrase the artist, is that computers only give answers. From these answers a clear picture of quantum chaos has yet to be developed.

REFERENCES

1.1 E.N. Lorenz, J. Atmos. Sci. 20, 130 (1963); Tellus 16, 1 (1964).

1.2 E.R. Buley and F.W. Cummings, Phys. Rev. A1454, 134 (1964).

3.1 M.J. Feigenbaum, J. Stat. Phys. 19, 25 (1978); 21, 669 (1979).

4.1 B.A. Huberman and J. Rudnick, Phys. Rev. Lett. 45, 154 (1980).

5.1 C. Corduneanu, Almost Periodic Functions (Interscience Publishers, N.Y., 1968).

5.2 E.O. Brigham, The Fast Fourier Transform (Prentice-Hall, Englewood Cliffs, N.J., 1974).

7.1 R.M. May, Nature 261, 459 (1976).

7.2 R.P. Feynman, R.B. Leighton, and M. Sands, The Feynman Lectures on Physics (Addison-Wesley, Reading, Mass., 1964), volume II, p. 41-11.

8.1 P. Collet and J.-P. Eckmann, Iterated Maps on the Interval as Dynamical Systems, Progress in Physics, volume I (Birkhauser Verlag, Basel, 1980).

8.2 R.H.G. Helleman, in Fundamental Problems in Statistical Mechanics, volume V, edited by E.G.D. Cohen (North-Holland, Amsterdam, 1980).

9.1 P. Collet, J.-P. Eckmann, and O.E. Lanford, Commun. Math. Phys. 76, 211 (1980).

9.2 D. Singer, SIAM J. Appl. Math. 35, 260 (1978).

10.1 B. Derrida, A. Gervois, and Y. Pomeau, J. Phys. A12, 269 (1979).

10.2 A.N. Sarkovskii, Ukranian Math. J. 16, 61 (1964); P. Stefan, Commun. Math. Phys. 54, 237 (1977).

10.3 T.-Y. Li and J.A. Yorke, Am. Math. Monthly 82, 985 (1975).

10.4 N. Metropolis, M.L. Stein, and P.R. Stein, J. Comb. Theory A15, 25 (1973).

11.1 B.A. Huberman and J.P. Crutchfield, Phys. Rev. Lett. 43, 1743 (1979).

11.2 S. Novak and R.G. Frehlich, Phys. Rev. A26, 3660 (1982).

11.3 J.N. Elgin, D. Forster, and S. Sarkar, Phys. Lett. 94A, 195 (1983).

11.4 J.R. Ackerhalt, H.W. Galbraith, and P.W. Milonni, in Coherence and Quantum Optics, edited by L. Mandel and E. Wolf (Plenum, N.Y.,1984).

12.1 F.T. Arecchi, R. Meucci, G. Puccioni, and J. Tredicce, Phys. Rev. Lett. 49, 1217 (1982).

14.1 P. Manneville and Y. Pomeau, Phys. Lett. 75A, 1 (1979); Y. Pomeau and P. Manneville, Commun. Math. Phys. 74, 189 (1980).

14.2 J.E. Hirsch, B.A. Huberman, and D.J. Scalapino, Phys. Rev. A25, 519 (1982).

14.3 J.E. Hirsch, M. Nauenberg, and D.J. Scalapino, Phys. Lett. 87A, 391 (1982).

14.4 B. Hu and J. Rudnick, Phys. Rev. Lett. 48, 1645 (1982).

15.1 L.D. Landau, Akad. Nauk. Doklady 44, 339 (1944); English translation in Collected Papers of L.D. Landau, edited by D. ter Haar, p. 387 (1965).

15.2 D. Ruelle and F. Takens, Commun. Math. Phys. 20, 167 (1971).

15.3 S.E. Newhouse, D. Ruelle, and F. Takens, Commun. Math. Phys. 64, 35 (1978).

15.4 C. Grebogi, E. Ott, and J.A. Yorke, Physica 15D, 354 (1985).

15.5 V.I. Arnold, Am. Math. Soc. Transl. Ser. (2) 46, 213 (1965).

15.6 S.J. Shenker, Physica 5D, 405 (1982).

15.7 M.J. Feigenbaum, L.P. Kadanoff, and S.J. Shenker, Physica 5D, 370 (1982).

15.8 D. Rand, S. Ostlund, J. Sethna, and E.D. Siggia, Phys. Rev. Lett. 49, 387 (1982); S. Ostlund. D. Rand, J. Sethna, and E.D. Siggia, Physica 8D, 303 (1983).

16.1 A.P. Fein, M.S. Heutmaker, and J.P. Gollub, Phys. Scr. T9, 79 (1985).

16.2 H.L. Swinney and J.P. Gollub, Phys. Today 31, 41 (1978).

16.3 J.P. Gollub and S.V. Benson, J. Fluid Mech. 100, 449 (1980).

16.4 A. Libchaber, S. Fauve, and C. Laroche, Physica 7D, 73 (1983).

16.5 S. Martin, H. Leber, and W. Martienssen, Phys. Rev. Lett. 53, 303 (1984).

16.6 C. Grebogi, E. Ott, and J.A. Yorke, Phys. Rev. Lett. 51, 339

(1983).

17.1 O.E. Lanford, in Hydrodynamical Instabilities and the Transition to Chaos, edited by H.L. Swinney and J.P. Golub (Springer, N.Y., (1981).

17.2 B. Mandelbrot, Fractals (W.H. Freeman, San Francisco, 1977).

17.3 M. Hénon, Commun. Math. Phys. 50, 69 (1976).

17.4 D.A. Russell, J.D. Hanson, and E. Ott, Phys. Rev. Lett. 45, 1175 (1980).

17.5 H.G.E. Hentschel and I. Procaccia, Physica 8D, 435 (1983).

18.1 G. Benettin, L. Galgani, and J.-M. Strelcyn, Phys. Rev. A14, 2338 (1976); G. Benettin, L. Galgani, A. Giorgilli, and J.-M. Strelcyn, Meccanica 15, 9 (1980); 15, 21 (1980).

18.2 A. Wolf, J.B. Swift, H.L. Swinney, and J.A. Vastano, Physica 16D, 285 (1985).

18.3 J.P. Eckmann and D. Ruelle, Rev. Mod. Phys. 57, 617 (1985).

18.4 H. Haken, Phys. Lett. 94A, 71 (1983).

18.5 F. Takens, in Proceedings on Dynamical Systems and Turbulence, Lecture Notes in Mathematics 898, edited by D.A. Rand and L.S. Young (Springer, Berlin, 1980).

18.6 J.A. Vastano and E.J. Kostelich, in Dimension and Entropies in Chaotic Systems, edited by G. Mayer-Kress (Springer-Verlag, Berlin, 1986).

19.1 P. Grassberger and I. Procaccia, Phys. Rev. Lett. 50, 346 (1983); Phys. Rev. A29, 2591 (1983).

19.2 M.C. Mackey and L. Glass, Science 197, 287 (1977).

19.3 R.S. Shaw, Z. Naturforsch 36a, 80 (1981).

20.1 J.D. Farmer, Z. Naturforsch 37a, 1304 (1982).

20.2 Ya. B. Pesin, Usp. Mat. Nauk 32, 55 (1977); Math. Surveys 32, 55 (1977).

21.1 G. Mayer-Kress and H. Haken, J. Stat. Phys. 26, 149 (1981).

21.2 J.P. Crutchfield and B.A. Huberman, Phys. Lett. 77A, 407 (1980).

21.3 A. Zardecki, Phys. Lett. 90A, 274 (1982).

22.1 L. Allen and J.H. Eberly, Optical Resonance and Two-Level Atoms (Wiley, N.Y., 1975).

23.1 H. Haken, Phys. Lett. 53A, 77 (1975).
23.2 C. Sparrow, The Lorenz Equations: Bifurcations, Chaos, and Strange Attractors (Springer-Verlag, N.Y., 1982).
23.3 J.L. Kaplan and J.A. Yorke, Commun. Math. Phys. 67, 93 (1979).
23.4 E.D. Yorke and J.A. Yorke, J. Stat. Phys. 21, 263 (1979).
23.5 K. Robbins, SIAM J. Appl. Math. 36, 457 (1979).
23.6 I. Shimada and T. Nagashima, Prog. Theor. Phys. 61, 1605 (1979).
24.1 C.O. Weiss and W. Klische, Opt. Commun. 51, 47 (1984).
24.2 T.Y. Chang, T.J. Bridges, and E.G. Burkhardt, Appl. Phys. Lett. 17, 357 (1970).
24.3 J.P. Gordon, H.J. Zeiger, and C.H. Townes, Phys. Rev. 99, 1264 (1955).
24.4 C.O. Weiss, J. Opt. Soc. Am. B2, 137 (1985).
24.5 C.O. Weiss, W. Klische, P.S. Ering, and M. Cooper, Opt. Commun. 52, 405 (1985).
24.6 H. Zeghlache and P. Mandel, J. Opt. Soc. Am. B2, 18 (1985).
24.7 R.G. Harrison and D.J. Biswas, Prog. Quantum Electron. 10, 147 (1985).
25.1 L.W. Casperson and A. Yariv, Appl. Phys. Lett. 17, 259 (1970).
26.1 L.W. Casperson, IEEE J. Quantum Electron. QE-14, 756 (1978).
26.2 L.W. Casperson, in Laser Physics, ed. by J.D. Harvey and D.F. Walls, Lecture Notes in Physics 182 (Springer-Verlag, Berlin, 1983).
26.3 M.-L. Shih, Ph.D thesis (University of Arkansas, 1984).
26.4 M.-L. Shih, P.W. Milonni, and J.R. Ackerhalt, J. Opt. Soc. Am. B2, 130 (1985); P.W. Milonni, J.R. Ackerhalt, and M.-L. Shih, Opt. Commun. 49, 155 (1984).
26.5 N.B. Abraham, T. Chyba, M. Coleman, R.S. Gioggia, N.J. Halas, L.M. Hoffer, S.-N. Liu, M. Maeda, and J.C. Wesson, in Laser Physics, ed. by J.D. Harvey and D.F. Walls, Lecture Notes in Physics 182 (Springer-Verlag, Berlin, 1983).
26.6 R.S. Gioggia and N.B. Abraham, Phys. Rev. Lett. 51, 650 (1983).
26.7 L.W. Casperson, Phys. Rev. A21, 911 (1980).
26.8 P. Mandel, Opt. Commun. 44, 400 (1983).

26.9 D.K. Bandy, L.M. Narducci, L.A. Lugiato, and N.B. Abraham, J. Opt. Soc. Am. B2, 56 (1985).

27.1 L.E. Urbach, S.-N. Liu, and N.B. Abraham, in Coherence and Quantum Optics V, ed. by L. Mandel and E. Wolf (Plenum, N.Y., 1984).

27.2 L.M. Hoffer, T.H. Chyba, and N.B. Abraham, J. Opt. Soc. Am. B2, 102 (1985).

28.1 H. Risken and K. Nummedal, J. Appl. Phys. 39, 4662 (1968).

28.2 R. Graham and H. Haken, Z. Physik 213, 420 (1968).

28.3 R. Graham, Phys. Lett. 58A, 440 (1976).

28.4 P.W. Milonni, Appl. Opt. 16, 2794 (1977).

28.5 M. Lax, G.P. Agrawal, M. Belic, B.J. Coffey, and W.H. Louisell, J. Opt. Soc. Am. A2, 731 (1985).

29.1 M.L. Minden and L.W. Casperson, J. Opt. Soc. Am. B2, 120 (1985).

29.2 S.T. Hendow and M. Sargent III, J. Opt. Soc. Am. B2, 84 (1985), and references therein.

29.3 L.A. Lugiato and L.M. Narducci, Phys. Rev. A32, 1576 (1985).

30.1 J.A. Fleck and R.E. Kidder, J. Appl. Phys. 35, 2825 (1964); 36, 2327 (1965).

30.2 M.-L. Shih and P.W. Milonni, Opt. Commun. 49, 155 (1984).

30.3 D. Ross, Lasers, Light Amplifiers and Resonators (Academic Press, N.Y., 1969), p. 320.

30.4 R. Hauck, F. Hollinger, and H. Weber, Opt. Commun. 47, 141 (1983); F. Hollinger and Chr. Jung, J. Opt. Soc. Am. B2, 218 (1985).

30.5 D.J. Biswas and R.G. Harrison, Phys. Rev. A32, 3835 (1985).

31.1 W.W. Rigrod, J. Appl. Phys. 36, 2487 (1965).

32.1 A.M. Albano, J. Abounadi, T.H. Chyba, C.E. Searle, S. Yong, R.S. Gioggia, and N.B. Abraham, J. Opt. Soc. Am. B2, 47 (1985).

32.2 N.B. Abraham, L.A. Lugiato, and L.M. Narducci, J. Opt. Soc. Am. B2, 7 (1985).

32.3 T. Midavaine, D. Dangoisse, and P. Glorieux, Phy. Rev. Lett. 55, 1989 (1985).

32.4 W. Klische, H.R. Telle, and C.O. Weiss, Opt. Lett. 9, 561

(1984).
32.5 Y.C. Chen, H.G. Winful, and J.M. Liu, Appl. Phys. Lett. 47, 208 (1985).
32.6 F.T. Arecchi, G.L. Lippi, G.P. Puccioni, and J.R. Tredicce, Opt. Commun. 51, 308 (1984); J.R. Tredicce, F.T. Arecchi, G.L. Lippi, and G.P. Puccioni, J. Opt. Soc. Am. B2, 173 (1985).
32.7 D.K. Bandy, L.M. Narducci, and L.A. Lugiato, J. Opt. Soc. Am. B2, 148 (1985).
32.8 C.O. Weiss and H. King, Opt. Commun. 44, 59 (1982); C.O. Weiss, A. Godone, and A. Olafsson, Phys. Rev. A28, 892 (1983).
32.9 H.G. Winful, Y.C. Chen, and J.M. Liu, Appl. Phys. Lett. 48, 616 (1986).
32.10 D.G. Aronson, R.P. McGehee, I. G. Kevrekidis, and R. Aris, Phys. Rev. A33, 2190 (1986).
32.11 T. Kai and K. Tomita, Prog. Theor. Phys. 61, 54 (1979).
32.12 I.G. Kevrekidis, private communication.
32.13 L.W. Hillman, J. Krasinski, R.W. Boyd, and C.R. Stroud, Jr., Phys. Rev. Lett. 52, 1605 (1984); L.W. Hillman, J. Krasinski, K. Koch, and C.R. Stroud, Jr., J. Opt. Soc. Am. B2, 211 (1985).
33.1 H.M. Gibbs, Optical Bistability: Controlling Light with Light, (Academic Press, N.Y., 1985).
33.2 E. Abraham and S.D. Smith, Rep. Prog. Phys. 45, 815 (1982).
33.3 A. Szöke, V. Daneu, J. Goldhar, and N.A. Kurnit, Appl. Phys. Lett. 15, 376 (1969).
33.4 S.L. McCall, Phys. Rev. A9, 1515 (1974).
33.5 K. Ikeda, Opt. Commun. 30, 257 (1979).
33.6 R. Bonifacio and L.A. Lugiato, Lett. Nouvo Cim. 21, 505 (1978).
33.7 H.M. Gibbs, S.L. McCall, and T.N.C. Venkatesan, Phys. Rev. Lett. 36, 1135 (1976).
34.1 K. Ikeda, H. Daido, and O. Akimoto, Phys. Rev. Lett. 45, 709 (1980).
34.2 H. Nakatsuka, S. Asaka, H. Itoh, K. Ikeda, and M. Matsuoka, Phys. Rev. Lett. 50, 109 (1983).
34.3 H.J. Carmichael, R.R. Snapp, and W.C. Schieve, Phys. Rev. A26,

3408 (1982).

34.4 H.M. Gibbs, F.A. Hopf, D.L. Kaplan, and R.L. Shoemaker, Phys. Rev. Lett. 46, 474 (1981).

34.5 R.G. Harrison, W.J. Firth, C.A. Emshary, and I.A. Al-Saidi, Phys. Rev. Lett. 51, 562 (1983).

34.6 K.J. Blow and N.J. Doran, Phys. Rev. Lett. 52, 526 (1984).

35.1 H. Goldstein, Classical Mechanics (Addison-Wesley, Reading, Mass., 1980).

35.2 D. ter Haar, Elements of Hamiltonian Mechanics (North-Holland, Amsterdam, 1961).

36.1 L.D. Landau and E.M. Lifshitz, Mechanics (Pergamon, Oxford, 1969), pp. 155-6.

37.1 M.J. Ablowitz, A. Ramani, and H. Segur, J. Math. Phys. 21, 715 (1980).

38.1 V.I. Arnold and A. Avez, Ergodic Problems of Classical Mechanics (Benjamin, N.Y., 1968).

38.2 J. Ford, in Fundamental Problems in Statistical Mechanics III, ed. by E.G.D. Cohen (North-Holland, Amsterdam, 1975).

38.3 A.J. Lichtenberg and M.A. Lieberman, Regular and Stochastic Motion (Springer-Verlag, N.Y., 1983).

39.1 See Collected Works of Enrico Fermi (University of Chicago Press, 1965), paper 266 and comments by S. Ulam.

39.2 A.N. Kolmogorov, Dokl. Akad. Nauk. SSSR 98, 527 (1954). An English translation may be found in Hao Bai-Lin, Chaos (World Scientific, 1984).

39.3 N.J. Zabusky and M.D. Kruskal, Phys. Rev. Lett. 15, 240 (1965).

39.4 M.D. Kruskal, lecture at Los Alamos National Laboratory, 13 August 1984.

39.5 R.K. Dodd, J.C. Eilbeck, J.D. Gibbon, and H.C. Morris, Solitons and Nonlinear Wave Equations (Academic Press, London, 1984).

40.1 V.I. Arnold, Russ. Math. Surveys 18, 9 (1963).

40.2 M.V. Berry, American Institute of Physics Conference Proceedings, no. 46 (A.I.P., N.Y., 1978).

40.3 J. Ford and J. Waters, J. Math. Phys. 4, 1293 (1963).

41.1 G.H. Walker and J. Ford, Phys. Rev. 188, 416 (1969).

41.2 B.V. Chirikov, Phys. Rep. 52, 263 (1979).

42.1 M. Hénon and C. Heiles, Astron. J. 69, 73 (1964).

43.1 E. Thiele and D.J. Wilson, J. Chem Phys. 35, 1256 (1961).

43.2 D.W. Oxtoby and S.A. Rice, J. Chem. Phys. 65, 1676 (1976).

43.3 P. Brumer, Adv. Chem. Phys. 47, 201 (1981).

45.1 H. Margenau, Physics and Philosophy: Selected Essays (D. Reidel, Dordrecht, 1978), p.41.

45.2 M. Born, Dan. Mat. Fys. Medd. 30, no. 7 (1955).

45.3 P. Rannou, Astron. Astrophys. 31, 289 (1974).

45.4 Nature 300, 311 (1982).

46.1 G. Casati, B.V. Chirikov, F.M. Izrailev, and J. Ford, in Stochastic Behavior in Classical and Quantum Hamiltonian Systems, ed. by G. Casati and J. Ford (Springer-Verlag, N.Y., 1979).

47.1 N.R. Isoner and M.C. Richardson, Appl. Phys. Lett. 18, 224 (1971); Opt. Commun. 3, 360 (1971).

47.2 See, for instance, J.L. Lyman, G.P. Quigley, and O.P. Judd, in Multiple-Photon Excitation and Dissociation of Polyatomic Molecules, ed. by C.D. Cantrell (Springer-Verlag, N.Y., 1987); O.P. Judd, J. Chem. Phys. 71, 4515 (1979).

47.3 T.B. Simpson, J.G. Black, I. Burak, E. Yablonovitch, and N. Bloembergen, J. Chem. Phys. 83, 628 (1985).

47.4 R.V. Ambartzumyan, Yu. A. Gorokhov, V.S. Letokhov, G.N. Makarov, and A.A. Puretskii, Sov. Phys. JETP Lett. 23, 22 (1976); R.V. Ambartsumyan, Yu. A. Gorokhov, V.S. Letokhov, and G.N. Makarov, Sov. Phys. JETP 42, 993 (1976).

47.5 D.M. Larsen and N. Bloembergen, Opt. Commun. 17, 254 (1976); D.M. Larsen, ibid. 17, 250 (1976).

47.6 J.R. Ackerhalt and P.W. Milonni, Phys. Rev. A34, 1211 (1986); J.R Ackerhalt, H.W. Galbraith, and P.W. Milonni, Phys. Rev. Lett. 51, 1259 (1983).

47.7 M. Bixon and J. Jortner, J. Chem. Phys. 48, 715 (1968).

47.8 P.W. Milonni, J.R. Ackerhalt, H.W. Galbraith, and M.-L. Shih,

Phys. Rev. A28, 32 (1983).

47.9 See, for instance, M.J. Lighthill, Introduction to Fourier Analysis and Generalized Functions (Cambridge University Press, Cambridge, 1970), p. 67.

48.1 L.C. Biedenharn and J.D. Louck, Angular Momentum in Quantum Physics, Encyclopedia of Mathematics and Its Applications, ed. by G.-C. Rota, Volume 8 (Addison-Wesley, Reading, Mass., 1981), Chapter 7, Section 10.

48.2 H.W. Galbraith, J.R. Ackerhalt, and P.W. Milonni, J. Chem. Phys. 79, 5345 (1983).

49.1 E.V. Shuryak, Sov. Phys. JETP 44, 1070 (1976).

49.2 P.I. Belobrov, G.P. Berman, G.M. Zaslavskii, and A.P. Slivinskii, Sov. Phys. JETP 49, 993 (1979).

49.3 D.A. Jones and I.C. Percival, J. Phys. B: At. Mol. Phys. 16, 2981 (1983).

49.4 W.E. Lamb, Jr., in Laser Spectroscopy III, ed. by J.L. Hall and J.L. Carlsten (Springer-Verlag, N.Y., 1977); Laser Spectroscopy IV, ed. by H. Walther and K. Rothe (Springer-Verlag, N.Y., 1979).

49.5 N. Bloembergen, Opt. Commun. 15, 416 (1975).

49.6 E.B. Alterman, C.T. Tahk, and D.J. Wilson, J. Chem. Phys. 44, 451 (1966).

49.7 R.C. Baetzold, C.T. Tahk, and D.J. Wilson, J. Chem. Phys. 45, 4209 (1966).

49.8 P.F. Endres and D.J. Wilson, J. Chem. Phys. 46, 425 (1967).

49.9 R.B. Walker and R.K. Preston, J. Chem. Phys. 67, 2017 (1977).

50.1 P. Bocchieri and A. Loinger, Phys. Rev. 107, 337 (1957).

50.2 A. Hobson, Concepts in Statistical Mechanics (Gordon and Breach, N.Y., 1971).

50.3 T. Hogg and B.A. Huberman, Phys. Rev. Lett. 48, 711 (1982).

50.4 A. Peres, Phys. Rev. Lett. 49, 1118 (1982).

50.5 I.C. Percival, J. Phys. B6, 1229 (1973).

50.6 K.S.J. Nordholm and S.A. Rice, J. Chem. Phys. 61, 203 (1974); 61, 768 (1974); 62, 157 (1975).

50.7 D.W. Noid, M.L. Koszykowski, and R.A. Marcus, J. Chem. Phys. 71, 2864 (1971).
50.8 J.S. Hutchinson and R.E. Wyatt, Phys. Rev. A23, 1567 (1981).
50.9 P. Brumer and M. Shapiro, Chem. Phys. Lett. 72, 528 (1980).
50.10 E.J. Heller, J. Chem. Phys. 72, 1337 (1980).
50.11 M.D. Feit and J.A. Fleck, Jr., J. Chem. Phys. 80, 2578 (1984).
50.12 W.P. Reinhardt, J. Phys. Chem. 86, 2158 (1982).
50.13 A. Peres, Phys. Rev. A30, 504 (1984); M. Feingold, N. Moiseyev, and A. Peres, Phys. Rev. A30, 509 (1984).
51.1 J.B. Keller, Ann. Phys. (N.Y.) 4, 180 (1958).
51.2 I.C. Percival, Adv. Chem. Phys. 36, 1 (1977).
51.3 N. Pomphrey, J. Phys. B7, 1909 (1974).
51.4 M.J. Davis and R.E. Wyatt, Chem. Phys. Lett. 86, 235 (1982).
51.5 M.V. Berry and M. Tabor, Proc. Roy. Soc. A356, 375 (1977).
51.6 M.V. Berry, in The Wave-Particle Dualism, ed. by S. Diner, D. Fargue, G. Lochak, and F. Selleri (Reidel, Dordrecht, 1984).
51.7 P. Pechukas, Phys. Rev. Lett. 51, 943 (1983).
51.8 O. Bohigas, M.J. Giannoni, and C. Schmit, Phys. Rev. Lett. 52, 1 (1984).
51.9 M.V. Berry, Philos. Trans. Roy. Soc. London, A287, 237 (1977).
51.10 M.V. Berry, J. Phys. A10, 2083 (1977).
51.11 G.M. Zaslavskii, Sov. Phys. JETP 46, 1094 (1977).
51.12 S.W. McDonald and A.N. Kaufman, Phys. Rev. Lett. 42, 1189 (1979).
52.1 P.W. Milonni, J.R. Ackerhalt, and M.E. Goggin, Phys. Rev. A35, 1714 (1987).
52.2 Y. Pomeau, B. Dorizzi, and B. Grammaticos, Phys. Rev. Lett. 56, 681 (1986).
53.1 P.W. Milonni, Phys. Rep. 25, 1 (1976).
53.2 E.T. Jaynes and F.W. Cummings, Proc. IEEE 51, 89 (1963).
53.3 M. Tavis and F.W. Cummings, Phys. Rev. 170, 379 (1968).
53.4 P.I. Belobrov, G.M. Zaslavskii, and G. Th. Tartakovskii, Sov. Phys. JETP 44, 945 (1976).
53.5 P.W. Milonni, J.R. Ackerhalt, and H.W. Galbraith, Phys. Rev.

Lett. 50, 966 (1983); 51, 1108 (E) (1983).
53.6 J.R. Ackerhalt and P.W. Milonni, J. Opt. Soc. Am. B1, 116 (1984).
53.7 R. Graham and H. Höhnerbach, Phys. Lett. A101, 61 (1984).
53.8 M. Kuś, Phys. Rev. Lett. 54, 1343 (1985).
53.9 R.F. Fox and J. Eidson, Phys. Rev. A34, 482 (1986); J. Eidson and R.F. Fox, Phys. Rev. A34, 3288 (1986).
53.10 J.H. Eberly, N.B. Narozhny, and J.J. Sanchez-Mondragon, Phys. Rev. Lett. 44, 1323 (1980); H.I. Yoo, J.J. Sanchez-Mondragon, and J.H. Eberly, J. Phys. A14, 1383 (1981).
53.11 G. Rempe, H. Walther, and N. Klein, Phys. Rev. Lett. 58, 353 (1987).
54.1 C.F.F. Karney, A.B. Rechester, and R.B. White, Physica 4D, 425 (1982).
54.2 J.D. Hanson, E. Ott, and T.M. Antonsen, Jr., Phys. Rev. A29, 819 (1984).
54.3 F.M. Izrailev and D.L. Shepelyansky, Theor. Mat. Fiz. 43, 417 (1980).
55.1 Ya. G. Sinai, Theor. Prob. Appl. 27, 256 (1982).
55.2 P.W. Anderson, Phys. Rev. 109, 1492 (1958).
55.3 K. Ishii, Prog. Theor. Phys. (Suppl.) 53, 77 (1973).
55.4 D.J. Thouless, in Ill-Condensed Matter, ed. by R. Balian, R. Maynard, and G. Toulouse (North-Holland, Amsterdam, 1979).
55.5 D.R. Grempel, R.E. Prange, and S. Fishman, Phys. Rev. A29, 1639 (1984).
55.6 Ya. B. Zeldovich, Sov. Phys. JETP 24, 1006 (1967).
55.7 R.E. Wyatt, in "Lasers, Molecules, and Methods," Adv. Chem. Phys., ed. by J.O. Hirschfelder, R.E. Wyatt, and R.D. Coalson, to be published.
55.8 G. Casati and I. Guarneri, Phys. Rev. Lett. 50, 640 (1983).
55.9 H.G. Schuster, Phys. Rev. B28, 443 (1983).
55.10 D.L. Shepelyansky, Physica 8D, 208 (1983).
56.1 J.E. Bayfield and P.M. Koch, Phys. Rev. Lett. 33, 258 (1974).
56.2 J.G. Leopold and I.C. Percival, Phys. Rev. Lett. 41, 944 (1978).

56.3 J.G. Leopold and I.C. Percival, J. Phys. B12, 709 (1979).

56.4 D.A. Jones, J.G. Leopold, and I.C. Percival, J. Phys. B13, 31 (1980).

56.5 K.A.H. van Leeuwen, G. v. Oppen, S. Renwick, J.B. Bowlin, P.M. Koch, R.V. Jensen, O. Rath, D. Richards, and J.G. Leopold, Phys. Rev. Lett. 55, 2231 (1985).

56.6 R.V. Jensen, Phys. Rev. A30, 386 (1984).

56.7 C.C. Grimes, T.R. Brown, M.L. Burns, and C.L. Zipfel, Phys. Rev. B13, 140 (1976).

56.8 D.K. Lambert and P.L. Richards, Phys. Rev. B23, 3282 (1981).

56.9 G. Casati, B.V. Chirikov, and D.L. Shepelyansky, Phys. Rev. Lett. 53, 2525 (1984).

